D0065405

INSTRUMENTATION IN ANALYTICAL CHEMISTRY

INSTRUMENTATION IN ANALYTICAL CHEMISTRY

Articles from Volumes 41-44 of
ANALYTICAL CHEMISTRY

Collected by **Alan J. Senzel,**
Associate Editor of ANALYTICAL CHEMISTRY

An ACS Reprint Collection

American Chemical Society
Washington, D.C.   1973

Library of Congress Catalog Card 72-95641
ISBN 8412-0159-5

PRINTED IN THE UNITED STATES OF AMERICA

CONTENTS

Chromatography

Electrochemistry

Other Related Areas

PREFACE

This volume consists of all the monthly Instrumentation articles which have appeared in ANALYTICAL CHEMISTRY from January 1969, when the feature assumed its current format of outside contributions by experts in various fields of instrumentation, to July 1972. The authors have been given the opportunity to update their articles and to make corrections as they saw fit, and many of them have done so.

ANALYTICAL CHEMISTRY recognizes the vital importance of high quality, innovative instrumentation to scientific research and development, and we believe that our dedication to leadership in this field is reflected in the articles contained herein. Articles included in the book have been grouped for reader convenience into four broad areas of analytical chemistry—spectrometry, chromatography, electrochemistry, and combination and other techniques. Areas of interest include not only instrument design, but also specific applications, such as in biomedical instrumentation, pollution measurement devices, and computer applications.

These articles have contributed original thinking useful to the solution of a wide selection of analytical problems, ranging from those bordering on physics to those with a biomedical flavor. The Editors of ANALYTICAL CHEMISTRY sincerely hope that this permanent collection of Instrumentation features will prove to be of great value as both a research and teaching aid.

Most importantly, we wish to thank the authors who devoted many hours to the preparation of these manuscripts and the members of our Instrumentation Advisory Panel who reviewed them. All of these people served with little compensation other than the satisfaction of providing our readers with the latest advances in still-developing fields of instrumentation.

Alan J. Senzel

Instrumentation of a Spectrophotometric System Designed for Kinetic Methods of Analyses

Theodore E. Weichselbaum and William H. Plumpe, Jr.
Sherwood Medical Industries, Inc.
Subsidiary of Brunswick Corp.
St. Louis, Mo. 63103

Harry B. Mark, Jr.
Department of Chemistry
University of Michigan
Ann Arbor, Mich. 48104

THE CONCEPT of kinetic-based analytical methods has been of interest to analytical chemists for many years. Initially, catalytic reactions, enzyme reactions especially, were recognized as useful analytical tools because of their sensitivity and selectivity. In recent years, the measurement of the rates of competitive reactions has been used for the *in situ* simultaneous determination of closely related mixtures. In spite of these and other advantages, the actual application of kinetic techniques to routine practical analysis has not been extensive. The reason for this is primarily instrumental. The commercial instrumentation readily available up to the present, although accurate, reliable, and sophisticated, has been designed primarily for equilibrium measurements. The first problem in kinetic analysis is that the measurements must be made with respect to an accurately known time reference. Also, the modes of taking and presenting data in useful forms for a technician are not convenient or built-in an equilibrium instrument. For practical routine kinetic analysis, instrument design must take into consideration the specific problems associated with measurements on a dynamic system.

The subject of the recent 21st Annual Summer Symposium on Analytical Chemistry (Pennsylvania State University, July, 1968) was the role of computers in analytical chemistry. Most of the discussion centered around the applications of computers for data reduction. However, as was pointed out, the computer itself can do little to improve poor data. This is especially true with kinetic methods. For example, on paper, one can do a simultaneous differential kinetic determination of a mixture of many unknowns. In practice, two or three components are the maximum, and this limitation is purely instrumental. Recently, the Digecon instrument (Sherwood Medical Industries, Inc., St. Louis, Mo.) has been designed and built, with modern computer electronic technology and components, to meet the requirements for use in kinetic-based analysis. This paper discusses the "philosophy" behind the criteria chosen as a requirement for practical kinetic measurements, the special instrumental problems, and the design solutions of these problems that were necessary to meet these criteria.

Design Criteria

The first general requirement was that this instrument be a very stable noise-free spectrophotometer. The transmittance range should be 0 to 100% (the actual instrument provides an overrun to 110%) with a linearity of 0.1% in the readout (this does not include non-linearity introduced by the photocell which would be added to this tolerance). The absorbance range should be 0 to 2.0

1

absorbance units. Again, the actual instrument with overrun can read to 2.1 absorbance units in two ranges with an accuracy of 1% of the reading plus 0.001 absorbance unit. The reason one increment on the readout must be added to the tolerance is that 1% of the reading becomes vanishingly small at low absorbances and leaves no room for the inevitable small variations in the servo potentiometer and gear train.

Secondly, because many of the important analytical kinetic reactions, such as enzyme assays, catalyst determinations, etc., are measured during their initial rate period (pseudo-zero order conditions), where the *initial reaction rate* is the parameter proportional to concentration, it was decided that the instrument should be capable of giving a direct readout of the derivative of the absorbance–time curve (which gives the rate of the reaction as a function of time). It was also decided that the derivative readout be continuous rather than a two (or even multiple) fixed-point method, because valuable information such as induction period variation with concentration, maximum reaction rates, etc., can be lost in the dead periods between measurements (actually a continuous readout can be considered to be an infinite fixed-point method) and,

because, multiple fixed-point sampling is electronically more complex than continuous measurement. It was felt that direct electronic differentiation of the signal should be employed rather than the more complex comparison-integrator technique. Both methods perform the identical operation on the signal and, hence, are equally sensitive to input noise. With careful design component choices, the stability of a direct differentiator is comparable with the comparison-integrator technique and has one less operational amplifier to serve as a possible source of trouble.

The range of reaction times of most kinetic methods over which the rates are to be measured is from 20 seconds to 10 minutes. Thus, the range of rates to be measured by this instrument must be between 0.001 to 1.0 absorbance unit per minute. No attempt was made to handle rates of very fast reactions, because these require specialized mixing and readout methods beyond the scope of a general purpose instrument. The rate function should be usable over the entire 0 to 2 absorbance-unit range of the instrument. The rate accuracy should be at least 1% + 0.001 unit from 0 to 1 absorbance unit; progressively deteriorating to about 5% at an absorbance of 2.

The last design criteria was digital

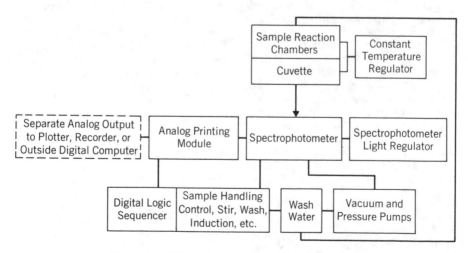

Figure 1.   Block diagram of the Digecon system

readout to 0.001 absorbance unit with a variable scale factor (or coefficient multiplier) which allows the readout to be obtained directly in concentration units. The instrument was designed so that the absorbance reading could be multiplied from 0.000 to 5.998 times and the rate reading could be multiplied from 0.000 to 11.996. As will be seen below, the digital readout is incorporated in a very simple and straightforward manner in the logarithm conversion operation of the circuit.

A permanent record should be provided, also. This system incorporates a digital printout and can give a continuous analog output for a pen recorder etc., by means of a repeater potentiometer on the servo.

Instrument Design

General Considerations. A general block diagram of the rate instrument is given in Figure 1. The important feature is the digital logic sequencer module which is programmed by the operator to control sequentially all operations (sample handling, stirring, washing, induction, measurement, etc.) of a particular analytical method. Once set for a particular analysis, the timing of the reaction and its measurement are automatically controlled and, hence, known to a high degree of accuracy. This module is constructed with conventional digital logic circuitry. The design and operation of the control sequencer will be discussed in detail in a subsequent paper and, hence will not be discussed here. It should be noted, however, that it is designed in a flexible manner with extra unused logic circuits so that it can be programmed for any future analysis sequence that is developed. Also, the details of sample handling, reaction chambers, and cuvette will be discussed elsewhere. Only the unique design features of accurate measurement of the reaction course will be discussed here.

For operation as a spectrophotometer in transmittance or absorbance (a nonlinear element converts transmittance to absorbance) as function of time mode, the basic operational amplifier circuitry is conventional for a single-beam spectrophotometer and the servo system acts only as a digital voltmeter. Extreme care must be taken to eliminate noise and drift in the light source and the photocell. This is especially important when operating in a rate mode, because the derivative of a signal containing noise and drift is impossible to handle. It was decided to use a single beam rather than dual beam, because excessive noise problems and both mechanical and electronic complications arise from the dual-beam system. Also, with care in design and the use of quality operational amplifiers, the long term drift of the single-beam system is less than 0.003 absorbance unit per hour. Thus, no significant improvement could be obtained with a dual-beam system.

Noise. To start with, 60 cycle line noise and transient pulses from switches, etc., must be eliminated by careful attention to shielding, line filtering, and ground paths. Light source, photocell, and amplifier noises are more difficult to deal with because they contain low frequency components, some of which have frequencies of the same magnitude as the period of the signals to be measured. This, of course, limits the use of filtering techniques, and these low frequency noises must be eliminated at their source.

First, the light source must be stabilized. Because typical line voltage fluctuations have periods on the same order of magnitude of the rates to be measured, a system with poor line voltage regulation might give the same light intensity at the beginning and end of an hour's operation, but will have short term fluctuations which seriously interfere with rate readings. For example, if a light intensity fluctuation caused a 0.001 absorbance unit fluctuation every 15 seconds (Figure 2), a variation which would be considered negligible in straight absorbance measurement, the average rate observed between points A and B would be + 0.008 unit per min, and a corresponding negative rate would be observed

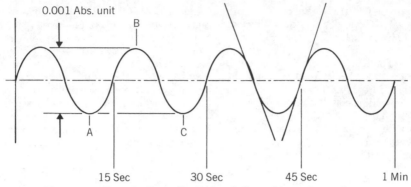

Figure 2. Hypothetical **effect of variation of lamp line voltage**

between B and C. Thus, the reading would show a rate variation in excess of 0.016 unit per min, which is a higher value than some of the rates measured in certain determinations. These requirements virtually dictate a high gain–low noise electronic regulator. A constant voltage regulator such as a cell storage battery is not satisfactory because a tungsten light bulb will give a decreasing light output for at least one hour after being turned on (apparently

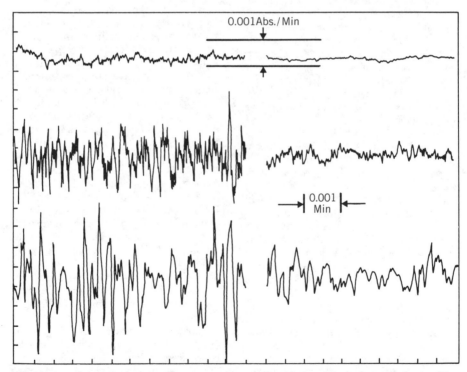

Figure 3.   Noise level of lamp output.   Recording of the differentiator output with filter time constants of 1 second (left hand side) and 10 seconds (right hand side). Bottom trace: Lamp in chimney; Middle trace: Forced circulation of air around lamp; Upper trace: Baffled lamp

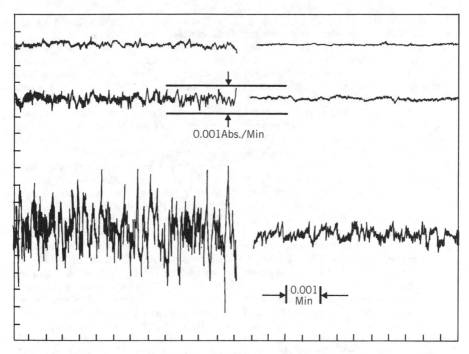

0.001Abs./Min

0.001 Min

Figure 4. Noise level of phototube output. Recording of the differentiator output with filter time constants of 1 second (left hand side) and 10 seconds (right hand side). Light adjusted for equal anode currents. Bottom trace: 5 dynode photomultiplier; Middle trace: Gas diode phototube; Upper: Vacuum diode phototube

the resistance changes on warming of the bulb and socket). It is necessary to use a regulator which samples both voltage and current, in effect monitoring the bulb resistance and compensating the applied voltage accordingly.

Even with a perfectly regulated and compensated lamp power supply used with a mechanically rigid optical system, serious light level variation was still found. These variations arise from heat convection flow around the light bulb itself. Instruments with the lamp in a chimney with vents at both the top and bottom are especially bad in this respect. The bottom trace in Figure 3 is a noise level recording of the output (at the differentiator output) of an old style Beckman DU lamp in its original form. The middle trace shows the effect of blowing air across the bulb. The top trace shows the same unit with baffling installed to prevent convection currents across the light path. Note the acceptable rate noise level of 0.001 unit per min indicated on the figure.

The proper choice of photocells is equally important from a noise point of view. Some solid state photocells exhibit good signal-to-noise ratios and allow low impedance circuitry to be used. However, to date no suitable solid state device has been found for operation at wavelengths under 4000 Å. Some photomultipliers have good wavelength response and high sensitivity, but all have signal-to-noise ratios of 1 to 2 orders of magnitude greater than a photodiode tube as shown in Figure 4. Cooling photomultipliers primarily reduces only the dark current noise which is only a small part of the total current and noise and, consequently, has little value. Reducing the number of dynodes used gives an improvement, but even connecting typical photomultipliers to utilize only

5

one dynode still gives a much poorer signal-to-noise ratio than two element tubes, without giving a significant improvement in sensitivity. Gas diodes were found to give poorer signal-to-noise ratios than vacuum tubes (see Figure 4). Statistical theory states that the signal-to-noise ratio will vary with the square root of the number of photoelectrons captured. As the use of gas or multiplier electrodes may give larger electrical outputs but do not change the number originally emitted from the cathode, it is generally better to use photodiodes and depend upon the excellent solid state amplifiers available to supply the required gain. However, it was found that it is necessary to use a gas photocell when operating below 3800 Å because a vacuum photocell has too low a signal in this range.

Photocell amplifiers and load resistors must be carefully chosen. Wirewound resistors would be desirable but are practically unobtainable in the light values required here. Vapor deposited resistors work well but must be tested for noise level. Field effect transistor input stages are used in all operational amplifiers and the finished amplifiers are tested and graded to utilize only the best in the photocell circuits.

Differentiator Circuit (Rate Mode). The output of the absorbance circuit (a nonlinear element is used in the direct ab-

sorbance readout mode) could simply be coupled to an active differentiator to give rate readings, but this approach tends to magnify errors in the absorbance circuit. The nonlinear element does not have a sufficiently accurate log output to be used in this way. Figure 5 shows the characteristics of a nonlinear element absorbance conversion circuit. Although the deviations from ideal behavior stayed within a 1% envelope, exaggerated here for clarity, this error would give much more than 1% error in rate reading. In this example, the output is 1% high at an absorbance of 0.5, at A and 1% low at 0.84 at C. With a reaction changing between A and C the output would appear to swing 0.327 absorbance unit instead of 0.340, an error of nearly 4%. Therefore, the requirements for a log converter are much more stringent when used in rate determinations than for equilibrium methods or direct absorbance measurements.

Fortunately this problem can be avoided. The derivative of absorbance with respect to time is equivalent to

$$\frac{d}{dT}(-\log_{10} \text{Trans})$$

$$= -\log_{10}e\left(\frac{d \text{ Trans}}{dT}\right)/\text{Trans} \quad (1)$$

This can be accomplished electrically by differentiating transmittance and reading the result with a servo system with transmittance as a reference rather than a fixed voltage as shown in Figure 6. Since this is a true electrical analog of the above mathematical function, the accuracy and absorbance range are now primarily limited only by noise level and the limits of the photocell and amplifiers. Three decades have been covered experimentally but the range was reduced to two decades to allow the use of concentration multipliers at up to twelve times the numerical rate value, while staying within the voltage range of readily available amplifiers.

The blank rate control shown provides an offsetting 0 to allow direct concentration readings on reactions whose concentration *versus* rate lines do not intercept zero.

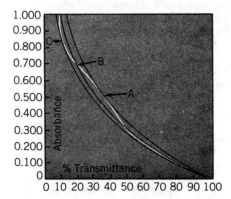

Figure 5. Simulated characteristics of a nonlinear element used in a log converter

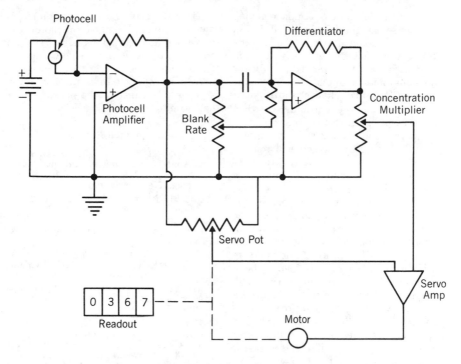

Figure 6. Basic diagram of the absorbance derivative circuit

The servo operates as a null seeker, driving the servo potentiometer to equal the differentiator output and displaying the potentiometer setting on a mechanical counter dial (the digital printout also arises directly from the position of this mechanical counter dial). As the transmittance value varies, the voltage per increment on the potentiometer changes proportionately, and over a two decade range will vary over a 100:1 ratio.

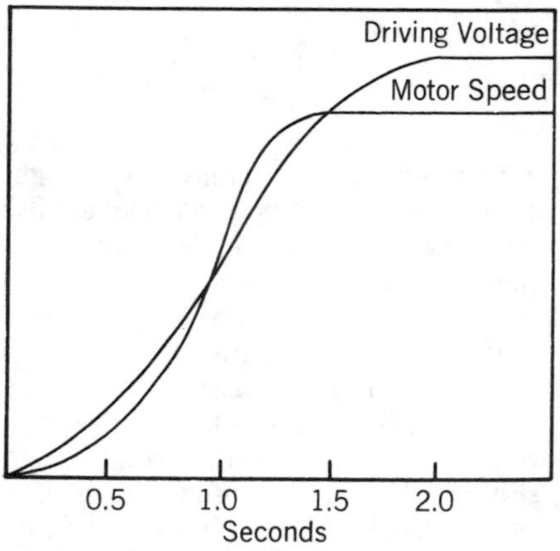

To maintain good servo action, the servo amplifier gain should also vary accordingly. Several automatic gain control systems were tested. The obvious approach would be to attempt to maintain the motor driving power proportional to the rotational distance off null. These systems all introduced additional problems such as disturbance of the amplifier null point, so a method was devised in which the motor drive was proportional to the length of time off null, regardless of voltage. This is shown in Figure 7. When an error is introduced the drive to the motor will slowly increase and the motor will start to move. After about two seconds the drive will reach maximum and the motor

Figure 7. Servo startup response to a step input

will run at full speed. Upon reaching the new setting the drive level will remain at full for about 100 ms to allow the application of full reverse drive to stop overshoot and then be reduced below the minimum required to move the motor. If a small change is introduced such as would be encountered when tracking a changing rate, the servo will reach the new reading while the motor is still turning slowly and will remove drive almost instantaneously. In practice this gives smooth following of a varying reading regardless of the voltage applied to the servo pot as long as the servo amplifier gain is sufficient to sense the smallest voltage increments required. Short duration noise peaks are attenuated since the servo will not build up drive rapidly enough to follow them. When resting on a reading with a high voltage applied to the servo potentiometer, as when working at low absorbances, the exact null will often lie between two adjacent wires on the potentiometer winding. The servo will then hunt back and forth between the two wires, but since this movement will be made as soon as the drive exceeds the minimum amount required to move the motor, the drive will not be high enough to sustain the type of oscillation usually encountered under these conditions. Since this movement will be a small fraction of the accuracy tolerance of the instrument it has no practical significance.

Suggested Reading

W. J. Blaedel and G. P. Hicks, "Advances in Analytical Chemistry and Instrumentation," Vol. 3. C. N. Reilley, Ed., Interscience, New York, 1964.

H. B. Mark, Jr., L. J. Papa, and C. N. Reilley, "Advances in Analytical Chemistry and Instrumentation," Vol. 2. C. N. Reilley, Ed., Interscience, New York, 1963.

H. B. Mark, Jr., G. A. Rechnitz, and R. A. Greinke, "Kinetics in Analytical Chemistry," Interscience, New York, 1968.

K. B. Yatsimerskii, "Kinetic Methods of Analysis," Pergamon Press, New York, 1966.

"Model 1011 Digecon System, A Semi-Automatic Spectrophotometric System for Chemical Analysis," Bulletin No. 10V-87, Sherwood Medical Industries, Inc., Brunswick Corp., 1831 Olive St., St. Louis, Mo. 63103.

COMMENTARY

by Ralph H. Müller

This stimulating article is an outstanding example of the extent to which deceptively small experimental factors can determine the reliability of a measurement. The system designed by the authors may well lead to a happy state of affairs in which we shall have fewer and better papers on reaction kinetics.

We find the authors' discussion on the stability of light sources most interesting and provocative. It is a perennial problem and one which deserves more attention than it usually receives.

One reason why highly stable light sources have not been developed has been the successful use of flicker, chopper, or double-beam systems which make a rapid intercomparison of intensities and, therefore, are independent of all but short-term changes in source intensity. The physics of the incandescent lamp is a vast and complicated subject. It has been under investigation throughout the century. During the thirties when we were studying precision photoelectric chemistry, our research students learned many simple but important facts, such as: 1) The best of lamp sockets are an abomination. Always solder stout leads to the lamp terminals and to the storage battery terminals as well. A heavy knife switch, preferably double pole, can be used to open the circuit. At low voltages, contact resistance variations of 0.01 ohm or less can account for large changes in intensity. 2) Always include a heat-absorbing filter in the optical system. The near infrared, which constitutes the major fraction of the total emitted radiation, is an efficient means of heating the photocathode and producing thermionic emission. 3) Perform the simple experiment [R. H. Müller, R. L. Garman, and M. R. Droz, "Experimental Electronics," Prentice-Hall, Inc., New York City, 1947. p. 70 (out of print, out of date)] of measuring the total phototube output as a function of lamp voltages. Over a considerable range, the data are accurately represented by the equation: $I = kV^n$ and a simple log–log plot of I $vs.$ V yields a straight line, the slope of which is n. For tungsten lamps, n lies between 3 and 4. For lamp operation at 6 volts, the attainment of 0.1% photometric precision requires voltage constancy to within a few millivolts. In electrical engineering handbooks, extensive tabulations are to be found listing coefficient and exponent in an expression of this sort for the calculation of numerous criteria, such as lamp life, lumens per watt, spectral distribution, etc. Voltages are expressed as $\left(\dfrac{V_o}{V_r}\right)^n$ in which V_o = operating voltage and V_r = recommended voltage.

For protracted operation, all lamp bulbs darken as a consequence of tungsten evaporation. This results in diminished transmittance and is only partially compensated by increased brilliance caused by the slight diminution of filament diameter. In the case of Hg arcs and glow tubes, the exponent n is much smaller and can ap-

proach unity and this is readily understood because one is now dealing with direct electronic excitation of atoms, whereas in the incandescent lamp the radiation is thermally excited.

The authors' observations on the effect of the bulb temperature are interesting and important and show the need for monitoring both current and voltage of the lamp. Also the use of baffles to minimize thermal convection currents is most useful. No doubt, many users of instruments have praised the manufacturers for efficiently ventilating lamp housings to keep things cool and comfortable without realizing this effect on lamp stability.

We hope no one will miss the full implications of the authors' observations on photomultipliers $versus$ photodiodes because the subject has come around full circle in the past thirty years. When Zworykin and his associates at RCA developed the photomultiplier, the claim was made that its signal to noise ratio was much higher than that of a high vacuum phototube followed by an amplifier providing equivalent gain. This claim was completely justified for the amplifiers of that era. Toward the end of World War II and for some years thereafter, a rash of papers appeared on performance improvement obtained by progressive decrease in dynode gain leading to the almost absurd limit in which the photomultiplier output was not much greater than that of a photodiode. Some of these papers could have been entitled "Gain in a Photomultiplier when it aint Multiplying." This series of Pyrrhic victories was similar to the interionic attraction theory of strong electrolytes in which one can, for example, calculate the equivalent conductance of potassium chloride at various concentrations and yet approach the true value asymptotically at infinite dilution—where there is no KCl. No one in his right mind disputes the intellectual achievement of calculating values in terms of fundamental constants and with few or no empirical factors, but in a transducer or instrument we want to know the answer while it is still working. As the authors point out, modern solid state amplifiers with field effect transistors (FET) in the input stage have changed the picture completely. As they indicate, two photocathodes of identical area and sensitivity exposed to the same level of radiation will emit the same number of photoelectrons. The respective outputs, in magnitude and signal to noise ratio, will depend upon the relative perfection of amplification —i.e., whether it is obtained by secondary electron emission at dynodes or by a conventional amplifier. The authors give data to show that photomultipliers show a higher noise level than a photodiode. In their special application, this was very important because, as they state, "in operating in a rate mode the derivative of a signal containing noise and drift is impossible to handle." This special requirement should put the question in proper perspective because the authors certainly do not imply any fundamental defect in photomultipliers. Their role in the measurement of extremely small levels of illumination is unique, especially if one is working under equilibrium conditions.

Further comment on this article seems superfluous; it speaks for itself and amply illustrates the importance of giving careful attention to the exact behavior of primary elements in an instrumental array. In many respects, there is a continuing need for the reexamination of our many transducers and primary elements. In our early studies in photoelectric photometry, we encountered two arguments. The first, "how can these complicated photoelectric-electronic methods compete with the simple Dubosque colorimeter?" That argument we handled by ignoring it. The Dubosque colorimeter has been a museum piece for many years. The second, "can one depend upon the linearity of response of a phototube?" We answered, to our own satisfaction at least, in the affirmative [R. H. Müller, R. L. Garman, and M. R. Droz. "Experimental Electronics," Prentice-Hall, Inc., New York City. 1947. p. 255 (out of print, out of date.)] It turned out that the use of a high vacuum phototube feeding a vacuum tube voltmeter with input compensation provided by a precision potentiometer and, therefore, using the electronics merely as a sensitive balance indicator could check linearity to better than 0.1%. The real problem was to reproduce levels of illumination with comparable precision. That is still true today. There is some promise in nuclear light sources such as the promethium-147 phosphor combination. These are widely used as railroad track and yard signals. In the scientific use, the random nature of radioactive decay is of no consequence because such large quantities of isotope are used that the statistics are excellent. The gradual decay of the isotope is slow; easily calculated or checked periodically and the same holds true for the slow destruction of the phosphor. The few which we have been able to examine turned out to be "hot as a pistol" and even affected nearby electronic circuitry. These are not insurmountable difficulties and these sources may become useful reference standards of illumination, eventually with NBS certification like a standard cell. Who knows? Let us hope that the story of precise standards of illumination will not remain a continuing paraphrase of Kipling's "The Light That Failed."

In his December editorial, Dr. Laitinen, in speaking of our new instrumentation offering, said "We are eagerly looking forward to an exciting new feature which we confidently expect will play an increasingly significant role in keeping analytical chemists abreast of developments in this most important area in years to come." As a consequence of the first contributions, we are sure that his fingers are to be found uncrossed and the future looks bright. Let us hope that some of our young investigators who are curious and genuinely inquisitive will direct their attention to pure instrumental problems and lend a hand to those gifted but overworked people on the research staffs of the large instrument companies. This is no chore for the investigator who feels compelled to write a paper every five or six weeks to insure tenure or promotion or funds. It requires an urge to do something new and better and not all are suited for the task. *Chacun a son métier.*

Infrared Fourier Transform Spectroscopy

New technology will greatly increase the capabilities of Fourier Transform

spectrometers so that the impact of these high resolution instruments will

be felt in all areas and applications of infrared analysis

By M. J. D. Low, New York University, New York, N. Y. 10453

IT IS A TRUISM that infrared spectroscopy has flourished within the past two decades and has found applications in a wide variety of analytical situations. A glance at the literature makes this as obvious as the connection between the spread of infrared spectroscopic techniques and the increased availability and improved quality of commercial spectrometers. It is also obvious and pertinent that the spread and success of "infrared" have so far been predominantly associated with spectrometers operating on the principle of dispersion. Such instruments, using prisms and/or gratings to disperse radiation mixtures, at present make up the bulk of the world's spectrometers.

It is important to note that dispersion spectrometers constitute just *one* general type of instrument system which can be used to measure spectra. Alternate, nondispersive systems based on filters or on interferometers have been available for quite some time, but these have not been used very much because the technology required to build adequate instruments was not available. However, this situation appears to be changing rapidly and, with the use of improved electronics and data handling techniques, there have been significant advances in the use of scanning interferometers for the measurement of infrared spectra. It is the purpose of the present article to outline the basis of interferometric or Fourier Transform spectroscopy, and to point out the advantages and disadvantages as well as the potential of this nondispersive method of analysis.

Optical Transformation

The end results produced by a Fourier Transform spectrometer and a conventional dispersion spectrometer are the same. Each type of instrument yields a spectrum—a plot of intensity *versus* frequency—but there are drastic differences in the way the spectrum is produced. In the conventional spectrometer, the polychromatic radiation must be fanned out or separated into bundles of (almost) monochromatic radiation because the frequencies are so high that a detector cannot discriminate between them; the detector can only respond to intensity, and not frequency. Then, by some suitable mechanical arrangement such as a tilting mirror, the series of bundles is slowly swept past a suitable detector. Each bundle can be termed a resolution element. The detector of the dispersion instrument then produces a simple electrical signal which is proportional to the intensity of the (almost) monochromatic radiation striking it. The spectrum produced in this fashion is thus the result of a series of radiometric measurements.

Conversely, the optical system of the Fourier Transform spectrometer does not disperse polychromatic radiation, but performs a frequency transform. The incoming signal is uniquely encoded so that the frequencies of the optically transformed signal fall within the time range of response of detectors, which can then pick up both the frequency and intensity information present in the signal. The "transformed" or coded polychromatic radiation falls on the detector throughout the entire observation period, and a complex signal termed an interferogram is produced. The original signal (the incoming polychromatic radiation) and the interferogram are complementary and constitute a Fourier pair. Consequently, by "transforming the interferogram,"— i.e., subjecting it to Fourier analysis (hence the term Fourier Transform spectroscopy)—it is possible to obtain intensity–frequency information about the original signal.

A precise description of the transforms requires fairly involved mathematics (see Selected Reading), but the way interferograms are produced and processed can be readily understood if one considers monochromatic radiation or a mixture containing just a few wavelengths.

Suppose a beam of monochromatic radiation of wavelength w enters a Michelson interferometer, shown schematically in Figure 1, and is divided into two equal components at a beamsplitter. Each is reflected at a mirror and returned to the beamsplitter, where their amplitudes will add. If the rays a_1 and a_2 arrive at the beamsplitter in phase, they will interfere constructively and a signal proportional to the sum of their amplitudes will be produced by the detector. The rays a_1 and a_2 will be in phase when the mirrors are equi-

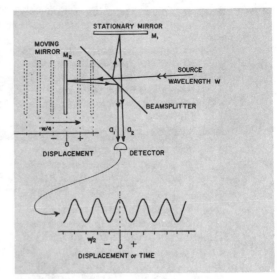

Figure 1. Optical system of Fourier Transform spectrometer

A simplified diagram of a Michelson interferometer (only the rays leading to the detector are shown, and a compensator plate was omitted). The mirror, M_1, is mounted on a suitable carriage and is mechanically or electrically moved. If monochromatic radiation enters the interferometer, the detector produces the signal shown below the optical system

distant from the beamsplitter. If M_1 is displaced by a distance $w/4$, the path of a_1 is changed by $w/2$, so that, on recombination at the beamsplitter, a_1 and a_2 will be 180° out of phase. There will then be destructive interference, and the detector output will be zero. This will also happen for all displacements which are odd multiples of $w/4$— i.e., for retardations of $\pm w/2$, $\pm 3w/2$, $\pm 5w/2$, and so on—where the plus and minus signs denote displacements on either side of the zero position. Similarly, constructive interference will occur at displacements of even multiples of $w/4$, with partial destructive interference occurring at in-between displacements. If M_1 were moved incrementally, the signal produced by the detector would fluctuate rhythmically and would, in fact, be a simple cosine wave.

Suppose that, instead of being displaced step-wise, the mirror is moved smoothly with a velocity V. As the cosine wave produced by the detector goes through one cycle if M_1 is displaced by a distance $w/2$, the frequency of the detector signal is then,

$$f = V/(w/2) = 2V\nu$$

Or, for a constant mirror velocity, there is a linear relation between the frequency ν (wave numbers) of the incoming monochromatic radiation and the frequency of the detector signal. For example, with a mirror velocity of 0.5 mm/sec, monochromatic radiation of 10 micron wavelength (1000 cm^{-1}, frequency 3×10^{14} Hz) will produce a detector signal of 50 Hz; for 5 micron radiation, $f = 100$ Hz. The amplitude of the low frequency signal is proportional to the intensity of the incoming monochromatic radiation. The extension to a radiation mixture follows from this: each frequency component of polychromatic radiation is made to undergo such a transformation in frequency and produces a detector wave of unique frequency. The signal or interferogram produced by the detector is then the summation of all such waves and is a complex signal like that shown in Figure 2. The center peak in the interferogram occurs when M_1 and M_2 are equidistant from the beamsplitter, so that all components reaching the detector have the same phase. The peak amplitude of the interferogram is proportional to the total energy in the incident beam. The peaks of smaller amplitude along each side of the central spike carry intensity and frequency information.

Data Reduction

Such an interferogram does not at all resemble a spectrum. It is, in fact, not a spectrum, but the Fourier Transform

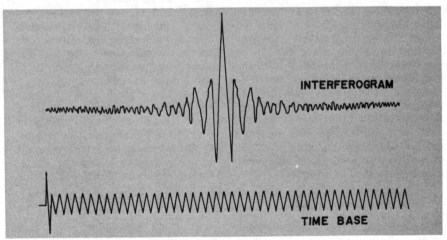

Figure 2. Interferogram of polychromatic source
The central spike corresponds to the position where the two mirrors are equidistant from the beamsplitter. The abscissa denotes time and, if the mirror velocity is constant, the mirror position. An electronically-generated time base is produced at the same time as the interferogram. The single sharp spike on the extreme left of the lower trace marks the start of the interferogram; the rest of the regular signal is used as time base. With fringe-referenced systems (see text), the time base is produced by the interferometer itself

of a spectrum, and as such carries the desired intensity–frequency information within it. In principle, all the spectral information may be extracted by performing the appropriate inverse Fourier transformation. In practice, reducing an interferogram has not been easily accomplished until recently. The computations involved in data reduction are so complex, lengthy, and tedious that manual computation is out of the question. This difficulty, as well as the absence of suitable analog conversion devices, was one of the factors which had kept Fourier Transform spectroscopy in obscurity. However, digital computer techniques and analog analysis are now readily available.

The interferogram can be recorded on magnetic tape, punched cards, paper tape, or directly in the core memory of a suitable computer. The Fourier Transform is then performed by digital means. This has the advantage that a number of corrections can be made, spectra can be ratioed, and so on. Fast computer programs are available for these computations, and the result is a normal spectrum. Alternately, the interferogram can be scanned by a wave analyzer. The latter is, essentially, an audio-frequency spectrometer, and is used to extract the intensity–frequency information from the interferogram so that a spectrum results. A variety of such devices ranging from "real-time" to slow scanning ones are available.

Analog data reduction is not as precise and nowhere near as flexible as the digital method, but has the advantage that data reduction and readout are rapid. The time delay between the experiment and the reception of the results, frequently encountered if somebody else's computer is used, can be avoided. In general, however, data reduction presents few problems and, with the increasing availability of small in-house or on-line computers, difficulties with data reduction are rapidly shrinking to the vanishing point (or at least, to a quite tolerable level).

Advantages

As outlined, Fourier Transform spectroscopy is a complex, indirect, and occasionally quite disagreeable way to measure spectra. One may wonder why one should bother with something so complicated. . . . However, there are advantages.

Remember that the interferometer brings about an optical transform. Essentially, radiation with frequencies of the order of 10^{14} Hz is processed and heterodyned to yield signals in the low audio-frequency range. The original signal is shifted "down stream" in frequency by about 11 orders of magnitude, thus drastically decreasing the demands on detector frequency response. The optical transformation moves the signal from a very high frequency range (one in which detectors cannot respond to frequency changes but respond only to intensity) to a low frequency range where response to both amplitude and frequency is possible. This leads to significant advantages in terms of signal-to-noise, based on the amount of signal which can be processed, and the time used to process it.

Dispersion or filtering is not required, so that energy-wasting slits are not needed. An interferometer has a relatively large, circular entrance aperture and relatively large mirrors (for example, in the mid-infrared range, a mirror diameter of 15 mm; in the far-infrared, several inches). The throughput, the amount of radiation which can enter the optics of the Fourier Transform spectrometer, is consequently quite large in comparison to that of a conventional spectrometer.

The second major advantage is not quite as obvious as the throughput gain, and also arises from the absence of the need to disperse or filter. In the conventional spectrometer, each radiation bundle or resolution element of the spectrum is scanned across the detector. Consequently, if there are M resolution elements, the intensity of each element is measured for only a fraction T/M of the total scan time, T. The signal proper (the intensity of an element) is directly proportional to the time spent observing it, while noise, being random, is proportional to the square root of the observation time. The signal-to-noise ratio (S/N) is then proportional to $(T/M)^{\frac{1}{2}}$. With the interferometer, however, the entering radiation falls on the detector, so that each resolution element is observed throughout the entire scan period, with the result that S/N is proportional to $T^{\frac{1}{2}}$. The improvement by the factor of $M^{\frac{1}{2}}$ for the case of the interferometer, termed Fellgett's Advantage, can be quite large under conditions of relatively high, or high resolution. Fellgett's Advantage is realized with detectors which are detector-noise limited —*i.e.*, as the signal level increases there is no increase in detector noise.

The advantage in S/N can be traded off for rapid response. The scan time can be decreased. A spectrum can be measured with a Fourier Transform spectrometer in the same time as with a conventional spectrometer, but with a better S/N, or in a much shorter time with an equivalent S/N.

Disadvantages

The sources of disadvantages range from the economical (at present most Fourier Transform spectrometers are fairly expensive in comparison to conventional instruments) to the psychological (Fourier Transform spectroscopy is new and therefore suspect; only the term spectroscopy sounds sort of familiar). One disadvantage, the necessity of a fairly complex data reduction procedure, has already been noted. Others stem from the generally complex nature of the instrumentation.

The optical system of an interferometer is simple, but it must always be correctly and precisely adjusted. Unlike a dispersion spectrometer, which will yield poor but still usable spectra if the instrument is misadjusted or mistreated, a misaligned Fourier Trans-

form spectrometer yields nothing. However, the available commercial instruments appear to be quite stable and trouble from this source does not appear to be at all serious. The performance of the Fourier Transform spectrometer is also more dependent on the quality and performance of the electronic components than is the conventional spectrometer.

Instrumentation

A rather large variety of spectrometer systems have been devised, mostly based on the Michelson interferometer. They differ little in principle, but vary greatly as far as the optical, mechanical, and electronic components are concerned. The spectral ranges covered depend on the nature of the beamsplitter and detector (and, of course, on the associated electronics). Commercially available instruments now cover the range from 40,000 cm^{-1} to 10 cm^{-1} (0.25 to 1000 microns). Scan times vary from about 1/10th second to several hours. Some instruments incorporate a small computer which is used for recording the interferogram and performing the necessary data reduction; others have a time-averaging computer used for multiple-scanning in order to enhance the signal-to-noise ratio when very weak sources are observed; still others are equipped with an analog data conversion device. All are quite sensitive.

Applications

There is little point in using a Fourier Transform spectrometer to make a measurement which can be handled quite well with a conventional spectrometer. Using a Fourier Transform spectrometer must be worthwhile. Consequently, much of the literature describing applications of Fourier Transform spectrometers deals with measurements which are either very difficult or impossible to make with conventional dispersion instruments. Generally, energy-limited situations have been involved, so far—*i.e.*, where the amount of radiation emitted by the. source or transmitted, reflected, or emitted by the sample is very small. Fourier Transform spectrometers have consequently performed very well in the far infrared. They have also been very useful in astronomy for recording the spectra of stars and planetary atmospheres; on a more mundane level, such remote sensing was used for the detection of SO_2 from smoke stacks, and promises to be useful in air pollution studies. The multiple-scanning instruments also appear to have a great potential utility in the fingerprint region for a number of diverse applications including gas chromatography, surface effects, and the examination of micro samples. One example will suffice to point out the relatively high sensitivity, speed, and versatility as well as the potential of such instruments.

During a study of diborane–trimethylboron equilibria it became desirable to characterize various compounds isolated from the equilibrium mixtures. Consequently, spectra were run with a Perkin-Elmer Model 521 instrument; spectra obtained with 30-minute scans are shown in Figure 3.

However, the changes in the spectra with time indicated that the sample was decomposing and suggested the use of a Fourier Transform spectrometer. The simple, mainly self-explanatory setup shown in Figure 4 was used in conjunction with a Block Engineering Co. mutiple scanning spectrometer. A carefully prepared, pure sample of the order of 10^{-7} mole was separated from the equilibrium mixture and immediately frozen out in the cell. The dewar was removed, and 50 consecutive scans, each of 1-second duration, were made when the sample had vaporized. The cumulative signal yielded spectrum A of Figure 5. Spectra B to G were obtained with a second sample of the same compound by similar methods, after the sample had been at or near room temperature for various periods of time.

The changes in the spectra speak for themselves. By the time the dispersion instrument had recorded the spectrum, the composition of the sample had already changed. There was even relatively little correspondence between the first and last portions of the spectrum. This and other work suggested that much of the relatively voluminous in-

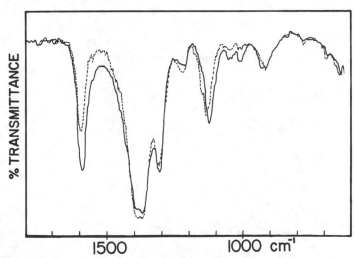

Figure 3. Spectra of cis-1,2-dimethyldiborane
Solid line: first scan. Dashed line: second scan of same sample. Each scan was about 30 min, measured with a Perkin-Elmer Model 521 spectrophotometer

17

frared literature on alkyldiboranes was suspect, because it was probable that many of the spectra on which band assignments were based had been run with impure, partially decomposed compounds. The Fourier Transform spectrometer was thus extremely useful in this study. The other side of the coin is that it was not feasible to make new band assignments because the resolution of the available instrumentation was not good enough.

Potential

Infrared dispersion spectrometers are well entrenched and, in view of some of the disadvantages of the Fourier Transform method, one may well wonder, Where do we go from here? Rapidly changing techniques point to an answer.

For example, interferometers are generally used as single-beam spectrometers, and consequently the spectra have the inherent bad qualities obtained from single-beam operation, such as bands superimposed on a sloping background, and consequently must be corrected. However, dual-beam operation is also possible, and methods for using a single interferometer to yield spectra in which the background has been removed are developing rapidly. Rapid changes in data handling procedures have already been mentioned; further improvements seem likely. Of greater importance is the matter of spectral resolution, particularly for the type of multiple-scanning instruments used in the fingerprint region.

The resolution is mainly a function of the length of the sweep of the mirror and of the precision to which the position of the mirror is known as function of time—*i.e.*, with a constant velocity of mirror motion—the mirror should move as far and as smoothly as possible. If the velocity is precise, the mirror position is known, and consequently one can make an electronically generated time coordinate for the interferogram. These requirements lead to severe mechanical problems, with the result that the resolution has been relatively poor, for example, about 18 cm^{-1} for an instrument covering the 2500–250 cm^{-1} range. However, new ex-

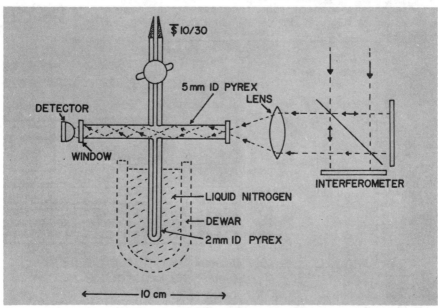

Figure 4. Experimental arrangement

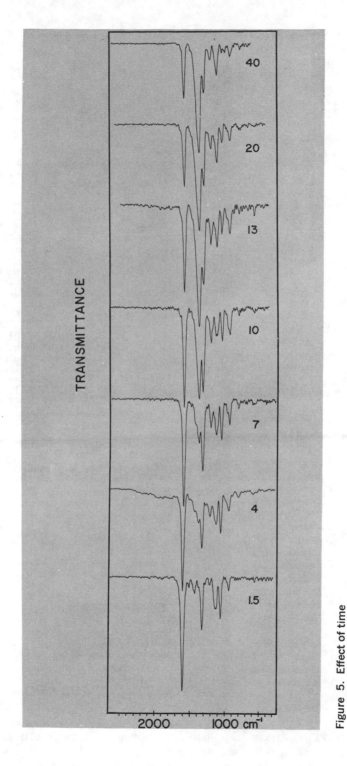

Figure 5. Effect of time

A trapped sample of cis-1,2-dimethyldiborane was allowed to vaporize, and spectra were recorded with a Block Engineering Co. Model 200 Fourier Transform spectrometer. The numbers indicate the time in minutes the sample was at or near room temperature. Spectrum A resulted from 50 1-second scans; 100 scans were taken for other spectra

perimental mirror drives have been developed, including drives in which the sweep length has been greatly increased but the linearity of the velocity has not been very much improved. In such a case, the interferometer itself is used to generate its own time scale: in addition to processing infrared radiation, the interferometer is used to process a very narrow, almost monochromatic line from a source such as a neon lamp or a laser. This produces a discrete signal which is time-locked to the mirror motion and hence to the interferogram. If the mirror velocity changes, the time-base signal changes, so that it is possible to derive a precise time coordinate for the interferogram and consequently improve the resolution. Such "fringe-reference" systems have been built, and will probably become commercial in the very near future. Fourier Transform spectrometers with 1 cm^{-1} and even 0.1 cm^{-1} resolution in the fingerprint range will become available. In view of the new technology, it is probable that their cost will be of the same order of magnitude as that of high quality dispersion instruments. What then?

The impact of the high resolution Fourier Transform spectrometers will be felt in all areas of infrared analysis. The extrapolation from existing experimental prototypes is not great: the restriction of interferometers to low energy situations will be greatly decreased, and the emphasis of applications will shift to include all types of infrared spectral measurements. The inherently high sensitivity of the Fourier Transform spectrometer will make it possible to record high quality infrared spectra in a matter of seconds; at worst, in a matter of minutes. Fourier Transform spectroscopy will therefore become profitable in all manner of scientific and analytical infrared applications.

Selected Reading

The principles of Fourier Transform spectroscopy have been treated in detail by G. A. Vanasse and H. Sakai, in Chapter 7 of "Progress in Optics," Vol. VI, E. Wolf, Ed., John Wiley & Sons, Inc., New York, 1967.

Short, general articles

M. J. D. Low and I. Coleman, Spectrochim. Acta, 22, 396 (1966). M. J. D. Low, J. Chem. Ed., 43, 637 (1966). W. J. Hurley, J. Chem. Ed., 43, 236 (1966). G. Horlick, Appl. Spectr., 22, 617 (1968).

Remote Sensing

M. J. D. Low and F. K. Clancy, Environ. Sci. Techn., 1, 73 (1967); R. Beer, Physics Teacher, 6, No. 4 (1968); D. M. Hunten, Science, 162, 3B (1968).

Miscellaneous Applications

M. J. D. Low, Anal. Letters, 1, 819 (1968); Appl. Spectr., 22, 463 (1968); Appl. Optics., 6, 1503 (1967); M. J. D. Low and I. Coleman, Appl. Optics., 5, 1453 (1966); M. J. D. Low and S. K. Freeman, Anal. Chem., 39, 194 (1967); J. Ag. Food Chem., 16, 525 (1968); M. J. D. Low and J. C. McManus, Chem. Commun., 1967, 1166; M. J. D. Low, R. Epstein, and A. C. Bond, Chem. Commun., 1967, 226; J. Chem. Phys., 48, 2386 (1968). C. H. Perry et al., Appl. Optics, 5, 1171 (1966); J. Appl. Phys., 37, 1994 (1966); Spectrochim. Acta, 23A, 1137 (1967); A Anderson and H. A. Gebbie, Spectrochim. Acta, 21, 883 (1965); D. M. Adams and H. A. Gebbie, Spectrochim. Acta, 19, 925 (1963).

References

(1) M. J. D. Low, J. Chem. Educ., 47, A163, A255, A349, A415 (1970).

(2) M. J. D. Low, Naturwissenschaften, 57, 280 (1970).

(3) S. J. Dunn and M. J. Block, Amer. Lab. (October 1969).

(4) M. J. D. Low, J. Agr. Food Chem., 19, 1124 (1971).

(5) M. J. D. Low, A. J. Goodsel, and N. Takezawa, Environ. Sci. Technol., 5, 1191 (1971).

(6) A. J. Goodsel, M. J. D. Low, and N. Takezawa, ibid., 6, 268 (1972).

(7) M. J. D. Low and S. K. Freeman, J. Agr. Food Chem., 16, 525 (1968).

(8) M. J. D. Low, A. J. Goodsel, and H. Mark, in "Molecular Spectroscopy 1971," p 383, Institute of Petroleum, London, England, 1972.

(9) M. J. D. Low, H. Mark, and A. J. Goodsel, J. Paint Technol., 43, 49 (1971).

(10) R. Yang and M. J. D. Low, J. Colloid Interface Sci., in press (1972).

Since the article was written, almost four years ago, technology has advanced rapidly, and the optimistic note of the last part of the article has not soured. Triglycine pyroelectric detectors, which have a flat response into the far ir, have become available, mirror drives have improved, and dedicated minicomputers have been mated to interferometers. The result has been the marketing of highly sophisticated (if somewhat expensive) Fourier transform spectrometer systems (1, 2). Of particular interest is the Digilab FTS-14 system which, with one of several available beamsplitters, operates in the fingerprint region (1–4). The fully computerized FTS-14 has been used for some exotic measurements but has found ready application to analytical situations such as the measurement of spectra of gas chromatography effluent (4), the interaction of air pollutants with solids (5, 6), small samples (7–9), coatings (8), and adsorbed layers by use of internal reflection techniques (10). The high sensitivity and speed of the instrument, as well as its resolution to 0.5 cm^{-1} (0.1 cm^{-1} with a special drive system), make it a "natural" for analytical applications. It seems, indeed, that as more of such instruments become available, Fourier transform spectroscopy will be profitable in all manner of scientific and analytical situations.

COMMENTARY

by Ralph H. Müller

Analytical chemists will be grateful for Professor Low's outline and discussion of infrared Fourier Transform spectroscopy. In so carefully outlining the advantages as well as the present limitations of the interferometric approach to infrared spectroscopy, we believe the conclusion is inescapable that these techniques will seriously rival spectrometers operating on the principle of dispersion, however elegant and highly developed the latter class of instruments happens to be. We are happy to note Dr. Low's observation of the "connection between the spread of infrared spectroscopic techniques and the increased availability and improved quality of commercial spectrometers." We think it is clearly demonstrable that in all ages, the amount of scientific information was at all times governed by the techniques, resources, and instruments available for the problem. At present, the vast amount of publications is ample proof that most information has been obtained with elegant instruments, most of them automatic or semi-automatic. It is probably demonstrable that a smaller proportion of all investigators who publish today are concerned with the development of new methods, new techniques, and new instruments than some thirty years ago. The information explosion does not run parallel to the population explosion—it exceeds it and not by working overtime at time-and-a-half remuneration. We suspect that sharp inflections in the growth curve arise about once every twenty years.

Someone then gets a Nobel Prize and during the new cycle publications increase at a rate proportional to the improvement in equipment developed for the new phenomenon. We are reluctant to dwell upon the boredom which sets in after the 1961st paper on the phenomenon has appeared.

Considering the question of improved resolution, we note that this is a function of the length of the sweep of the interferometer mirror and the constancy of its velocity. To achieve the highest precision is an undertaking requiring the best resources of physics and engineering, and the effort has been going on for about three quarters of a century. When Henry Augustus Rowland, at Johns Hopkins, built his first engine to rule diffraction gratings, he spent several years in perfecting the lead screw used to advance the diamond cutting tool. This was done by running a split-nut, lapped with fine abrasive, back and forth along the screw in order to eliminate periodic errors in pitch. It is not quite certain who first suggested the use of an interferometer (counting of interference fringes) as a criterion of perfection. As a rough benchmark of the dimensions which are involved, the distance between fringes for the median wave length of the sodium D lines (5893 Å) is about 11 microinches. Several decades ago, Dean George Harrison at MIT developed the Compander, an ingenious system for the automatic and continuous correction of tool position for second order correction of residual screw errors. With an interferometer mirror placed on the cutting tool carriage, the moving fringes were detected photoelectrically

—giving rise to a sinusoidal voltage which was constantly compared with a similar voltage derived from the rotation of the feed screw. The two sinusoidal voltages were adjusted to equal amplitude but for corrections, the phase relations were utilized. A slight error in pitch would cause a phase shift and this difference, after amplification, was used to advance or retard the cutting tool position momentarily. One hundredth of a fringe displacement could be measured with this system, which corresponds to a mechanical displacement of $\lambda/200$. The source of light for the interferometric monitor was the 5461 Hg line obtained from an arc using the ^{208}Hg isotope. This had been developed by Meggers, and in other aspects, the line is of such high spectral purity that it may well serve as an absolute alternate standard of length. It would not surprise the chemist to be told that the achievement and maintenance of such high metrical precision required the control of the partial pressure of CO_2 and water vapor in the air surrounding the dividing engine because these determine the refractive index of the air and, therefore, the velocity of light in the measuring arm. These factors, along with temperature fluctuations and vibration, determine the precision which can be attained. Present day resources in electronics, computing, and in engineering are vastly better than when these requirements were established. Probably the technique will be competitive or superior to conventional methods when it can provide equal or greater precision per unit time, cost, and operation ease. The principles of interferometry have other, including unexplored, uses.

A New Detector for Spectrometry

The use of this device in the vacuum ultraviolet is shown. However, the technique can feasibly be applied in other spectral regions and in mass spectrometry

by A. Boksenberg
Department of Physics, University College London

This article discusses a new technique for reading spectral information in the form of a spatial distribution of charge, at present under development in the Physics Department of University College London [A. Boksenberg, R. L. F. Boyd, J. C. Jones, *Nature*, **220**, 556 (1968)], that promises to be useful in ultraviolet spectrometry, mass spectrometry, and possibly other applications too.

A charge image is first recorded on the surface of an insulating layer by direct photoemission, photoconduction, or other means, according to the character of the information to be detected.

For example, to record a spectrum in the vacuum ultraviolet, an insulating photoemitter can be conveniently used in an otherwise conventional spectrograph; then the unbalanced positive charges that remain after the photon-induced surface emission of electrons, constitute the recorded image and this is a direct analog of the spectral pattern to which the layer is exposed.

To extract the information thus recorded, a reading process is employed that is basically similar to the charge measuring technique used in the vibrating reed (dynamic capacitor) electrometer, but here added mechanical sophistication is required for the measurement of charge images.

In the vibrating reed electrometer, a quantity of charge is determined by mechanically varying the value of an associated capacitor and measuring the resulting alternating current with a sensitive amplifier. In the new application described here, a charge image residing on the surface of a thin insulating layer backed by a grounded conducting plate is read out, element by element, by means of a fine, closely scanning, vibrating, conducting probe connected to a current amplifier, the direction of vibration being normal to the surface.

During the entire reading operation the surface of the layer is never touched by the probe, so the image remains intact and may repeatedly be read without detriment. But a more important consequence of nondestructive reading is that the probe can remain over each effective image element sufficiently long to enable the inherent charge level to be measured as accurately as desired by use of the narrow-band lock-in amplifier technique. In other words, the image can be read out without substantially degrading it with noise introduced in the reading process.

This is in direct contrast to the conditions obtaining in a conventional television camera tube—which uses electron beam readout—where the image is necessarily destroyed as it is read; a normal wide-band amplifier must then be used so the signal-to-noise advantage of a lock-in amplifier is lost. Nevertheless, this loss can be effectively reduced by intensifying the image before recording, although this is usually at the cost of considerable complexity and a severe limitation in dynamic range.

However, the real value of the new image detecting technique follows from the total avoidance of electron optics both in the reading and recording processes. This allows the image area to assume almost any desired extent and contour—as is true for photographic film—and makes the system particularly appropriate for use in spectrographic instruments where focal surfaces are often long and curved. In comparison, the need for accurately defined electric and magnetic fields for electron acceleration and focussing in image intensifiers and television camera tubes severely constrain the permissible geometrical extent of their sensitive areas.

Figure 1 shows, in schematic form, the various practical steps of the detection procedure for use in the vacuum ultraviolet. First, the image is recorded by exposing a highly insulating photoemitting layer through an open mesh held at a high positive potential relative to the conducting backplate. This mesh collects the emitted photoelectrons while recording and is subsequently drawn aside, or otherwise removed, for reading. In general, the storage layer is not homogeneous but consists of a dielectric spacer layer with a thin surface film of photoemitter, as indicated. Such a surface layer is not difficult to find for detection in the vacuum ultraviolet since many insulators are also efficient photoemitters in this region, notably the alkali halides. Probably the best of these is cesium iodide, having near unit quantum efficiency (the ratio of the numbers of emitted electrons to incident photons) over the range 10–1000 Å.

In comparison, the appropriate photographic emulsions have efficiencies between one and two orders of magnitude lower.

Next, the reading process: since spectra essentially contain information in one dimension only, the scanning action can be correspondingly one-dimensional and the probe knife-edged in form and held normal to the dispersion direction.

To get a better understanding of the reading process, we define the image density in the vicinity of the probe at any instant as σ and the respective capacitances of the probe and the conducting backplate to an area a of surface beneath the probe $C_1(t)$ and C_2, the former being a function of time because of the impressed vibration. For simplicity, we neglect fringe effects and assume the potential difference between probe and backplate is zero. Then, if the instantaneous charge on $C_1(t)$ is $q_1(t)$ and the induced signal current is $i_1(t)$:

$$q_1(t) = \frac{C_1(t)\sigma a}{C_1(t) + C_2}$$

$$i_1(t) = \frac{d}{dt}[q_1(t)] = \frac{C_2 \sigma a \dfrac{dC_1(t)}{dt}}{[C_1(t) + C_2]^2}$$

Thus, the signal current is periodic with the probe vibration and its amplitude is proportional to the surface charge density and to the rate of change of $C_1(t)$—i.e., to the frequency of vibration. Also, it can be shown that optimum signal generation occurs when C_2 and the mean value of $C_1(t)$ over one cycle are approximately equal.

The current is passed to a sensitive current pre-amplifier, then to a lock-in amplifier where phasing reference is provided by the vibrator oscillator, and finally to a recorder. The sensitivity of the system is such that a charge equivalent to a few tens of electrons can be detected with certainty using a vibration frequency of 10 kHz

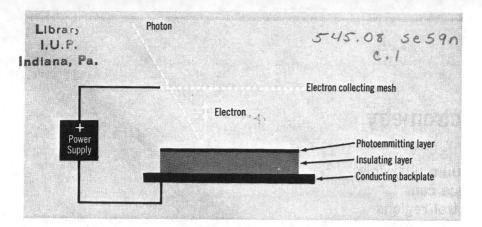

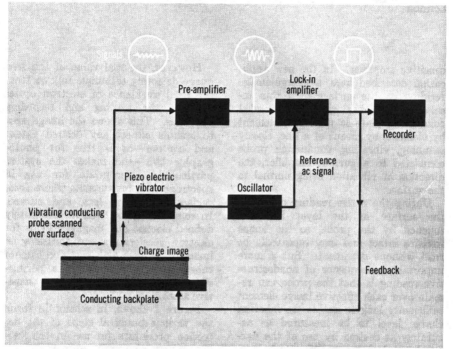

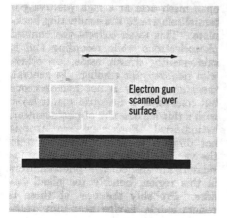

Figure 1. Schematic showing the practical steps in the detection procedure in the vacuum ultraviolet. Top: Recording the image; Middle: Reading the image; Bottom: Erasing the image

25

and a reading rate of ten image elements a second.

In practice, the probe is vibrated by means of a piezo-electric crystal operating near resonance. Scanning may be achieved by mounting the probe unit on a movable carriage which is borne against a reference surface accurately related to the surface of the storage layer. Alternatively, the probe unit may be fixed and the storage layer moved.

The spatial resolution of the system is naturally dependent both on the geometrical configuration of the probe and its mean separation from the surface. The probe width and mean separation both need to be in the region of 10 microns to achieve a resolution comparable to that obtainable photographically. Furthermore, the second harmonic component of the signal gives a better resolution than the first, since it is generated predominantly when the probe is closest to the surface during each vibration cycle. For optimum resolution performance, the probe must also be electrically blinkered by means of carefully designed guard electrodes—not shown in the diagram—in order to better confine its "view" to the region of the storage surface in its immediate vicinity.

To maintain the resolution—and therefore, the probe mean separation—constant over the whole of the scan for a nominal separation of only a few microns is a difficult mechanical undertaking but can be adequately achieved if aided by servo control. This also has the effect of stabilizing sensitivity, a feature which likewise becomes increasingly necessary with decreasing separation. Such servo action may be realized simply by applying a correcting bias to the piezo-electric vibrator to stabilize the mean separation and, if necessary, also controlling the vibrator drive oscillator to stabilize amplitude. One way of measuring the two parameters to be controlled is by the use of a conducting strip deposited along the edge

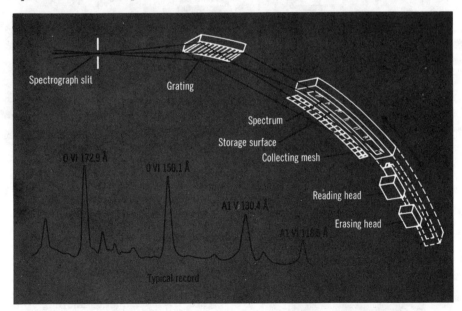

Figure 2. Vacuum spectrograph with image detector in place of photographic plateholder

of the storage surface and an overlaying subsidiary probe mounted on the vibrator. By maintaining a constant potential difference between them and measuring the first and second harmonic components of the resulting signal current, the mean separation and vibration amplitude can be uniquely determined. However, in applications where the passive mechanical performance of the system is adequate for resolution but not for maintaining a sufficiently constant sensitivity, the latter may be acceptably stabilized by the usual method of negative feedback, as shown in the diagram.

After reading, the image is erased for a fresh exposure by scanning it with a low velocity electron beam, a procedure that effectively replaces the electrons that were emitted from the surface during recording.

As a first commercial application of the image detecting system, work is proceeding to apply it in place of the photographic plateholder in a Hilger and Watts two-metre grazing incidence vacuum spectrograph (normally working in the range 5–950 Å), in collaboration with the U.K. Atomic Energy Authority, Hilger and Watts Ltd., and the U.K. National Research Development Corporation. The main advantages here to be gained over the use of film are: a high detecting efficiency, linearity of response, one to two orders of magnitude increase in dynamic range and an electronic readout that obviates the need to change exposed plates—obviously a great operational convenience for a vacuum instrument.

The basic components of the experimental system are illustrated in Figure 2. As described above, the sequence of operations consists of (a) conversion of the ultraviolet spectrum image to a static charge image, (b) electromechanical readout and (c) erasure. In the current trials, the collecting mesh, reading and erasing heads are fixed and the storage layer, which is coated on to a large curved glass block,

is moved to the collecting mesh for recording and then slowly scanned past the reading and erase heads at the completion of the exposure. The wavelength range covered in this trial instrument is 5–250 Å. A portion of a typical record obtained in preliminary and as yet unoptimized experiments is included in the figure and shows spectral features in the region 100–200 Å obtained with an air-filled flash tube source. The quality of the storage layer is such that it is possible to retain spectra as charge images for many days after exposure.

As an additional function, the exposure can be electronically controlled by appropriately gating the potential of the collecting mesh relative to the backplate to alternately inhibit and encourage photoemission as required (the backplate is a metallic film deposited on the glass support block before applying the storage layer). Although this in itself is probably an unnecessary complication, it does lead to an interesting extension of the basic detection technique, namely, in time-resolving transient spectral phenomena. This can be achieved by constructing a collecting mesh in the form of a number of longitudinal sections insulated from one another. During exposure, the applied potential is gated from negative to positive and back to negative for each in turn, thus recording the spectral information as a stack of adjacent time-staggered spectrum strips. To read the composite image, it is envisaged that the piezo-electric vibrator would contain a separate probe system for each spectrum strip.

The use of the reading technique for detection in other spectral regions is inherently more difficult than in the vacuum ultraviolet, largely because of problems associated with the use of suitable layers. In the near ultraviolet and optical regions, a photoconductive rather than photoemissive approach is currently being studied, the former being preferred since it avoids the inconvenience of a vacuum envelope not

27

now needed optically.

Mass spectrometry is another possible area of interest for application of the technique. Here, ions conventionally separated and then received on an insulating layer could be directly read out without further processing. Alternatively, charge multiplication by secondary emission at the storage surface could provide a useful degree of intensification to aid in the detection of very low levels of impurities.

Radiative Detection of Nuclear Magnetic Resonance

NMR/RD uses nuclear decay products—beta or

gamma radiation— to detect resonances. This technique is

extremely sensitive as it is a microscopic detection scheme.

Chemical applications are yet to be explored

by D. A. Shirley
Department of Chemistry and Lawrence Radiation Laboratory
University of California, Berkeley, Calif.

An exciting development taking place in chemistry today is the widespread evolution of techniques for studying matter on a truly atomic scale. Chemists are naturally motivated to explore new, high-sensitivity techniques because chemical thought and discussion is most often directed toward matter at the atomic level, while classical measurements usually deal with macroscopic properties. While statistical mechanics help to resolve this problem, it seems clear that the acquisition of much information about the behavior of matter on an atomic scale awaits the development of experimental methods that are sensitive to small numbers of atoms.

Nuclear magnetic resonance (NMR) is an area for which these comments are especially appropriate. Its very high precision and wide applicability make NMR a powerful method. Unfortunately conventional NMR has relatively low sensitivity, and a sample of macroscopic dimensions is required to obtain an observable signal. This is true only because resonance is detected through a change in the macroscopic magnetization of nuclei in the sample.

Nuclear magnetic moments are small; therefore many nuclei are required in order to produce a sizable magnetization. The resulting low sensitivity is a major weakness of conventional NMR.

This article describes a small, new area of very unconventional NMR in which the sensitivity problem is avoided by utilizing a *microscopic* detection scheme. The area is termed "radiative detection of nuclear magnetic resonance," or NMR/RD (1). Applicable to metastable nuclear states, NMR/RD employs nuclear decay products—beta or gamma radiation—to detect resonances. Because nuclear counters are sensitive enough to record individual nuclear events, a resonance can be confirmed whenever it causes a statistically significant change in the counting rate. Only about 10^6 nuclei are usually required to achieve this result, making NMR/RD the most sensitive type of NMR by many orders of magnitude. Unfortunately the range of application of NMR/RD is apparently very limited.

In the next section the distribution of radiation from oriented nuclei is discussed, and three methods for orienting

29

nuclei are discussed briefly. The third section deals with the power requirements for producing observable resonances in oriented nuclear spin systems. Finally in the fourth section the results to date are summarized and the outlook for NMR/RD is surveyed.

Radiation from Oriented Nuclei

The only way in which NMR can appreciably affect the radiations from oriented nuclei is to alter their angular distributions. In an ordinary assembly of radioactive nuclei, however, the distribution of radiation is spherically symmetrical, and an impressed radiofrequency field can have no effect. For an NMR/RD experiment to be feasible a very special kind of sample is required: a sample whose nuclei are *oriented*. By this we mean that the nuclear spins must be "pointed" along a certain direction in space. The nuclei may be *polarized*, with their spins parallel, as in a ferromagnet, or they may be only *aligned*, with spins oriented in direction but not in sense, as in an antiferromagnet.

Each nucleus by itself has a radiation probability pattern which is spatially anisotropic, analogous to that of an antenna. For an assembly of unoriented nuclei the individual patterns add incoherently, but when the nuclei become oriented the intensities of these individual probability patterns add coherently, and the entire assembly exhibits an anisotropic radiation pattern. An illustrative example is shown in Figure 1. We consider a nucleus with a spin-zero ground state and a metastable state of spin 1. Application of a magnetic field H resolves the degeneracy of the metastable state, giving three magnetic substates with magnetic quantum numbers $M = 1, 0,$ and -1. A nucleus in the metastable state can emit a dipole γ ray and decay to the ground state, and the probability for this process is unaffected by H. The substates have different γ-ray angular distributions, however. Suppose that a

counter placed near the sample registered A counts/sec before H was applied. With H present the total counting rate is still A, but the contribution from each substate depends on θ, the angle between the γ-ray direction and H. The $M = 0$ substate contributes $(A/2)\sin^2\theta$, and, while each of the $M = \pm 1$ states contributes $(A/4)(1 + \cos^2\theta)$. Note that the total counting rate is still isotropic, as $\sin^2\theta + \cos^2\theta = 1$.

Now suppose that the nuclei are aligned along H. The quantum-mechanical description of this situation is simple: more nuclei are in each of the two states $M = \pm 1$ than are in the $M = 0$ state. A $\cos^2\theta$ term appears in the radiation intensity, which is thereby enhanced along the H direction relative to perpendicular directions. If radiofrequency radiation is subsequently used to equalize the substate populations, the radiation anisotropy will decrease or disappear: this is the essence of NMR/RD. Most real cases are more complicated, but the principles are still the same: oriented nuclei radiate anisotropically, and NMR decreases this anisotropy. Let us now consider the three common methods for orienting nuclei.

The simplest method is to apply a strong magnetic field H, splitting the M substates by γH (where γ is the nuclear gyromagnetic ratio), and to lower the temperature. Populations of adjacent substates are related by the Boltzmann factor $e^{-\gamma H/kT}$. For $kT \cong \gamma H$ appreciable population differences, and nuclear orientation, can occur. Very low temperatures ($\sim 10^{-2}$ °K) are required to orient nuclei in this way, but the method can be applied to nearly every element in the periodic table. Since it takes time to reach such low temperatures, only metastable states with lifetimes of 10^3 sec or more can usually be studied in this way. If magnetic resonance is performed, the method is termed "NMR in oriented nuclei" or NMR/ON.

The second technique for obtaining

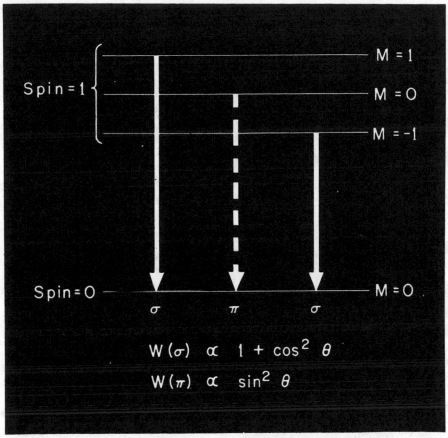

Figure 1. Level scheme for a nucleus with two levels, of spins 1 and 0, connected by a dipole transition. The σ, π notation is taken from atomic spectroscopy

oriented nuclei is more subtle. It is based on the *angular correlation* that exists between two radiations emitted successively from the same nucleus. If the two radiations (γ_1 and γ_2 in Figure 2) are detected in two separate counters, and recorded only when they are in coincidence, then this coincidence counting rate $W_{12}(\theta)$ will vary with the angle θ between the propagation directions of the two γ quanta. This result may be understood by analogy to the thermal nuclear orientation method described above. After a nucleus has emitted a γ_1 quantum in a certain direction it is oriented with respect to that direction. Radiations emitted sub-sequently will be distributed anisotropically about that direction—i.e., the distribution will depend on θ. Coincidence counting is employed only to select nuclei oriented in the γ_1 direction. When a magnetic field H is applied and NMR is performed on the intermediate state, the method is termed "NMR detected by perturbed angular correlations," or NMR/PAC. It is applicable to metastable states with lifetimes in the 10^{-8}–10^{-6} sec range, and these states must be preceded by radiative transitions.

The third way to orient nuclei is by producing them in an oriented configuration, through nuclear reactions at an

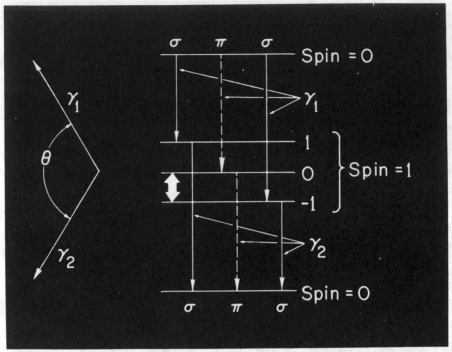

Figure 2. Correlation angle (left) and level diagram for an angular correlation experiment. For $\theta = 180°$ only σ–σ coincidences appear. An NMR transition (heavy arrow) in intermediate state can alter substrate populations and affect correlations

accelerator. When a particle in the beam collides with a target nucleus in a nuclear reaction, there is most often a sizable orbital angular momentum between the two particles. The angular-momentum vector is perpendicular to the beam direction, so the reaction-product nuclei are oriented with their spins perpendicular to the beam, and they radiate anisotropically. Nuclei thus produced in nuclear reactions are candidates for NMR if their lifetimes lie in the range 10^{-6}–10^3 sec. This method is abbreviated NMR/NR.

Power Requirements for NMR/RD

Turning now to the other side of the NMR/RD methods, we need a description of the magnetic resonance phenomenon that will allow us to determine the requirements for a successful

NMR/RD experiment. Conventional NMR is often discussed using perturbation theory, but that approach is not suitable for NMR/RD because these experiments require that nearly every metastable nucleus undergo a transition in a very short time. The approach sketched below is ideal for NMR/RD experiments, and it is actually simpler than the perturbation-theory method.

A particle with magnetic moment $\vec{\mu}$ in a magnetic field H will precess about H according to the equation

$$\frac{d\vec{\mu}}{dt} = \gamma\vec{\mu} \times \vec{H}.$$

The precession frequency v is given by $hv = \gamma H$, where h is Planck's constant. Resonance is brought about at frequency v by applying a radiofrequency field $H_1(t) = 2H_1\cos 2\pi v t$ perpendicular to H. The problem is greatly simplified by transforming into a coordinate frame that rotates about H with frequency v. In this "rotating frame" the influence of H is removed and μ is fixed in space. From $H_1(t)$ only a component H_1 survives the transformation. This component is time-independent and is perpendicular to the original H axis. Now we can write another equation

$$\frac{d\vec{\mu}}{dt} = \gamma\vec{\mu} \times \vec{H_1}$$

for the motion of μ in the rotating frame. Thus μ precesses about H_1 in this frame with frequency v_1 given by $hv_1 = \gamma H_1$. When this precession has carried μ through a sizable angle—about $90°$ or $\pi/2$ radians—the probability is high that magnetic resonance absorption will have taken place, because the orientation of μ in the laboratory frame would now be given by a different magnetic quantum number M.

The "power" requirement for NMR/RD is easily obtained from this picture of NMR. We may simply regard NMR as one of two competing processes. The other process is either radioactive decay, with time constant τ (the nuclear lifetime) or relaxation, with time constant T_1 (the spin-lattice relaxation time). For NMR to compete favorably with decay, for example, we require $2\pi v_1\tau \geq \pi/2$, or $v_1 \geq 1/4\tau$. This requirement is met for $H_1 \geq h/4\gamma\tau$. Unusually large radiofrequency field strengths are required to satisfy this condition—for $\tau = 10^{-6}$ sec, an H_1 of 100 gauss is needed—but these fields can be attained with some effort.

Results to Date

The NMR/RD field started in the early 1950's. It isn't practical to give a complete review of the early work here with a discussion of the value of each contribution, but we can mention a few important papers. A more complete set of references is available in a recent review article (1).

In 1952 Deutsch and Brown reported the radiative detection of magnetic resonance in positronium (2): this work was the precursor of the NMR/RD field. Later, in 1953, Bloembergen and Temmer pointed out the possibility of NMR/ON (3) and Abragam and Pound suggested NMR/PAC the same year (4). During the next 12 years several experiments were reported that were individually very elegant. Among these were an NMR/ON study of dynamically oriented ^{76}As, by Pipkin and Culvahouse (5) and an NMR/NR experiment on ^{8}Li in LiF, by Connor (6). Unfortunately none of the experiments done in this period showed promise of being applicable to any but a few nuclear states. As recently as 1965 no NMR/PAC experiment had been reported, nor had an NMR/ON experiment been successfully carried out on *thermally* oriented nuclei. This latter fact was particularly important because thermal nuclear orientation has a very wide range of application. The earlier theoretical papers (3, 4) had discussed the necessary conditions for these NMR/RD experiments, but had not suggested practical ways to achieve these conditions. There was widespread opinion that the experiments were not feasible.

Early in 1966 a group at our laboratory in Berkeley tried an NMR/PAC experiment based on "hyperfine enhancement" of the applied field in ferromagnetic nickel. This effect, discovered in 1959 by Gossard and Portis (7), has the effect of amplifying the H_1 field by a factor of about 10^3, thereby making the crucial condition $v_1 \geq 1/4\tau$ attainable for even very short-lived nuclear states. The first NMR/PAC result, on ^{100}Rh ($\tau = 3 \times 10^{-7}$ sec) in nickel, was reported by Matthias *et al.* (8), and within nuclei) has also emerged from this work.

Applications to solid-state phenomena seem promising. A number of hyperfine magnetic fields, Knight shifts, and relaxation times have already been obtained by NMR/RD. It seems to be

later that year another NMR/PAC resonance, on ^{100}Rh in iron, was observed with the same apparatus.

After the NMR/PAC experiments had been carried out with the aid of hyperfine enhancement, extension of NMR/ON was a natural next step. NMR/ON experiments in ferromagnetic lattices were especially appropriate because the "universal" method of thermal nuclear orientation employs ferromagnets (9). In fact the NMR/ON experiments appear somewhat less difficult, and far more widely applicable, than the NMR/PAC method. Matthias and Holliday at Berkeley reported the first such NMR/ON result, on ^{60}Co in iron, later in 1966 (10). Subsequent work in several laboratories, notably in Oxford and Leiden, has confirmed the generality of the NMR/ON method and has extended it considerably.

With these two methods it was thus possible to perform NMR/RD studies on a wide variety of long-lived ($\tau >$ 10^3 sec) and short-lived ($\tau < 10^{-6}$ sec) nuclear states, but the large intermediate range with $10^{-6} < \tau < 10^3$ sec was still not generally accessible. It was clear that only the NMR/NR methods would be applicable here.

Nuclear states with lifetime in the upper half of this range—i.e., for $10^{-2} < \tau < 10^2$ sec—had been studied by Connor and Tsang, by Sugimoto and coworkers, and by others (1), using NMR/RD and employing the asymmetry of β particles for detection. In 1967 Sugimoto and coworkers extended their work to metallic hosts (11), thereby substantially generalizing the NMR/RD method for these longer-lived cases. There still remained a lifetime range $10^{-6} < \tau < 10^{-3}$ sec in which no NMR work had been done, however, and many interesting γ-emitting nuclear isomers lay in this range.

In 1968 Christiansen and coworkers oriented nuclear isomers in liquid gallium (12), and early this year Quitmann and Jaklevic used a liquid gallium host to observe NMR in a 6-microsecond isomer of ^{73}As (13).

With the success of the nuclear reactions-NMR work, NMR/RD can be applied to a great many nuclear states. Resonances have already been observed in about 20 states, and a fair diversity of information has been derived from NMR/RD studies. Several nuclear magnetic moments have been measured accurately. Other nuclear information, particularly about "hyperfine anomalies" (the distribution of magnetism uniquely appropriate for very dilute systems or very small samples, where conventional NMR isn't sensitive enough. It also holds promise for studies of ion-implantation and other dynamic processes.

Like many physical techniques NMR/RD has quickly proved its value in physics, while its potential (if any) in chemistry is as yet unexplored. This is understandable, because several barriers must be overcome if NMR/RD is to enjoy widespread use in any area of chemistry. First of all its feasibility in gaseous molecules, or liquids has not yet been demonstrated. Then too, serious restrictions are intrinsic in all three NMR/RD methods. NMR/ON requires very low temperatures, NMR/PAC usually involves nuclear transmutation, and in NMR/NR the nuclei recoil immediately before the resonance is observed. Overcoming these restrictions will require breakthroughs as substantial as anything described above. The incentive for attacking these problems is provided by the extremely high sensitivity available with NMR/RD—a sensitivity that makes NMR a truly microscopic method.

Literature Cited

(1) D. A. Shirley, "The Detection of NMR by Nuclear Radiation," University of California Lawrence Radiation Laboratory Report UCRL-18315 Rev, (July 1968), to be published in Proceedings of the XV° Colloque A.M.P.E.R.E., Grenoble, France, 16–21 September 1968.
(2) M. Deutsch and S. C. Brown, *Phys. Rev.*, **85**, 1047 (1952).
(3) N. Bloembergen and G. R. Temmer,

Phys. Rev., **89**, 883 (1953).

(4) A. Abragam and R. V. Pound, *Phys. Rev.*, **92**, 943 (1953).

(5) F. M. Pipkin and J. W. Culvahouse, *Phys. Rev.*, **106**, 1102 (1957).

(6) Donald Connor, *Phys. Rev. Letters*, **3**, 429 (1959).

(7) A. C. Gossard and A. M. Portis, *Phys. Rev. Letters*, **3**, 164 (1959).

(8) E. Matthias, D. A. Shirley, M. P. Klein, and N. Edelstein, *Phys. Rev. Letters*, **16**, 974 (1966).

(9) David A. Shirley, "Annual Reviews of Nuclear Physics," Vol. 16, (Annual Reviews, Inc., Palo Alto, Calif., 1966) pp. 89–118.

(10) E. Matthias and R. J. Holliday, *Phys. Rev. Letters*, **17**, 897 (1966).

(11) K. Sugimoto, K. Nakai, K. Matuda, and T. Minamisono, *Phys. Letters*, **25B**, 130 (1967).

(12) J. Christiansen, H.-E. Mahuke, E. Recknagel, D. Riegel, G. Weyer, and W. Witthuhn, *Phys. Rev. Letters*, **21**, 554 (1968).

(13) D. Quitmann, J. M. Jaklevic, and D. A. Shirley, Lawrence Radiation Laboratory Report **UCRL-18871**, April 1969.

COMMENTARY

by Ralph H. Müller

D R. SHIRLEY'S DISCUSSION of the Radioactive Detection of Nuclear Magnetic Resonance will arouse the interest of people in many fields and, hopefully, that of analytical chemists. Even with lack of comprehension of what is really going on in these nuclear phenomena, their symmetric and asymmetric distribution in angle of emitted beta or gamma radiation, the analyst will sit up and take notice of the fantastic sensitivity which these techniques permit. If only 10^6 nuclei are required to get a good statistical answer, this is of the order of 10^{-17} mole and superb trace analysis by any criterion. They will not be too discouraged by the author's statement that "its potential (if any) in chemistry is as yet unexplored." And also "its feasibility in gaseous molecules, or liquids has not yet been demonstrated." It is enough to know that NMR/RD has quickly proved its value in physics. It will command increasing interest in solid state physics where purposely introduced trace impurities can provide such a wide variety of electrical properties and functions. One thinks instinctively of alternative methods, which over the years have developed into highly useful and widely used techniques. Neutron activation analysis, which is applicable to a large number of elements, provides exact identification in terms of the decay rate of the radionuclides which are formed, and quantitative results as well. In the case of nuclei which have very large cross section values, the sensitivity is enormous.

The electron probe technique of Castaing, Gunier, and others affords detailed analysis of minute specimens and definite identification of their nature.

Many years ago, we heard a fascinating lecture by an eminent physiologist who used an unusual approach for delineating the distribution of calcium in striated muscle. A microtome section of muscle was deposited on a quartz slide and then carefully incinerated in a furnace. The slide containing the ash residue was then heated to something less than 800 °C and thermionic electrons emitted by the ash were then accelerated and focused on a fluorescent screen. The image was a precise replica of the gross morphological struc-

35

ture of the muscle structure as revealed by ordinary microscopic examination. This was no mean feat in those early days of electron optics. The method was obviously of limited use, but, presumably would work for any residue exhibiting thermionic emission at moderate temperatures. It had been known for a long time that calcium oxide as well as other alkaline earth oxides, are excellent thermionic emitters. Today, the electron probe by a scanning technique would yield the same information but it is doubtful whether the special delineation would be as precise or, for comparable precision, could be obtained as quickly.

Another example is striking. Newton Harvey at Princeton, who revealed the nature of the chemiluminescence of cyprodina, showed that it arose from the oxidation of luciferon, catalyzed by the enzyme luciferase. In one experiment, he chose to furnish oxygen by its liberation at a platinum electrode immersed in the lucifer-in-luciferase system. By observation through a low power microscope, the least perceptible flash of light (dark accommodated eyes) appeared at an electrolysis current corresponding to the order of one oxygen atom.

If one begins to wonder about the feasibility of detection at the single atom level, it is merely necessary to recall Rutherford's first demonstration of the transmutation of matter by the bombardment of nitrogen with alpha particles in which high speed protons were one of the products. These were detected by their very long range by visually observed scintillation on a fluorescent screen. To be sure, many such protons were observed but each flash represented a single proton. The occasional forked track in a cloud chamber again represents a single nuclear event and detailed analysis of the track reveals the components resulting from the event.

If the nuclear physicist is bewildered and embarassed by the vast array of nuclear particles, some of an extremely fugitive nature, we analysts are confronted with the same difficulties when we inquire into the nature of an allegedly pure sample. Even emission spectroscopy, when applied to the examination of a very pure sample of metal will reveal the unmistakable presence of an appreciable fraction of all the known elements.

Trace analysis when it approaches the single atom level is something like throwing a party for a dozen congenial friends. If one has earned even a modest reputation for hospitality, it is astonishing how many people show up.

Laser-Excited Fluorescence of Matrix-Isolated Molecules

James S. Shirk[1] and Arnold M. Bass

National Bureau of Standards, Washington, D. C. 20234

Matrix isolation and laser-excited fluorescence combined form highly specific, sensitive analytical techniques. The methods have great potential in solving analytical problems

THERE is a great deal of information in molecular emission spectra. For heavy diatomic or polyatomic molecules, the spectra are often very complex and difficult to interpret. The complexity arises because the emission is usually between many different vibrational, rotational, and electronic energy levels. Also, in heavy diatomic molecules and polyatomic molecules, the rotational energy levels are closely spaced and the rotational fine structure leads to broad bands that can be resolved only by very large spectrographs.

There are several recently developed techniques for studying these complex spectra. We shall discuss here the use of the matrix isolation technique in conjunction with laser-excited resonance fluorescence. This combination is especially useful since the matrix isolation technique not only provides a means of simplifying the spectra by freezing out the rotational fine structure, but also allows one to, in effect, "tune" an absorption band into coincidence with some known intense laser line. The laser emission can then be used to excite fluorescence spectra. The commercial availability of intense lasers at several

[1] Current address, Dept. of Chemistry, Illinois Institute of Technology, Chicago, Ill. 60616

different wavelengths has made such studies readily possible. The more recent availability of ultraviolet lasers and frequency doublers, one of which (the frequency doubled argon-ion laser) operates at 2579Å, has vastly increased the number of molecules that can be studied with laser light sources.

The intense, highly monochromatic light characteristic of lasers is especially useful for fluorescence studies. Excitation with a laser source makes it possible to obtain a significant population in a single, or at most, a few upper states. Since the fluorescence occurs only from a few well-defined upper states, the spectra are simple and easier to analyze than typical emission or even absorption spectra. One can thus obtain information on the molecular constants for molecules whose spectra are otherwise quite complex. This technique has already been used successfully to study the diatomic species K_2, I_2, Rb_2, and Cs_2 in the gas phase. Resonance fluorescence studies of this sort have also been done in the infrared on several molecules, including CH_4, CO_2, and SF_6. The aim in the latter cases has been to study energy transfer and relaxation processes. Unfortunately, in the gas phase only selected molecules can be studied since it is necessary to have a molecular absorption exactly overlap a laser line. Tuning a laser over a spectroscopically significant range of wavelengths is usually not possible (except for the pulsed dye lasers), thus in the gas phase one is restricted to studying only the molecules for which a discrete absorption fortu-

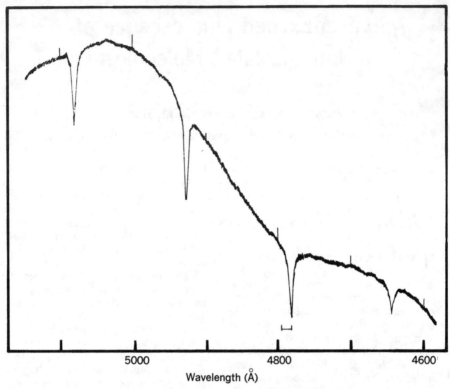

Figure 1. Absorption spectrum of CuO isolated in xenon at 15°K

itously overlaps a laser line.

Matrix isolation is another technique which is useful for the study of complex spectra. In this technique the sample is diluted with a large excess of inert gas, usually 200 to 1000 or more parts of inert matrix to 1 part of sample. The mixture is then frozen onto a transparent window maintained at 15 or 4 °K. The sample is thus trapped in a gas-like environment where the molecules cannot react, diffuse, or usually even rotate. Since the sample is at very low temperatures, essentially all the molecules are present in their electronic and vibrational ground state. The absorption spectra are typically sharp bands with little fine structure. Since the absorptions arise from ground state molecules and there is no rotational structure, the spectra in the visible ultraviolet region are typically sim-

ple progressions in the upper state vibration. In the infrared region the spectra are sharp bands corresponding to the vibrations of the molecule. This technique has proved especially useful in obtaining spectra of free radicals, high temperature species and reactive species in general, as well as in the analysis of complex spectra. Recently, techniques for rapid deposition of the sample has made it feasible to use this method for the qualitative and quantitative analysis of complex multicomponent gas mixtures. All these applications benefit from the simplicity inherent in matrix isolation spectra.

While the rare gas matrix does provide a gas-like environment for the trapped molecules, there are small shifts in the absorption frequencies, typically on the order of a few per cent from the gas phase absorption maxima.

This "matrix shift" is dependent on the particular rare gas chosen. As an example, Table I gives the shifts we have observed for the well-known $A^1\Pi \leftarrow X^1\Sigma^+$ transition of SiO trapped in various matrices. The absorption frequency can be shifted up to about 2% from the gas phase by choosing an appropriate matrix. Furthermore, by using a mixture of two rare gases for the matrix, —e.g., 75% krypton and 25% xenon, shifts in frequency intermediate between those for pure matrices can be obtained. An absorption can thus be "tuned" over a range of 100 or 200 Å by the simple device of choosing an appropriate matrix. Using the matrix isolation technique to tune an absorption band onto a laser line, it is possible to excite resonance fluorescence from many molecules. Such fluorescence spectra have all the virtues that matrix isolation provides—i.e., they are simple, sharp, and easy to interpret.

In our experiments the sample is deposited on a transparent cold window maintained at 15°K with liquid hydrogen in a double-walled Dewar. Alternatively it is possible to use liquid helium or a Joule-Thompson refrigerator;

these are all commercially available. A laser of an appropriate wavelength is defocused through a lens to illuminate the entire sample. The resulting fluorescence from the sample is collected with a lens and focused on the slit of a large aperture spectrograph. A filter is used to remove scattered laser radiation. Alternatively, we focus the light on the slit of a scanning spectrometer with photoelectric detection. A photoelectric detection system can be used with the sophisticated signal processing electronics now available to increase the sensitivity of the method. By employing Q-switching techniques which enable one to obtain nanosecond pulses of laser light of very high power, and observing the fluorescent lifetimes, it is possible to study the lifetimes of the states excited by the laser.

After an appropriate matrix gas has been chosen, the sample preparation depends upon the molecule under study. Stable gas samples can be premixed with the matrix gas and then deposited. Photolysis of the solid matrix isolated sample has been the most effective method for producing free radicals. Refractory materials and high tempera-

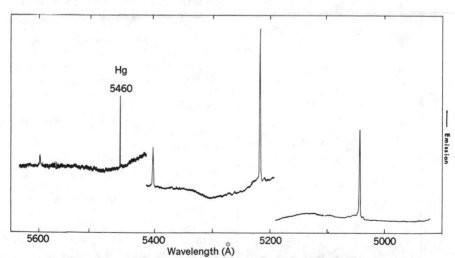

Figure 2. Resonance fluorescence from CuO isolated in argon at 15°K
Excitation with the 4880 Å line of an Argon—Ion Laser. The emission is from the $v' = 0$ vibration level of the upper electronic state. The Hg line at 5460 Å, arising from fluorescent room lights provides a convenient calibration point

Table I. Observed Frequencies for SiO in Various Matrices

Gas, cm⁻¹	Argon		Krypton		Krypton/Xenon = 3/1		Xenon	
	Freq., cm⁻¹	% Shift	Freq., cm⁻¹	% Shift	Freq., cm⁻¹	% Shift	Freq., cm⁻¹	% Shift
42,657	42,766	+0.25	42,425	−0.55	42,240	−0.99	41,855	−1.84
43,499	43,563	+0.15	43,254	−0.56	43,126	−0.86	42,673	−1.93
44,328	44,407	+0.18	44,103	−0.51	43,991	−0.77	43,478	−1.95
45,139	45,208	+0.15	44,903	−0.53	44,783	−0.79	44,277	−1.95
45,943	45,983	+0.09	45,704	−0.52			44,998	−2.10
46,694	46,792	+0.21	46,443	−0.54				

ture species can be deposited either out of a Knudsen effusion cell or sputtered from a solid sample by bombarding the sample with accelerated ions.

We have successfully used resonance fluorescence to study the spectra of matrix isolated CuO, CCl_2, and C_2^-. For example, the gas phase emission spectrum of CuO is exceedingly complex and extends over much of the visible spectrum. Despite much effort over almost 50 years, this spectrum had not been successfully interpreted. Using the matrix isolation technique, we were able to trap CuO in solid xenon and take the absorption spectrum shown in Figure 1. This absorption spectrum is very much simpler than the gas phase spectrum. It consists of a progression in the upper state vibrational frequency. If the CuO is trapped in argon instead of xenon, the $v' = 0 \leftarrow v'' = 0$ absorption to the upper electronic state, which was at 5080 Å in xenon, is now at 4880 Å, in exact coincidence with an intense argon-ion laser line. Figure 2 shows a photoelectric recording of the fluorescence spectrum obtained when a sample of CuO in solid argon was excited with an argon-ion laser. Again a simple progression is observed; the separation of the lines gives the ground state vibrational frequency directly. Further work using the resonance fluorescence technique with photographic recording showed another electronic state of CuO lying about 4000 cm⁻¹ above the ground state. Thus it was possible to determine the ordering of the three low-lying states of CuO and to determine their vibrational constants.

If we trapped CuO in a matrix that was 85% Xe and 15% Kr, we were able to tune the $v' = 2 \leftarrow v'' = 0$ absorption, which appeared at 4781Å in pure xenon, onto the 4765Å line of an argon-ion laser. When the sample was excited with the laser, fluorescence was observed from the $v' = 1$ and $v' = 0$ vibrational levels of the upper state. Observation of emission from the $v' = 1$ level leads one to the surprising conclusion that the vibrational lifetime of the upper state must be quite long—*i.e.,* CuO can undergo about 10^5 vibrations in a solid matrix without being completely vibrationally relaxed.

The analytical applications of matrix isolation spectroscopy are beginning to be explored by Rochkind and coworkers. It has been shown to be useful for the analysis of gas mixtures and isotopically substituted compounds. New sample preparation techniques could widen its usefulness. For example, a focused laser could be used to vaporize a refractory sample and the vapors trapped in a rare gas matrix for analysis.

Laser-excited fluorescence, of either gaseous or matrix isolated compounds, could be developed into a highly specific analytical method, since only a compound which has an absorption overlapping the laser line will fluoresce. Laser-excited fluorescence would be quite sensitive as well. The very high intensities possible from laser excitation sources make it possible to detect fluorescence from samples in the range of 10^{-9} mole or smaller. Analysis of air pollutants is a possible application of laser excited fluorescence now that lasers exist that can excite into intense absorptions of NO_2, SO_2, and O_3, as well as many unsaturated hydrocarbons.

Combining the techniques of laser excited fluorescence and matrix isolation has been useful or the study of molecular spectra. We are currently working on extending this technique to other complex molecular spectra and to a direct measurement of the vibrational lifetimes of trapped molecules. Laser excited fluorescence of matrix isolated or gas phase samples will hopefully prove equally useful in solving analytical problems.

References

Gas Phase Resonance Fluorescence; W. J. Tango, J. K. Link, and R. N. Zare, *J. Chem. Phys.,* **49,** 4264 (1968).

Matrix Isolation Spectra; J. S. Shirk and A. M. Bass, *J. Chem. Phys.,* **49,** 5156 (1968); M. M. Rochkind, ANAL. CHEM., **40,** 762 (1968); A paper, "The Absorption and Laser Excited Fluorescence Spectrum of Matrix Isolated CuO," by J. S. Shirk and A. M. Bass has been submitted to *J. Chem. Phys.*

Infrared Detectors

Users of currently available infrared detectors have
considerable choice. Sensitive areas may range anywhere
from 0.01 mm² to 1 cm² with thresholds from the visible
to the millimeter wave region and time constants
shorter than 1 nsec

Henry Levinstein
Syracuse University, Syracuse, N.Y. 13210

INFRARED DETECTORS may be characterized by three basic parameters:
their spectral response, speed of response, and the minimum amount of
radiant power they can detect. In addition to these basic parameters, the
user is generally interested in several
other characteristics, such as the temperature of operation, the magnitude
and type of signal produced by the incident radiation and the type of circuit
which best couples the detector to a
readout system.

Detectors may be placed into two
broad categories determined by the
physical principles underlying their operation: thermal detectors and photon
detectors. In thermal detectors, the incident energy is absorbed by a tempera-
ture-sensitive material or by an absorbing layer in contact with the tempera-
ture-sensitive material. The receiving
layer usually consists of materials, such
as "blacks," whose absorption is uniform over a wide spectral range. Tem-
perature-sensitive materials may be
either metal or semiconductor layers,
whose resistance is temperature dependent (bolometers); junctions of unlike

metals (thermocouples); ferroelectric
materials, where incident radiation affects the polarization of the sensing
element (pyroelectric detectors); or
gases, where pressures are temperature
dependent (pneumatic cells).

Photon detectors are usually semiconductor devices, where the incident
photon flux interacts with electrons in
bound states, exciting them to a free
(conducting) state. This may result
in a decreased resistance or a photo
voltage when charge carriers are separated by a built-in field. The different
principles of operation of the two
classes of detectors are evidenced by
differences in their properties. Thermal detectors usually have a spectral
response determined only by the characteristics of the absorber. For constant incident energy this might be
wavelength invariant over a wide spectral range. Ideal photon detectors
have a response which for constant incident energy rises linearly with wavelength, then drops abruptly. This is
based on the assumption that the number of charge carriers excited per photon is constant and independent of

photon energy until the photon energy is no longer large enough to excite a charge carrier. The speed of response of thermal detectors is determined by the rate at which the sensing layer heats and cools. This is partly a matter of construction detail involving the dimensions of the sensing element and its thermal contact with the environment and partly its specific heat. Since in photon detectors the speed of response is determined by the time required for the excited charge carriers to become immobilized through recombination, they usually respond faster. Furthermore, in the vicinity of the peak of their spectral response, photon detectors may be used to detect a considerably lower radiant flux than thermal detectors. To make possible a comparison among commercially available detectors, a more precise specification of measurement techniques and parameters is necessary.

Measurement Techniques

The minimum detectable power, noise equivalent power (NEP), is usually measured by placing a black body of known temperature (usually 500 °K) at a convenient distance from the detector. Its radiation is interrupted periodically by means of a rotating sectored disk, such that a detector signal of constant frequency is obtained. The radiant power received at the detector may be calculated. The detector noise is measured, usually with a narrow band amplifier, when the detector is shielded from the radiation source. The noise equivalent power (NEP) is calculated from the relation $NEP = P/(S/N)$ where P is the power received by the detector and S and N are signal and noise voltage, respectively.

It is often more convenient to specify the reciprocal of NEP, the detectivity, D, or in the comparison of detectors, a normalized quantity, D^*. D^* is the detectivity of a detector of a 1 cm² area whose noise is reduced to that obtained with an amplifier of 1 Hz bandwidth.

$$D^* = \frac{S/N}{P_D}\left(\frac{\Delta f}{A}\right)^{1/2}$$

where Δf is the amplifier bandwidth, A the detector area, and P_D the power density at the detector. The assumptions made in the normalization process require that detector noise varies as $A^{1/2}$ and $\Delta f^{1/2}$. The first assumption is valid for photon detectors as long as the detector areas do not vary by more than an order of magnitude. It may not be valid for certain thermal detectors. The second approximation requires either that noise is frequency invariant over the amplifier bandwidth, or that it is measured over such a narrow band that its variation is insignificant.

The spectral response of a detector is determined by measuring its signal, as it is exposed to radiation over a range of wavelengths obtained from a monochromator. In this process it is important that the energy received at the detector be held constant or that its variation be known. When its magnitude is measured by means of a calibrated detector, the absolute spectral response may be calculated directly by comparison. Otherwise, absolute response at any wavelength may be calculated from the response to a black body at known temperature and a knowledge of the shape of the relative response.

The speed of response of a detector is usually determined by exposing the detector to square radiation pulses and then observing the rise and decay of the signal on an oscilloscope. The radiation pulses are obtained either mechanically by rotating a sector disk in front of a line radiation source or optically by various modulating methods. Since the speed of response may be dependent on the wavelength and the intensity of incident radiation, conditions of measurement should approximate those under which the detector is actually being used.

Characteristics of Thermal Detectors

Uncooled thermal detectors have values of D^* in the 10^8–10^9-cm Hz$^{1/2}$-watt^{-1} range. Their time constant is usually no less than several milliseconds. Shorter time constants usually result in decreased detectivity. The most commonly used thermal detectors are thermocouples and thermistor bolometers. Thermocouples, because of their low impedance ($\sim 10\Omega$) are usually matched to an amplifier by means of specially designed preamplifiers or by transformers. Their time constant is of the order of 10 msec and responsivities vary from 2 to 7 V/W.

Thermistor bolometers have a resistance in the megohm range. They consist of thin flakes from oxides of manganese, nickel, and cobalt. Materials are selected on the basis of their high temperature coefficient of resistance. Another of the conventional thermal detectors is the pneumatic cell. It depends, in its operation, on mechanical rather than electrical properties of materials. As radiation strikes an absorbing layer in thermal contact with an enclosed gas, the increase in temperature of the layer produces an increase in gas pressure. This, in turn, causes a displacement of a diaphragm which may be monitored by various optical or electrical techniques. One of the versions, which employs optical monitoring of the diaphragm, is the Golay cell.

Most recently, two new types of thermal detectors have been introduced: pyroelectric detectors and cooled bolometers. The pyroelectric detector consists of a slice of a ferroelectric material, usually single crystal tryglycine sulfate (TGS), whose polarization is temperature dependent. The device is fabricated in the same manner as a capacitor. Two electrodes, one of them transparent, are formed on opposite sides of a TGS slice. Radiation is received through the transparent electrode. The voltage generated within the crystal is usually applied directly to a field effect transistor which is an integral part of the detector package. The cooled bolometer consists of a germanium crystal with impurities having low charge carrier activation energies. It is cooled to a temperature of between 1 and 4 °K, where its resistance variation with temperatures is large and its thermal capacity is low. Detectors of this type have detectivities several orders of magnitudes larger and time constants considerably shorter than the uncooled thermal detectors and, if coated with a good absorber, have a response which extends into the far infrared.

Characteristics of Photon Detectors

While thermal detectors have been in use in one form or another from the beginning of infrared in 1800, photon detectors are of relatively recent vintage, having been developed during the past 25 years. Their operation requires that the incident radiation signal liberate charge carriers from either the crystal lattice (intrinsic detectors) or from impurities which have been intentionally added to the host crystal (extrinsic detectors). There is thus associated with these detectors a wavelength threshold determined by the minimum energy required to free these charge carriers. Since charge carriers may be liberated also by thermal radiation from the background and by vibrations of the crystal lattice, as determined by the temperature of the crystal, detectivities depend on these factors.

The limiting noise source of a detector is G-R (generation-recombination) noise. It is associated with fluctuations in the density of free charge carriers, produced either by lattice vibrations, when the detector is not cooled sufficiently or by the random arrival of photons from the background. When the latter is the dominant source of noise, the detector is said to be background limited (BLIP). With no other source of noise, this condition is reached by cooling the detector to the point where the number of charge carriers freed by lattice vibrations is consider-

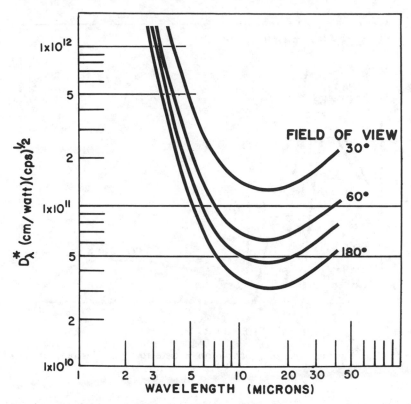

Figure 1. Detectivity at spectral peak as a function of long wavelength threshold for background limited (BLIP) detectors. Solid curves represent photoconductive detectors with 180°, 60° and 30° angular fields of view, dotted curve represents photovoltaic detector with 180° field of view

ably fewer than the number freed by the background.

A reduction in the amount of background radiation, obtained by a smaller field of view or the use of cooled narrow band filters, usually requires lower detector temperatures for BLIP operation. In general, detectors with a response to about 3μ and 180° field of view looking into thermal background (300 °K) require little if any cooling. Detectors prepared from materials with lower activation energy require cooling to temperatures as low as liquid helium for materials with a response beyond 30μ. Detectors with a response in the intermediate IR, 5–10μ, may be operated at liquid nitrogen temperature.

Because background-limited detectivities are dependent only on the amount of background reaching the detector, their detectivities may be calculated as a function of the number and energy distribution of background photons. Figure 1 shows BLIP detectivities as a function of detector threshold wavelength, for various fields of view, assuming unity quantum efficiency. Of course, actual values of detectivities are lower, since part of the incident energy is reflected from the surface of the detector element and some of it is transmitted through the element without being absorbed. Absorption may be increased by selecting optimum element thickness; reflection may be re-

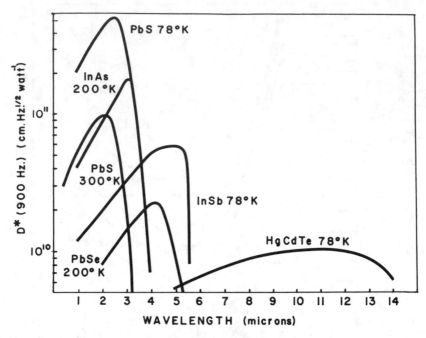

Figure 2. Spectral Response of typical intrinsic detectors (angular field of view is 180°)

duced by the use of nonreflecting coatings. Detectivity is also reduced by excess noise. The most commonly observed excess noise varies inversely as the square root of the frequency ($1/f$ noise). It has been made negligible above 100–1000 Hz for most detectors. With proper care in the construction of detectors, those most advanced reach their ideal detectivity within 10–20%.

When photon detectors are of the photoconductive variety, they are connected in series with a dc power source and a load resistor. Optimum supply voltage must be determined before the detector may be used effectively. Too high a voltage may produce excess noise and may occasionally damage the detector; too low a voltage may produce a noise level below that of the amplifier. Photovoltaic detectors, depending on their resistance, may require the use of a transformer or a specially designed preamplifier.

PbS, PbSe, and PbTe were among the earliest IR photon detectors developed. They consist of evaporated or chemically deposited layers sensitized by various techniques involving the use of oxygen in one form or another. Production of these detectors is more of an art than a science. PbS and PbSe are commercially available. PbS is operated at room temperature. It responds to 3μ, has a time constant of several hundred microseconds and a peak D^* of about 10^{11} cm Hz$^{1/2}$W^{-1}. When cooled to liquid nitrogen temperature, D^* for specially prepared detectors may increase by as much as an order of magnitude, but at the expense of time constants, which are lengthened considerably. PbSe has a response extending to 5μ, when cooled to dry ice temperature, a value of D^* of the order of 10^{10}, and a time constant of about 50μsec. Liquid nitrogen cooling shifts the response to slightly longer wavelengths and makes possible the use of material with element resistance of the order of 1 $M\Omega$ instead of 100 $M\Omega$ of PbSe materials designed for operation in the dry ice temperature range. The spectral curves of the lead salt detectors are shown in Figure 2.

The development of crystal growing and material preparation techniques has led to the preparation of InSb and InAs detectors. These are prepared by well-understood techniques from single crystals of the material. InSb can be prepared to operate in the photoconductive mode or by the formation of a p-n junction in the photovoltaic mode. These detectors have reached such a state of perfection that D^* approaches its theoretical limit (10^{11} at a spectral peak of about 5μ) closer than any other material. Time constants are of the order of 1μsec for photovoltaic detec-

tors and 10μsec for the photoconductive detectors. InAs, because its region of response overlaps that of PbS, but requires cooling to at least dry ice temperature, has not found as many applications as PbS. It is available only in the photovoltaic mode. When cooled to dry ice temperature, it has a detectivity which may exceed that of uncooled PbS, especially near the spectral cutoff wavelength, and a time constant considerably shorter than PbS.

Several approaches have been used in the construction of detectors with a response beyond 6μ. No single element or binary compound with charge carrier activation energy less than 0.2 eV has so far been found. One method for producing detectors with a threshold at longer wavelengths has consisted of using germanium with an activation energy of about 0.7 eV as the host lattice

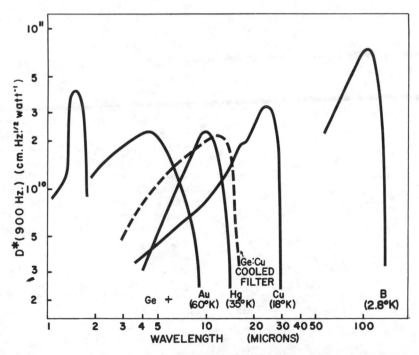

Figure 3. Spectral Response of typical impurity activated germanium detectors. All detectors with the exception of Ge:B have a 60° field of view; Ge:B has a 10° field of view

47

and adding selected impurities. The materials so prepared have a composite response, one due to excitation of charge carriers from the germanium with a response to about 1.7μ, the other due to excitation of charge carriers from a particular impurity. Thus Ge:Au has a response to about 9μ, Ge:Hg to 14μ, Ge:Cu to 30μ, Ge:Zn to 40μ, Ge:B to 120μ.

Of course, detectors responding to longer wavelengths require lower operating temperature. By properly controlling the concentration of compensating impurities, time constants of these detectors may be made as short as 10^{-9} sec. Figure 3 shows the spectral response of a series of impurity-activated Ge detectors. The other approach, which has been successful only recently consists of combining two compounds, one with a very small activation energy, the other with a larger activation energy. By using various fractions of each compound, detectors with a wide range of long wavelength thresholds have been prepared. The system HgTe–CdTe has been studied in most detail.

Of particular interest have been detectors with a response to 14μ and time constants of the order of 1μsec, prepared from the predetermined fractions of HgTe and CdTe. Other cutoff wavelengths, both shorter and longer than 14μ, can be obtained merely by changing the composition of the components, but those detectors have not, as yet, been equally well developed. These detectors have an advantage over impurity-activated germanium detectors in that they require less cooling for the same cutoff wavelength. In the 8–14-μ region HgCdTe, as shown in Figure 2, is competitive with Ge:Hg (Figure 3). It is to be preferred over Ge:Hg, when time constants in the microsecond-range

rather than the nanosecond-range are called for and when cooling temperatures below 78 °K are not easily attainable. It should be realized, however, that the lower element resistance (~ 100–1000Ω) of HgCdTe, as compared with a resistance in the megohm range for impurity-activated Ge, may require the use of specially designed low-noise amplifiers, when HgCdTe detectors are substituted for impurity-activated Ge.

Infrared detectors have now reached the stage of development where the user has considerable choice in obtaining them in a variety of styles and sizes, with sensitive areas ranging anywhere from 0.01 mm^2 to 1 cm^2, with thresholds from the visible to the millimeter-wave region and with time constants shorter than 1 nsec. A better understanding of the processes underlying their operation, as well as controlled techniques in the preparation of materials, should lead in the not too distant future to detectors tailormade to fit particular applications.

Acknowledgment

The author thanks the many detector manufacturers who have made available literature on detector characteristics and occasionally new and improved detectors for test purposes. This includes Texas Instruments, Santa Barbara Research Center, Barnes Engineering, Honeywell, Infrared Industries, Societe Anonyme de Telecommunications Mullard. A more detailed discussion on some of the aspects covered in this paper is contained in "Applied Optics and Optical Engineering," Vol. 2, Chapter 8, Academic Press, 1965.

COMMENTARY

by Ralph H. Müller

D R. LEVINSTEIN has given our readers an interesting and most useful account of the status of infrared detectors and factors which govern their range, sensitivity, and time response. The extraordinary properties of some photon detectors depend upon sensitization or "doping" and the nature of the additive determines spectral range and other properties. As he points out, the technique is largely an art. One wonders what might be accomplished by a systematic study of a wide variety of binary or ternary systems in the hope of obtaining some general principles governing such sensitization. It would, no doubt, be a dreary and tedious task, but present achievements already indicate how much can be gained even by the sporadic examples. The theoretical aspects of solid-state behavior are well understood and, in a sense, we are lacking sufficient experimental data to exploit this knowledge to the fullest extent.

Questions Raised

We raise several questions, purely in the sense of inquiry, about some unexplored phenomena. In most cases, the investigations would have to be highly empirical but always with the knowledge that some unusual and useful result could, no doubt, be explained by modern theory to restore the discoverer to a state of moderate respectability. After all, Michael Faraday did make thermistors by sintering mixtures of metallic oxides, but, because in testing them he never got the same answer twice, he lost interest in the matter. A century later, others found that by prolonged sintering at high temperatures and prolonged aging, a highly reproducible system could be obtained.

In an intermediate approach, one might ask what happens if one were to compact mixtures of metals, in the powder-metallurgy technique, obtaining rods or sheets somewhere in the wide range between "green" density and theoretical density? Would such a system resemble a multijunction thermocouple (with plenty of local short circuits), a good noise thermometer, or some illegal mixture of both? He who contemplates such an ill-defined mixture with undisguised horror should consider the carbon granules jiggling between carbon buttons in the carbon microphone. It responds nicely and faithfully to acoustical excitation. The main question is—is it useful, sensitive, and reproducible?

Multijunction thermocouples or thermopiles are ancient devices but their assembly usually petered out when the designer ran out of patience or was confronted by an awkward assembly of increasing complexity. There is no reasonable limit to the number of junctions which can be connected in series, provided the operation is done automatically. It should be possible to electroplate or vacuum evaporate two dissimilar metals alternately on a base or mandrel coated with a releasing agent (grease or aquadag) and thus stack up extremely thin layers. The composite film could be stripped off and cut into small detector elements. It should be possible to obtain an element of very small heat capacity and possibly a response as high as 50-100 mv/° C.

These are obviously automated extensions of known thermal detectors, but receivers of this class, when properly blackened have the great advantage of response—independent of wave length.

There may be a limited number of cases in which a differential measurement of the quenching of phosphorescence by infrared could be used. Early investigations showed that phosphors excited by uv or visible light could be quenched by infrared. In some systems, there is an initial burst of luminescence followed by quenching, in others, just quenching. Similarly, the effect is attributable to a mere heating effect in some cases but in others, only by selective IR absorption. Where such effects are feasible, the actual measurement could be made with visible light and conventional phototubes.

Early observations of Pohl with single crystals and others examining such items as zinc blende and the diamond showed that the inner photoelectric effect exhibited unit quantum efficiency. Illumination with ultraviolet light produced induction centers in the crystal and subsequent irradiation with infrared would liberate one electron per photon absorbed. Presumably, this effect has its counterpart in modern solid-state photon counters.

The gap between the longest infrared and the shortest electric waves was closed early in this century. R. W. Wood had isolated radiation longer than 400μ from the Hg arc and Mme. Arkadieva in Russia had produced electric waves shorter than this. Her technique involved a suspension of finely divided iron particles suspended in oil which was stirred violently. A heavy electric discharge was passed through the suspension resulting in the emission of a broad band of extremely short waves. The iron particles acted as short-gap Hertzian oscillators. They were detected by a screen coated with a random array of iron particles, groups of which acted as hertzian dipole receivers. It is intriguing but possibly useless, to contemplate how such a screen might be modified to act as a useful detector. Could the particles be imbedded in an electroluminescent material? At this point, we refrain from further speculation. The progress which Dr. Levinstein has reported is impressive but, if history is any guide, other phenomena may be discovered which offer an entirely new approach to this important problem.

Metastable Ions in Mass Spectra

J. H. Beynon, Department of Chemistry,
Purdue University, West Lafayette, Ind. 47907

MANY OF THE IONS in the ionization chamber of a mass spectrometer are produced with so much excitation energy that they undergo a series of competing unimolecular decompositions before leaving this chamber. The product ions of these decompositions are drawn out of the ionization chamber and many of them are sufficiently stable to be accelerated and to traverse the flight path of the mass spectrometer without further reaction. These "stable" ions form the conventional mass spectrum. Some of the ions, however, do break down before reaching the ion collector. These are the so-called "metastable" ions.

DOUBLE-FOCUSING MASS SPECTROMETER

In a double-focusing mass spectrometer, such as is shown schematically in Figure 1, consider ions which fragment in the field-free region in front of the magnetic sector. When an ion m_1^+ breaks into an ion m_2^+ and a neutral fragment $(m_1 - m_2)$, the mean velocity through the tube of m_2^+ and of $(m_1 - m_2)$ will be the same as that of the initial ion m_1^+. That is to say, the kinetic energy of the initial ion is shared between the two fragments formed in direct proportion to their masses. When these low energy m_2^+

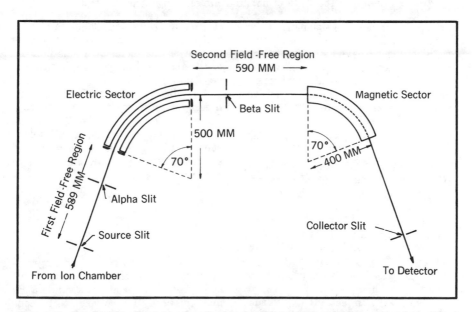

Figure 1. Schematic diagram of a double-focusing mass spectrometer of modified Nier-Johnson geometry (the diagram refers to Model RMH-2 made by Hitachi—Perkin Elmer)

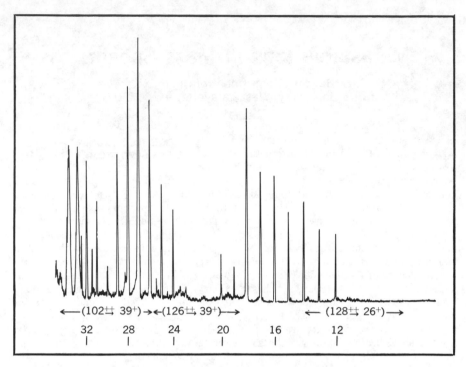

Figure 2. Partial mass spectrum of naphthalene showing weak "metastable peaks"

ions pass through the magnetic sector, they are deflected to a greater extent than m_2^+ ions having the full energy of acceleration, and it can be shown that they will appear as a weak, diffuse peak on the mass scale centered at an apparent mass m^* given by:

$$m^* = \frac{m_2^2}{m_1} \qquad (1)$$

The appearance of these "metastable peaks" and the difficulties of measuring their positions and intensities are illustrated in Figure 2.

Because the m_2^+ ions have less than the full energy of acceleration, eV, they can be prevented from reaching the collector if the potential of this electrode is raised to a value between $(m_2 \text{ eV})/m_1$ and eV. Early mass spectrometers were, in fact, fitted with such a "metastable peak suppressor" to avoid any interference with the normal peaks in the mass spectrum. Now, however,

such suppressors are seldom employed, and it is more usual to concentrate attention on the metastable ions because of the analytical information which can be obtained from them.

According to the quasi-equilibrium theory of mass spectra, the rate constant (k), for a particular decomposition may be related to the internal energy (E) of the dissociating ion, the activation energy (E_0), the effective number of oscillators (s), and a frequency factor (ν). The simple original expression connecting these quantities:

$$k = \nu \left(\frac{E - E_0}{E} \right)^{s-1}$$

has been found to be a poor approximation near threshold, but the modified expression relating k and E is mathematically too complex to be considered here. We shall use only the simple expression to suggest that for the slow decompositions of metastable ions, the excess energy $(E - E_0)$ is likely to be small. Thus, when a mass spectrometer is used to study the decompositions of metastable ions, the instrument is behaving as a "filter," studying the decompositions only of those ions which have little energy in excess of the minimum necessary for a particular reaction path, whatever the activation energy for reaction along that path may be. A low frequency factor will also lead to observation of a "metastable peak" in the mass spectrum and will be associated, for example, with reactions in which rearrangement of the atoms in the transition state may be necessary before reaction can occur.

Bombarding electrons in the source can transfer widely differing amounts of energy to molecules, and many ions are formed with large excess internal energies. The aim of much mass spectrometric work is the correlation of mass spectra with molecular structure. It seems likely that these correlations may be improved and minor differences in structure detected if the amount of excess energy $(E - E_0)$ can be kept small. Thus, a great deal of attention is now being directed toward ways of increasing the sensitivity of detection of "metastable peaks" and of observing them free from interference by the stable ions which make up the conventional mass spectrum.

A method of doing this is to study the decompositions of metastable ions which occur in the region preceding the electric sector in a double-focusing mass spectrometer. When this is done, the ions making up the "metastable peaks" will be more sharply focused by the combined effects of the electric and magnetic fields. The width of these peaks will thus be narrower with consequent increase in the peak heights. Indeed, peak heights two orders of magnitude greater than those due to metastable ion decompositions in front of the magnetic sector have been obtained when the product ions m_2^+ are made to pass through the electric sector.

OBSERVING STABLE IONS

When the mass spectrometer is set to observe stable ions, the field between the electric sector plates is such that ions which have received the full energy of acceleration will follow a central path through the electric sec-

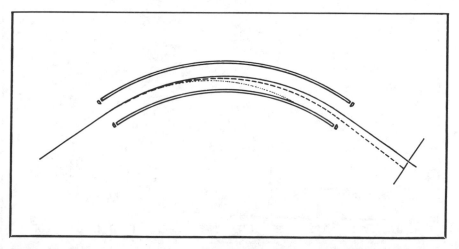

Figure 3. Paths of ions through the electric sector at normal voltage. The full line represents the path of the main beam of stable ions, the dotted lines the paths of the product ions from metastable ion decompositions

tor as shown by the full line in Figure 3. A product ion m_2^+ of a metastable ion decomposition will follow the dotted path through the electric sector and will not be transmitted because it possesses only a fraction m_2/m_1 of the necessary energy. It can be made to follow the central path either by increasing the ion accelerating voltage by a factor m_1/m_2 (keeping the electric sector voltage constant), or by reducing the electric sector voltage by a factor m_2/m_1 (keeping the ion-accelerating voltage constant). In either case, the main beam of stable ions is then not transmitted.

The first method has the disadvantages that as the ion-accelerating voltage is changed, the optimum "tuning" conditions within the ion source are altered owing to changes in field penetration and also that the voltage cannot be changed by more than a factor of about 4, limiting the ratios of m_1/m_2 which can be studied. It has the advantage that the mass scale

of the instrument is unchanged during the scan so that the magnet current can be set to observe m_2^+ ions, and as the accelerating voltage is changed, a series of peaks can be observed, each corresponding to a different metastable ion which decomposes into an m_2^+ ion.

The second method has the disadvantage that the mass scale changes during the scan of electric sector voltage, but the advantage that all the products of metastable ion decompositions (without limitation on the ratio m_1/m_2) can be made to follow the central path through the electric sector. A detector placed at the exit of the electric sector will then plot a complete ion kinetic energy (IKE) spectrum of the products of metastable ion decompositions. A typical IKE spectrum is shown in Figure 4 and provides a detailed "fingerprint" of the sample, which can be used for its identification. Mass analysis of the ions making up any peak in the IKE spec-

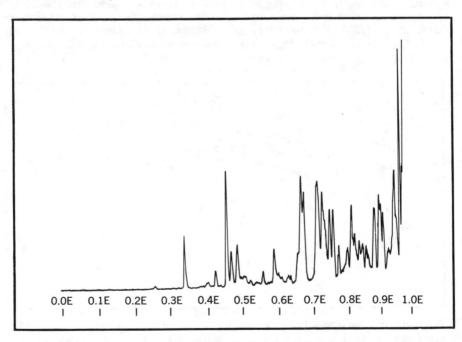

0.0E 0.1E 0.2E 0.3E 0.4E 0.5E 0.6E 0.7E 0.8E 0.9E 1.0E

Figure 4.   Partial IKE spectrum of N-carbobenzoxy-L-analyl-L-valine, methyl ester

trum can be effected by setting the electric sector to the appropriate value, raising the detector out of the ion beam and scanning the magnet current.

Whatever method is used to study the "metastable peaks" in a mass spectrum, the position, shape, and size of the peaks can be used to obtain useful structural information about the decomposing ions. For example, in the mass spectrum of aniline two of the "metastable peaks" occur at masses m^* of 46.88 and 45.96, respectively. These numbers are sufficiently accurate that it is not only possible to show that the transitions concerned are:

$$93+ \rightarrow 66+ + 27$$

and

$$92+ \rightarrow 65+ + 27$$

but that on the basis of more accurate masses which take account of the nuclear packing fractions of the elements in the decomposing ion, the decompositions can be written

$$C_6H_7N+\cdot \rightarrow C_5H_6+\cdot + HCN$$

and

$$C_6H_6N+ \rightarrow C_4H_3N+\cdot + C_2H_3\cdot$$

In this way, the reaction pathways along which many of the fragment ions are formed can be deduced. Also, when impure samples are being examined, the "metastable peaks" serve to link together various pairs of ions in the spectrum, thus showing they are due to a particular component present in the sample.

It is immediately apparent when mass spectra are examined that the "metastable peaks" are not so sharply focused as the peaks due to stable ions. It can be shown that this is a consequence of the release of a small amount of energy during the decomposition of the metastable ion, which appears as kinetic energy of separation of the fragments. Both the mean amount of energy and the range of energies released can be deduced from

the resultant broadening of the "metastable peak." The amount of energy released can be used to obtain information on ion structure because in a unimolecular decomposition, release of energy must correspond to an increased stability of the products formed. Thus, in the mass spectra of all three isomeric nitrophenols, a "metastable peak" is observed corresponding to loss of neutral NO• from the molecular ions. In the case of the o- and p-isomers, the peak is considerably broadened.

Stable quinonoid structures

can be visualized for fragment ions formed by loss of NO• from the o- and p-nitro phenols. This suggests that in the rearrangement reaction the remaining oxygen of the nitro group attaches itself to the same ring position as was previously occupied by the nitrogen atom, perhaps via isomerization to nitrite or by formation of a three-membered ring transition state. Correlation of the above structures with the kinetic energy released also lends weight to the hypothesis that charges in these positive ions are localized.

INTERCHARGE DISTANCE

Information on the intercharge distance in doubly charged ions can also be obtained from the widths of "metastable peaks." The first such system to be studied was benzene, the mass spectrum of which contains a very broad peak corresponding to the metastable transition:

$$C_6H_6{}^{2+} \rightarrow C_5H_3{}^+ + CH_3{}^+$$

The release of 2.7 eV of energy in this transition is mainly due to the potential energy released when the two positive charges are separated to infinity. Assuming all the energy released to come from this source, the initial charge separation can be calculated. If part of the energy is due to the increased stability of the fragments, the intercharge distance will be larger than the value calculated. The calculated minimum distance of 5.8 Å in benzene is much greater than the diameter of the benzene ring and shows that the $C_6H_6^{2+}$ ion has an open chain structure. The method might be used to determine, for a homologous series of molecules containing two oxygen or nitrogen atoms, whether the released energy correlates with the distance apart of the heteroatoms. This would confirm localization of charge on these atoms.

The heights of "metastable peaks" have also been used to study the slow fragmentation of ions even when much higher probability fast fragmentations are taking place. For example, the mass spectrum of benzoic acid labeled with deuterium on the carboxylic acid group shows a large peak due to loss of neutral $OD^\bullet$ and also a very small peak due to loss of $OH^\bullet$. Study of the "metastable peaks" shows that the peak corresponding to loss of $OD^\bullet$ in the slow decomposition of metastable molecular ions is only half as large as the peak corresponding to loss of $OH^\bullet$. This suggests that if the positive charge localizes on the carbonyl oxygen, this might enable transfer of an o-hydrogen to take place and that, ultimately, loss of a neutral $OH^\bullet$ or $OD^\bullet$ might involve either of the o-hydrogens as well as the deuterium atom. Structures such as

are therefore visualized for the molecular ion.

All the above kinds of studies can be carried out at maximum sensitivity by using the technique of IKE spectrometry. The sensitivities which can be achieved are so high that "metastable peaks" of intensity only a few parts per million of that of the largest "metastable peak" in the spectrum can be studied. This enables metastable ions to be used in analyses often with advantages over stable ions. For example, the mass spectra of many organic molecules include both a molecular ion peak and a peak due to loss of a single hydrogen atom from the molecular ion. Analyses of partially deuterated compounds are then rendered difficult because of the large, and often uncertain, corrections which have to be made for the presence of the

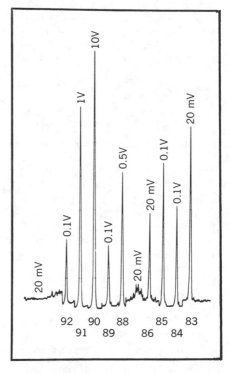

Figure 5. Focused "metastable peaks" in the spectrum of toluene. All the peaks arise from processes in which a single hydrogen atom has been lost

fragment ion. The method has been illustrated by the case of toluene which shows a "metastable peak" in its mass spectrum at an apparent mass of 90 due to the transition:

$$92^+ \rightarrow 91^+ + 1$$

To transmit the ion of mass 91 (to give the "metastable peak" at mass 90), the electric sector is "tuned" to a voltage equal to a fraction 91/92 of its normal value. If the "tuning" is successively altered to values of 90/91, 89/90, and so forth, other metastable transitions can be plotted in which ions of mass 91, 90, and so forth lose a single hydrogen. Such a scan is shown in Figure 5 and it can be seen that there are no peaks (except the isotope peak at mass 91) greater in intensity than ~2% of the peak at mass 90. If a monodeutero toluene is examined in the same way, the peak as mass 91 (corresponding to $93^+ \rightarrow 92^+ + 1$) is again by far the largest peak. The peaks at masses 90 and 91 (correcting for naturally occurring isotopes) are proportional in the mixture spectrum to the molar amounts of toluene and monodeutero toluene. The proportionality constants can be simply determined by adding a known amount of toluene to the mixture and remeasuring the peak heights of the "metastable peaks" at masses 90 and 91. This method is quick and accurate and is thought to be of wide application in the analysis of deuterated compounds.

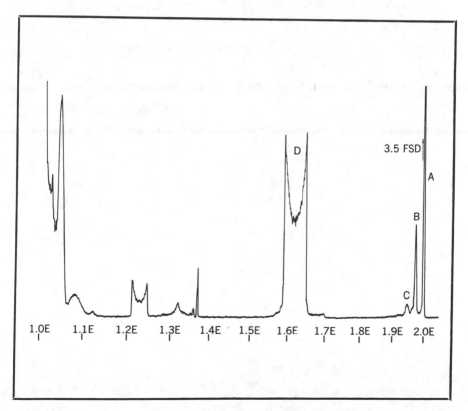

Figure 6. The IKE spectrum of benzene at electric sector voltages greater than the voltage *E* at which the main beam of stable ions is transmitted

OTHER IONIC REACTIONS

All of the ions discussed so far have been truly metastable and undergo fragmentation unimolecularly. Many other ionic reactions can be made to occur by causing the ion beam to interact with a gas in the field-free region in front of the electric sector. Such a "collision gas" can cause charge exchange between the ion beam and neutral gas molecules and can also lead to fragmentation of the ions in new ways. The study of all these product ions can be carried out using exactly the same methods as for metastable ions.

Consider Figure 6 which shows part of the IKE spectrum of benzene in the presence of nitrogen as a collision gas. All the peaks shown in this partial spectrum are transmitted at electric sector voltages *above* the voltage E at which the main beam of stable ions is transmitted. Peak A, for example, is due to all the species of doubly charged ions in the ion beam which, because of their two charges, have received twice the energy of acceleration of the singly charged ions. The ions have then lost a single charge by capturing an electron from the collision gas and consequently are transmitted through the electric sector at a voltage $2E$. The mass spectrum of this transmitted peak gives another "fingerprint" for benzene, equally as unique as the normal mass spectrum. Other "fingerprints" could be obtained by mass analysis of the other peaks such as B and C. Peak D is due to the fragment ion $C_5H_3^+$ from $C_6H_6^{2+}$ which has been discussed above. The measurement of kinetic energy release can be made much more easily in the IKE spectrum where there is no interference from stable ions.

Thus, the study of metastable ions has become of increasing interest and promises to lead to an improved understanding of fragmentation pathways, the energetics of ion decomposition and the structures and stabilities of positive ions. It has important applications in analysis, especially of isotopically enriched materials, and the wealth of detail in the IKE spectra and the various other "fingerprints" may lead to an improved ability to distinguish between structurally similar molecules on the basis of their spectra.

Thin-Layer Densitometry

Morton S. Lefar[1] and Arnold D. Lewis

Department of Analytical/Physical Chemistry, Warner-Lambert Research
Institute, Morris Plains, N. J. 07950

Precise and sensitive instrumentation for the quantitative evaluation of thin-layer chromatograms is now available

THE USE OF thin-layer chromatography (TLC) as a method of providing rapid separation of mixtures is well-known and documented. Since there are only microgram amounts on the TLC plate, usually of the order of 0.001 to 0.1 mg, the quantitative analysis of separated components has been difficult. The procedure of scraping and collecting the adsorbent followed by extraction with solvents is laborious and difficult especially when several components of low concentrations are to be quantitated. Many investigators have reported that analytical blanks are generally high and not reproducible owing to impurities in the adsorbent and to particles which remain in suspension in the eluate. Some substances such as certain steroids and amines may be decomposed during elution from the absorbent depending on the type of adsorbent and elution technique used.

These problems have been recognized by the scientific instrumentation industry, and equipment is now available which has both the desired precision and sensitivity. The purpose of this article is to discuss some of the aspects of thin-layer densitometry.

[1] Present address, Rhodia Inc., P.O. Box 111, New Brunswick, N. J. 08903

General Considerations

The results of investigators vary greatly as to the relationship between absorbance and the concentration of a material spotted on a TLC plate. (Readers should note that technically the term "absorbance" should be replaced by the mathematical equivalent "optical density" for scattering samples, where the photocurrent depends on the solid angle collected by the photometer. However, since the term "absorbance" is used by most manufacturers in describing their TLC densitometers, we have also done so in this article.) Several investigators have reported a direct relationship under suitable conditions between the integrated area under a densitometric peak, the absorbance value of the spot center, or the optical density of the entire spot and the compound concentration (where the term "concentration" actually means concentration × volume or quantity of the analyte). Other authors have reported a linear relationship between the logarithm of the weight of material spotted and the square root of the integrated absorbance reading. There are also reports of a linear relationship between the absorbance and the logarithm of the concentration of compound or between the curve area and the square root of the quantity of the substance applied. If the Beer-Lambert law were obeyed, one could predict a linear relationship between the logarithm of the absorbance area and the logarithm of the concentration of the compound spotted.

Unfortunately, this most often is not the case because of the translucence of the plate, the scattering of light, and other unknown factors. The behavior of a light beam when it impinges upon an absorbing material distributed on and in the surface of the gel is complex and not clearly understood. The optical phenomenon in a translucent material can be written as:

$$I_o = I_a + I_t + I_r$$

where I_o = incident light
I_a = absorbed light
I_t = transmitted light
I_r = reflected light

Shibata (3) reported at least six other types of resultant light, which may also be applicable to TLC densitometry. One of the criteria which affect the precision in the densitometry of colored materials is the use of reagents which will react with the compounds to be quantitated to afford a stable colored spot without diffusion of color to the surrounding areas. The color intensity must be reproducible and show a definite relationship between the amount of light absorbed by the compound and the amount of material spotted. The design of an instrument which meets these criteria is, therefore, not a small task.

Methods. There are primarily three modes of optical scanning of thin-layer plates. They are visible reflectance (or transmission), ultraviolet-excited fluorescence, and fluorescence quench.

Visible Reflectance. For the detection of some compounds, the developed chromatogram can be visualized by spraying with a reagent to develop a color. Charring with acid and heat is also used extensively to make the compound visible. However, with the current instruments available, this is per-

haps the most difficult technique with which to obtain quantitative data. It is critical that the developing reagents are sprayed evenly onto the plate. If this is not done (in practice difficult to perform), an uneven baseline will be obtained when the spot is scanned with a densitometer. The reagent must also be chosen and applied in a manner so that the spot intensity is constant during the scanning period and that compounds of high concentration have reacted completely. This is critical in the preparation of a calibration curve and in obtaining analytical precision. If the charring technique is used, then the oven temperature and time of reaction must be carefully and reproducibly controlled. Samples must be applied so that a reference lane is adjacent to each row of spots. If the plate has been uniformly charred or sprayed, it will permit the "cancelling out" of background absorption.

Ultraviolet Fluorescence. Those compounds that fluoresce visible radiation when excited with ultraviolet light can give good quantitative results since the area under the scan curve is usually linear with concentration over a wide range. Using this technique, nanogram amounts of compounds may be detected provided the plates are of good quality. The baseline is generally straight and not greatly affected by the thickness of the adsorbent. It is important to prescan developed plates which do not contain samples to assess the amount of fluorescence "noise" present in the adsorbent layer. This may be caused by the use of solvents containing fluorescent impurities. Since some compounds show a deterioration of fluorescence with time, calibration curves should be prepared for each substance being analyzed.

Fluorescence Quench. This is presently one of the most popular methods for visualization and quantitation of materials since neither a color-forming reagent nor charring is necessary. The number of compounds which can quench fluorescence is quite large when compared to those which exhibit fluorescence upon ultraviolet excitation. The most useful fluorescent phosphors are those which radiate 5220 Å energy when excited with 2537 Å ultraviolet light. Fluorescent plates should be excited first with uv light in the instrument for a few minutes before starting the analysis to "dark adapt." Quench systems usually show a nonlinear relationship between the integrated area under the scan curve and the sample concentration. A separate calibration curve must be prepared for each compound since different substances show different "extinction coefficients."

Transmission vs. Reflectance. With transmission scanning, the background absorbance varies with layer thickness. The adsorbent thickness must, therefore, be the same at all parts of the plate. The compound being analyzed should absorb the same amount of light regardless of the thickness of the layer. Most commercially coated plates give little problem with variations in thickness when the instrument used was operated so that the area adjacent to the spot being analyzed is used as a reference. Since the scanning beam must pass through a small amount of adsorbent layer before it can contact the spot, the amount of scattered light will vary. A compact spot will, therefore, permit the scatter to become relatively unimportant. The manner in which the compound is distributed on the adsorbent layer becomes of greater significance when using a reflectance method since this takes place mainly at the surface of the plate and detector response may decrease with increasing concentration of compound.

Layer Thickness. In an investigation utilizing the Joyce Loebl Chromoscan densitometer, Dallas (1) reported that the thickness of the adsorbent layer is a critical factor in densitometry. Peak height, width, and area all depended on layer thickness. As the layer thickness decreased, the peak height and area increased when scanning with reflected light. With transmitted light, both peak height and area decreased as the layer thickness decreased. No loss of peak resolution was observed as the layer thickness increased with transmitted light, but an appreciable loss of resolution occurred with reflected light.

Precision in Spotting. Although Dallas found a slightly higher standard deviation for scanning by transmitted light, consideration of the variance due to the method of evaluation and to the densitometry indicated that the standard deviation in the volume of different 2-μl capillary pipets was 2.2%. This error was greater than that due to the chromatography, the densitometry, and the method of determining the area under the curve combined. Other investigators have also reported that the application of compounds to a TLC plate is a major source of error.

Calculating Amount of Material. If there is a considerable variation in the area measured under the curve for the identical amount of material on the same plate and from different plates, it is desirable to use a calibration curve for each compound to determine the regression line equation. This can then be utilized for the calculation of the amount of material actually present. Shellard and Alam (2) reported that for two alkaloids, the mean square between readings on different plates is higher than the mean square between readings on the same plate. For these oxindole alkaloids, the average result obtained by a number of readings from one plate was associated with a lower variation than the average result obtained by a number of measurements from different plates. The average result obtained from a number of measurements from one plate may be higher or lower than the theoretical value. Better results can be obtained by the combination of measured values from different plates than by the combined values obtained from one plate.

Mobile Phase. A change in the stationary and mobile phase used for the separation affects the shape and size of the separated component. Different systems offer different rates of movements depending on the adsorption isotherm and (or) the partition coefficients of the component with respect to the solvent system. The solvent system that permits a low R_f usually gives a small, round compact spot. Systems that result in high R_f of the compound usually offer larger, often elongated and sometimes laterally diffuse spots. The changes in size and shape are reflected in a different distribution and deposition of compound on the adsorbent. Of importance is that the same quantity of substance may give a different response with regard to the densitometer readings after separation utilizing different solvent systems.

Instruments

The Nester-Faust Uniscan 900 (Nester-Faust Manufacturing Corp., Newark, Del.) utilizes the illumination and detection geometry used for many years in measuring colored solids and surfaces. The chromatogram is illuminated at a 45° angle to the surface of the plate by an incandescent source and then scanned at 90° to the surface. Two methods of operation are available in the visible reflectance mode. The instrument can be operated as a single-channel photometer where space is not available for use in scanning the plate surface along with the compound to be analyzed. The differential scan mode may be used which provides for a reference path adjacent to the separated sample. For uv fluorescence (native fluorescence or quenching mode) a second optical head is supplied as well as a plug-in front panel section with the necessary controls for amplification and power for the source and photomultipliers.

A separate module is used to afford regulated high-voltage power for the low-pressure mercury source and the photomultiplier. The uv fluorescence head contains a low-pressure mercury

source and filters to provide maximum emission at 2537 Å or 3654 Å. Two slits, 1 mm and 2 mm, are utilized in the instrument. The thin-layer plate is inserted in a carrier assembly. The appropriate detector head then moves over the plate at a rate of 8 in. in 2.5 min. The optional Summatic 1502 electronic digital integrator provides for automatic detection and digital printout of the area under the analog curve. A plug-in log-to-linear converter is available which can convert the logarithmic function output to a function linear with concentration. Front panel adjustments allow the transfer function log-to-linear to be modified to accommodate wide deviations from Beer's law.

The Schoeffel Model SD-3000 (Schoeffel Instrument Corp., Westwood, N. J.) double-beam spectrodensitometer utilizes a 150-watt Xenon or 200-watt Xenon-Mercury lamp as a light source. The radiated light is focused onto the entrance slit of a monochromator by means of a quartz condenser and surface reflector. The monochromator has a linearly calibrated wavelength selector dial, enabling the user to select monochromatic light in the range 200–700 nm. The light emitted from the monochromator exit slit is magnified by a quartz optic and a folded-beam mirror system, thus uniformly illuminating the sample and reference area on the TLC plate.

The beams of light are scattered by the TLC plate and appear to the detectors as a pair of secondary light sources. Since the illuminated areas can be varied in width as well as in length, maximum resolution (narrow slit width) or good background-noise integration (wide slit width) can be obtained. The scattered light (secondary light sources) can be detected either directly from underneath the plate or from above the plate by phototube interception of the scattered light at 45° to the optical normal (reflection). If necessary, interference wedge monochromators (continuously adjustable from 400–650 nm) can be inserted for emission analysis.

The sample and reference signals are balanced by means of a continuously variable gain balance adjustment on a plate location where reference and sample area have identical optical densities. Both signals are separately and continuously amplified and automatically computed into an optical density which equals log (reference signal)/(sample signal). On special request, a linear ratio of (sample signal)/(reference signal) can also be provided to facilitate fluorescence and reflection measurements. Signals are recorded by a strip chart recorder and integrated by a disk integrator.

The TLC stage drive is controllable and synchronized in a number of ratios to the recorder speeds of the system. For obtaining maximum quantitation accuracy, signal interference, caused by sample spillover into reference areas, is prevented by scoring the adsorbent layer into 10-mm width strips. The zones are spotted alternately, providing a blank reference strip between samples.

The Joyce Loebl Chromoscan densitometer (Technical Operations Inc., Burlington, Mass.) operates on the double beam principle. It is a null-point instrument in which a grey wedge is balanced against the reflection from or transmission through the sample. Measurements can be made by reflectance, transmission, fluorescence, or uv absorption through the use of an auxiliary high-resolution attachment. The light source is a quartz iodide tungsten lamp, with gelatin or interference filters. In the uv, mercury and deuterium lamps can be utilized. The uv lamp can be placed alongside the visible source in a dual lamphouse for rapid changing between sources. The parallel focused beam with a 90° incident angle can be adjusted to any shape within a 1-cm diameter circle.

For reflection measurements, only light scattered at a 45° angle to the incident light beam passes to the photomultiplier. With the standard instrument plates, up to 25-mm widths

Figure 1. Schematic block diagram of Farrand chromatogram analyzer

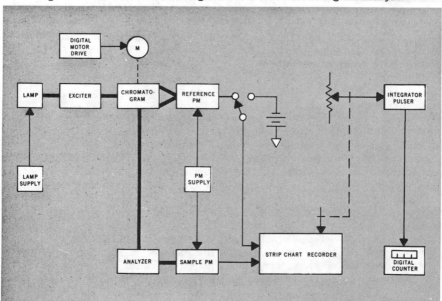

can be scanned. For plates up to 200 × 200 mm, an accessory is available. In transmission measurements, a separate photomultiplier is positioned several centimeters behind the sample so that scattered light is not measured. One of the unique features of this instrument is the use of plastic cams so that a correction factor may be built into the analysis affording linearity of results. Different angles of straight cams may be used for scale expansion when dealing with weak reflection or small extinction changes. A variable slit is available in the range of 0–1 mm in width and 0–3 mm in height. Filters are utilized to vary the spectral characteristics of the light. Sample movement and the recording drum can be rigidly linked, which in turn may be coupled through a gear box to give different expansion ratios of recorder/sample movement ≤ 9:1. An automatic six-digit readout integrator is provided on the instrument.

The Farrand VIS-UV Chromatogram Analyzer (Farrand Optical Co., Inc., Bronx, N. Y.) measures uv-absorption by either fluorescence quenching or by emission of native fluorescence. It also measures uv or visible absorption directly with or without staining of the plate or the spotted compounds. It can perform these functions by either double-beam ratio or single-beam operation (Figure 1).

The arc from a xenon lamp is imaged in magnified form at the entrance of the exciter monochromator. By optical techniques and slit masks, a sliver of light is produced which illuminates the surface of the plate. The sample carrier moves in such a manner that the length of the light beam is at right angles to the direction of scanning. In this manner, the spot to be analyzed and the lane on each side are illuminated. The two reference signals are then averaged. The double-beam procedure lessens the effects of lamp intensity variations and plate-background gradients. This instrument utilizes two monochromators, one in the excitation

portion and one in the analyzer leg with provisions for filters and polarizers. In the reference leg, a filter is used routinely. The spectral range of this instrument ranges from 200–785 nm. A built-in integrator/counter integrates the area under the curve by translating the pen position into a signal which is a function of scan time. After integration, the data are converted to digital form for a digital counter.

The Zeiss Chromatogram Spectrophotometer (Carl Zeiss, Inc., New York, N. Y.) is designed to determine quantitative and qualitative reflectance, transmission-absorption, fluorescence, and fluorescence quenching. Its major components consist of a dual light source (deuterium and tungsten), the M4Q III prism monochromator, an optical imaging assembly, the chromatogram scanning table, and an electronic detection system.

In addition to the standard sources, a xenon and mercury lamp can be furnished. The monochromator provides a spectral range of 200–2500 nm. Controls at the optical imaging assembly permit precise focusing of the monochromator slit or prism image. Slit width is continuously variable from 0.01–2.0 mm and slit height may be adjusted from 2–14 mm. For ideal coverage of circular spots a continuously variable circular aperture ranging from 6–30 mm is provided. The instrument can be operated in two basic arrangements as shown in Figure 2. Arrangement *M-Pr* facilitates monochromatic illumination normal to the sample surface while the detector is located at 45° for determination of diffuse reflection or emission. Arrangement *Pr-M* positions the source at 45° to the sample surface while diffuse reflection or emission is observed normal to the sample surface *via* the monochromator. In the transmission mode the detector is placed underneath the sample.

Intensities thus observed are indicated on linear and logarithmic scales of the photometer and can be docu-

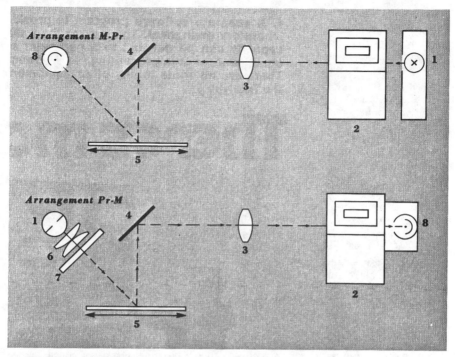

Figure 2. Simplified path of beam in Zeiss chromatogram spectrophotometer

Arrangement *M-Pr* (*M* = monochromator; *Pr* = sample): Radiation from light source first traverses the monochromator, then strikes sample. Arrangement *Pr-M:* Radiation first strikes sample then traverses the monochromator

1. Light source: in arrangement *M-Pr*, hydrogen lamp, incandescent lamp, or high pressure xenon lamp; in arrangement *Pr-M*, Hg lamp, hydrogen lamp, or incandescent lamp in special lamp unit
2. Monochromator

3. Intermediate optical system (simplified)
4. Deviating mirror
5. Chromatogram
6. Intermediate optical system
7. Filter
8. Detector

mented on recorders with or without integrators. For spectrophotometric analysis of a specific spot, the chromatogram table is held stationary while the monochromator is automatically scanned through a desired spectral range. An internal reflectance standard permits immediate reference to confirm instrument setting and stability.

The chromatogram adapter can also be supplied as an accessory to Carl Zeiss PMQ II users.

The Photovolt Densitometer Model-530 (Photovolt Corp., New York, N. Y.) operates on the principle of straight transmission measurement with a high degree of collimation-rejected scattered light. A variety of light sources can be used. For fluorescence measurements two mercury vapor lamps are utilized. One affords a sharp 254-nm emission and the other a broader uv range centered at about 350 nm. A uv-transmitting glass filter is used in front of the mercury arc to eliminate the visible radiation when utilized as an ultraviolet source. Most commonly used is a stabilized, small-filament tungsten lamp. This light source is essentially re-imaged on the thin-layer plate giving a small area of illumination. The upper collimating slit picks up only the central transmitted beam and rejects the scattered light.

The multiplier photometer permits high-resolution scanning with a 0.1-mm slit. The multiplier photometer has a large indicating meter calibrated in absorbance units. It has a four-decade switch in addition to a continuous sensitivity control. The photomultiplier tube is contained in a detachable housing, and a choice of tubes is available which covers the visible and ultraviolet region of the spectrum. Automatic scanning requires the use of a motor drive for the stage and for the variable response recorder. The synchronous motor advances the plate 1 in./min. The photometer output is then fed into the recorder. The key feature of the recorder is its variable recording capability. Twelve responses are available. The recorder can be set so that either linear or nonlinear responses result, so that the logarithm (absorbance) or steeper curves may be plotted.

Conclusion

There is now available suitable instrumentation for the quantitative evaluation of thin-layer chromatograms. The sources of error involved in the spotting, layer thickness, development, visualization procedures, and quality of the plates can be greater than that inherent in the use of any given instrument. Since this article is not intended to be an exhaustive survey of densitometers for thin-layer chromatography, the reader should refer to the American Chemical Society's 1969–1970 *Laboratory Guide to Instruments, Equipment and Chemicals* for further information on other instruments. Readers are also referred to E. J. Shellard's "Quantitative Paper and Thin-Layer Chromatography," Academic Press, 1968, and E. Stahl's "Thin-Layer Chromatography," (second edition), Academic Press, 1967.

Acknowledgment

The authors are grateful for the assistance and cooperation of the instrument manufacturers mentioned in the article.

References

(1) M. S. J. Dallas, *J. Chromatog.*, **33**, 337 (1968).
(2) E. J. Shellard and M. Z. Alam, *ibid.*, p. 347.
(3) K. Shibata, *Methods Biochem. Anal.*, **7**, 77 (1959).

COMMENTARY

by Ralph H. Müller

THIS DISCUSSION of thin-layer chromatography is timely and useful. It is at once apparent that these elegant instruments are far more reliable and reproducible than the complex system (chromatogram) which they are called upon to measure. The timeliness is evident if we consider Stahl's statement, "The golden era of paper chromatography began and by 1956 over ten thousand publications on the use of this 'universal' method had come out." In the case of TLC, by the end of 1965, over 4500 publications had appeared as well as a dozen monographs and reviews in 11 languages.

Paper and thin-layer chromatography have been applied to almost every conceivable class of substance, but in quantitative evaluation a half dozen or more functional relationships have been proposed and used, relating the optical measurement to concentration. It seems that further improvements must come from the chemical and manipulative factors leading to more reproducible matrix systems. For the present, the instrumental resources are more than adequate.

We continue to be confused by the term "quenching" as applied to that technique in which a chromatographic spot may, in one way or another, diminish the fluorescence of a substance incorporated in the thin layer or sprayed on the chromatogram. In most monographs or treatises, the term is used ambiguously. Obviously, there are two cases involved. In true fluorescence quenching, the "quencher" deactivates the excited fluorescent substance and diminishes or obliterates its emission of radiation. In the other case, by absorbing the exciting uv radia-

tion, it acts in a manner not much more subtle than interposing a piece of cardboard or a thick sheet of glass. Either phenomenon can be useful in locating or measuring a spot, but the functional relationship with concentration should be hyperbolic in the first case and logarithmic in the second. For true quenching, the simple explanation in terms of deactivation by "collision of the second kind" should suffice. As first pointed out by Klein and Rossland, Franck, and others, this leads to the equation by Stern-Volmer. The extinction is a hyperbolic function of quencher concentration, and a simple algebraic manipulation can furnish a linear plot. Chemists may prefer chemical explanations, such as complex formation with the appropriate equation, but both may be right. In any case, an exact formulation can be obtained from good data with the advantage of knowing something about what is really going on.

We think it might be useful to look into flying-spot scanners. At least 15 years ago, it was possible to buy scanners for a matter of a little over $100. These were designed for radio hams and TV hobbyists and consisted of a small cathode ray tube acting as a flying light source. The light was focused on a photographic negative (could be a chromatogram) and thence to a photomultiplier tube. The output could be displayed on a large oscilloscope or two-dimensionally on a TV screen. This would give an instantaneous and repetitive scan of the pattern. By more recent techniques, such as CAT (Computer Averaged Transients), feeble signals can be "lifted" out of noise, with a signal-to-noise ratio of $n/\sqrt{n}$, where n is the number of scans.

It would seem that the advantage of high-speed scanning systems is not trivial because it could take advantage of running a large number of calibrating standards along with an unknown. Aside from the time required for spotting a sample and standards, the time required for chromatogram development is the same.

Inherently, radioactive techniques provide an approach which is simpler to interpret than optical methods. One of the earliest applications in scanning labeled metabolites (2) gave improved precision and a saving of time of the order of 1800:1 compared with conventional radioautographic methods. Neutron activation methods are extraordinarily sensitive in many instances but depend upon accessibility to large irradiation facilities. Much more attractive is characteristic X-ray absorption using isotope excited X-ray sources. Tiny but acceptable sources can be prepared for X-rays of all elements between Z equals 22–93 (1). They are also suitable for taking advantage of the absorption edge technique. The nuclear techniques have the advantage of digital readout and integration if scalers are used instead of count meters.

References

(1) Müller, R. H., "Radioisotopes in the Physical Sciences and Industry," International Atomic Energy Agency, Vienna, 2, 65 (1965).
(2) Müller, R. H., and Wise, E. N., ANAL. CHEM., 23, 207 (1951).

Pulsed and Fourier Transform NMR Spectroscopy

T. C. Farrar

Institute for Materials Research
National Bureau of Standards
Washington, D. C. 20234

Certain advantages in pulsed nmr spectroscopy, especially increased sensitivity, make it an attractive technique. With the solution of many of the problems in the spectrometers and associated computer instrumentation, it is safe to predict it will rival continuous wave–nmr spectroscopy in usage

Pulsed nmr spectroscopy has been with us for almost as long as the more conventional, continuous wave (cw)–nmr spectroscopy; however, it is less familiar to most chemists. This situation is now changing rapidly since it has been pointed out that the induction decay signal obtained in a pulsed nmr experiment and the cw-nmr spectrum constitute a Fourier transform pair (1), provided that certain conditions are met.

Since potentially one of the most exciting and fruitful areas of research is ^{13}C Fourier-transform (Ft)–nmr spectroscopy, a large part of this article will be devoted to that subject. Other problems which are amenable to pulsed nmr methods will be mentioned briefly.

The primary purposes of this article are fourfold: (1) to point out the advantages of Ft-nmr spectroscopy, (2) to introduce the reader to some of the jargon used in pulsed and in Fourier-transform–nmr spectroscopy, (3) to attempt to explain in a simple way what pulsed and Ft-nmr spectroscopy are about and how they differ from "normal" or cw-nmr, and (4) to give the reader some information about the minimum requirements needed in the basic instrumentation of a Ft-nmr spectrometer.

IMPORTANCE AND LIMITATIONS OF CW-NMR

In the past six years or so a number of scientists have been working with great success in the area of ^{13}C, cw-nmr spectroscopy. In particular, the elegant work of Professor Grant and his coworkers at Utah, of Professor Lauterbur and his coworkers in New York, and of Professor Roberts and his coworkers in California attests to the vast wealth of information available from ^{13}C-nmr spectroscopy (2). As has been pointed out by these workers, ^{13}C-nmr spectroscopy is probably the single most important tool in determining the structure of organic and biological molecules and, as such, it is almost impossible to overestimate its importance. In an overwhelming number of cases where almost no information is available from proton nmr, ^{13}C-nmr spectra easily show differentiation between two complex molecules which differ only in the subtle rearrangement of a few carbon atoms. The reason for this lies partly in the fact that the range of ^{13}C chemical shifts (200 ppm or more) is much greater than for protons (about 20 ppm) and partly because in non-^{13}C-enriched molecules only one ^{13}C nucleus is in any given molecule.

The primary limitation of the technique has been a lack of sensitivity, due both to a relatively small magnetogyric ratio, γ_c, and to the fact that ^{13}C has a natural abundance of only 1%. By employing signal-averaging methods and by taking advantage of proton noise-decoupling techniques, great advances have already been made in increasing the signal-to-noise ratio (S/N) obtainable from ^{13}C-nmr spectrometers. By utilizing all the tricks available, good ^{13}C spectra have been obtained of relatively complex steroids (2). This has been accomplished by signal averaging for long periods of time and by using large amounts of material. Often, however, only small amounts of material are available. This is especially true in the case of samples of biological interest. The practical limit for signal averaging is about 20 to 30 hr, restricting the number of systems that can be investigated. Furthermore, in the area of biological research, the lifetime of a sample is often such that one has only a few hours in which to complete an investigation. In short, the sensitivity inherent in ^{13}C-cw-nmr spectroscopy, even using proton noise decoupling and long-term signal averaging, is woefully inadequate. Happily, Fourier-transform techniques, especially when used in multiple pulse experiments, afford one a significant increase (in principle, up to a factor of 100 or more) in this sensitivity.

It has been demonstrated (3) that for ^1H-Ft-nmr, one can obtain a substantial increase in sensitivity. Since one should realize even greater gains in the sensitivity of ^{13}C-Ft-nmr over ^{13}C-cw-nmr, it is surprising that more work has not been done in this area.

The reasons for this are probably due to the lack both of good pulsed nmr spectrometers and of good, reliable, and inexpensive data-recording and data-reduction devices (i.e., inexpensive, dedicated computers suitably interfaced to the spectrometer).

SOME ADVANTAGES OF FT-NMR

Up to this point we have attempted to give some evidence of the importance of ^{13}C-nmr spectroscopy, and we have stated that the main limitation in the technique is its lack of sensitivity. We have also stated that Ft-nmr offers a way of obtaining a substantial increase in sensitivity over that possible via cw-nmr spectroscopy. We now try to point out why this is so.

Basically, pulsed Ft-nmr spectroscopy is just a special form of interferometry and, as is usually the case, to interpret the results it is necessary to go to the conventional form of spectral representation by means of a Fourier transformation. This can be done either by an analog Fourier analyzer or it can be done by collecting the interferogram digitally and performing the Fourier transformation on a digital computer. The technique consists of the application to the sample of a short, intense pulse of radio-frequency (r.f.) energy and the measurement as a function of time of the resulting free induction decay signal from the nuclear spins in the sample. Fourier transformation of the free induction signal gives the ordinary high-resolution spectrum (3, 4).

The advantage of the Ft-nmr technique is that the free induction decay signal is obtained rapidly, so that in a given length of time it is possible to apply the pulse repetitively and add the free induction decay signals coherently in a digital computer or a time-averaging device. The time, T, required to record a spectrum is given by $T \simeq r^{-1}$ sec, where r is the resolution desired (usually about 1 Hz); T is independent of the width of the

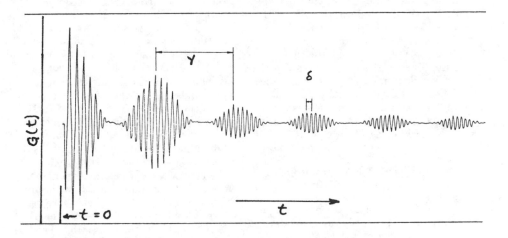

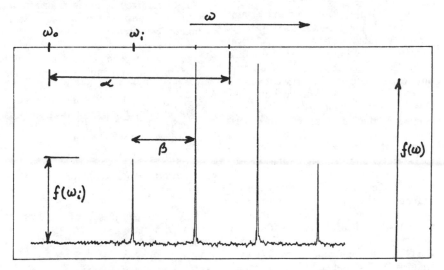

Figure 1. (a) depicts a "normal" ^{13}C cw-nmr spectrum of $H_3^{13}CCl$. The chemical shift from a given reference frequency, ω_o, is α Hz and the $^{13}C-H$ spin-coupling constant is β Hz. $f(\omega_i)$ is the intensity of the signal at the position ω_i ($= 2 \pi \nu_i$) in the spectrum. (b) depicts the interferogram of $H_3^{13}CCl$. The C—H spin-coupling constant is proportional to $1/\gamma$, and the chemical shift is proportional to $1/\delta$

spectrum, Δ. The consequence is that sensitivity is much higher. The sensitivity enhancement, factor ϵ_s, is given by $\epsilon_s \simeq (\Delta/r)^{1/2}$ or, alternatively, the time enhancement factor ϵ_t is given by $\epsilon_t \simeq \Delta/r$ (S/N improves only as the square root of the time, or as the number of scans in a time-averaged experi-

ment). Thus, the time required to obtain a spectrum of a given signal-to-noise ratio (S/N) is shorter using the Ft-nmr technique by a factor of ϵ_t. In ^{13}C-nmr studies of complex systems, Δ is typically 200 ppm and r is about 1 Hz, so at 15 MHz (at which frequency many current spectrometers operate), the time enhancement factor, ϵ_t, is $\epsilon_t \simeq$ (200 ppm · 15 MHz/1 Hz) = 3000. With the newer ^{13}C spectrometers which operate at 25 MHz or 55 MHz, the time saved is even greater. Also, one can achieve even greater time enhancements (preliminary theoretical and experimental work indicates that an additional factor of about 10 can be realized) by using multiple pulse techniques (5).

FURTHER COMPARISONS AND APPLICATIONS

The two methods, Ft-nmr and cw-nmr, differ primarily in the duration and magnitude of the r.f. field used in the experiments. In the pulsed Ft-nmr experiment, an intense (200 G) r.f. field is applied to the sample, but it is left on only for a very short time (typically, about 1 μsec). In the cw-nmr experiment, a rather weak (10^{-3} G or less) r.f. field is used, but is allowed to run continuously. The result of this is that in the cw-nmr experiment, one produces an r.f. field, H_1, at a single discrete frequency which perturbs only slightly that line in the spectrum which resonates at precisely this frequency. This frequency is then changed slowly at a rate comparable to the resolution desired. For example, for a resolution of 1 Hz the frequency is changed at a rate of about 1 Hz/sec. For ^{13}C-nmr one works usually at a frequency of about 15 MHz, and in a typical experiment this frequency would be changed slowly from a value of v_o to a value of $v_o + 3000$ Hz. The field, of course, is "locked" to a fixed value given by the expression 2 πv_o = $\omega_o = \gamma_c H_o$, where γ_c is the magnetogyric ratio for ^{13}C. The amplitude, $f(\omega)$, of the signal induced is recorded as a function of the frequency, ω, and the result of a sample of $H_3{}^{13}$CCl is the ^{13}C-nmr spectrum shown in Figure 1a. In the pulsed nmr experiment, the sample is perturbed very much indeed by the very large H_1 field. This has

the effect of irradiating the sample not with a single discrete frequency, as is the case with cw-nmr, but with a band of frequencies centered at 15 MHz and covering a range of about (1/4 τ) Hz where τ is the time duration (in seconds) of the pulse used. For a 1-μsec pulse, then, the pulse covers the frequency range 15 MHz $\pm$ 120 kHz. The result is that the nuclei associated with the various lines in the spectrum are induced to radiate their characteristic r.f. signals simultaneously and, as is the case whenever one mixes a number of signals simultaneously with different frequencies, an interference pattern is obtained. The result for H_3-^{13}CCl is shown in Figure 1b. This is usually referred to as an interferogram or the Ft-nmr spectrum or a "time-domain" spectrum.

Since the interferogram shown in Figure 1b, even in this simple example, is much more difficult to interpret than the frequency spectrum shown in Figure 1a, we desire some method of transforming Figure 1b into 1a. The transformation from 1b to 1a is given according to the formula

$$f(w_j) \rightarrow \sum_k G(t_k) \exp{(iw_j t_k)} \Delta t$$

It is, of course, possible to go from the form in Figure 1a to 1b. In this case the proper procedure is given by

$$G(t_j) \rightarrow \sum_k f(w_k) \exp{(iw_k t_j)} \Delta w$$

An alternative, at least in principle, to the pulse technique is a many-channel experiment. In this method one could, for example, apply 3000 transmitter frequencies from v_o to ($v_o + 3000$) Hz and use 3000 phase detectors to detect separately each one of the individually induced signals. This experiment, then, too, would give one a time-enhancement factor, ϵ_t, of 3000 since one is doing 3000 cw-nmr ex-

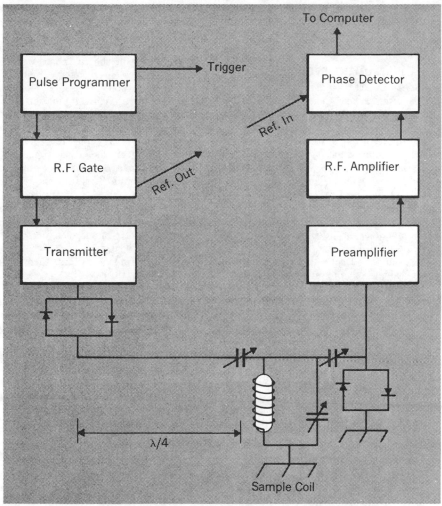

Figure 2. A simplified block diagram of a typical pulsed nmr spectrometer

periments simultaneously. In this experiment one could also use a single phase detector and obtain an interferogram just as in the usual pulsed Ft-nmr experiment.

In addition to the Ft-nmr application, there are a number of other problems amenable to pulsed nmr techniques. By measuring the spin-lattice relaxation time, T_1, and the spin-spin relaxation time, T_2, as a function of temperature, one can get a wealth of in-

formation about a given molecule system in the solid, liquid, or gaseous state, ranging from a determination of both activation energies and frequencies of motion of a given molecular species to the study of the "structure" of liquids and the determination of correlation times for angular momentum. A variety of multiple pulse techniques now exist for measuring self-diffusion coefficients and the Carr-Purcell-Meiboom-Gill pulse experiment can be used to study chemical kinetics over a wide dynamic range. An excellent recent review (6) gives a number of applications in these areas.

INSTRUMENT REQUIREMENTS

One remark and one question that we are often asked these days is as follows: "I would like to get started in the area of Ft-nmr. What are some of the more important considerations to keep in mind when buying equipment?" In the hope that it will serve a useful purpose, I would like to comment on what I believe are some of the essential requirements to look for in a pulsed nmr spectrometer and in a signal averager or "dedicated" computer to be used with it.

A simplified schematic diagram of a typical pulsed nmr system is shown in Figure 2. The pulse programmer determines the sequence of pulses in an experiment, the width of each pulse, and the repetition rate between sequences. It also provides a synchronizing pulse to start the time base drive in a recording instrument such as an oscilloscope, a gated boxcar integrator, or a signal averager. The programmer should be flexible enough so that almost any sequence of pulses can be generated. Since the timing in many experiments, for example, the DEFT (5) experiment, is extremely critical, a digital programmer is to be preferred. In this case it is preferable to have a clock frequency of 2 MHz or higher to minimize radio frequency interference (RFI) problems. It is important that the pulse widths be adjustable to a precision of 1% or better from 1 μsec up to 1 msec or longer. The longer pulses are needed for rotating reference frame experiments.

The r.f. gate unit and transmitter should be capable of generating enough power to give $\pi/2$ pulse widths of 1 μsec or less. This is an important requirement because if the pulse widths are too long, say much greater than 3 or 4 μsec for ^{13}C-nmr, the assumption that pulsed nmr and cw-nmr form a

Fourier-transform pair begins to break down. Of course, it is necessary that the r.f. gate unit supply a coherent r.f. reference signal to drive the ring demodulator in the r.f. phase detector. The total system recovery time at 15 MHz should be 15 μsec or less and the preamp-receiver combination should have a voltage gain of 120 db or more. The entire system should be extremely well shielded to avoid RFI problems arising from clocks in the signal averager, pulse programmer, or computer.

For maximum flexibility the digitizer in the A/D converter of the computer should have at least 10 bits or more and it should have a digitizing rate of at least 20 kHz or more. For 15 MHz ^{13}C-nmr, a writing speed of 6 kHz is required [from sampling theory one knows that to cover a spectrum of "x" Hz (in our case $x = 3000$ Hz), it is necessary to sample at a rate of "$2 x$" Hz]. If one samples for 1 sec (to obtain a resolution of 1 Hz), then 6000 data channels are required in which to store the data. To make use of the FFT (fast Fourier transform) computer programs available, one needs a signal averager and/or computer with a capacity of at least 8192 words for data storage; the word size should be at least 18 bits although 16 will do. Of course, another 4000 words of memory are needed in the computer to carry out the Fourier transformation. After the cw-nmr spectrum has been calculated, a D/A converter is needed to plot out the final cw spectrum.

In summary, the field of ^{13}C-Ft-nmr spectroscopy is developing at a very rapid rate. Although there have been many problems, they have, for the most part, been solved. With the rapid development and ever-decreasing cost of small computers, it seems safe to predict that the time when Ft-nmr spectrometers are as ubiquitous as the cw-nmr spectrometers is not far away.

References

(1) I. J. Lowe and R. E. Norberg, *Phys. Rev.*, **107**, 46 (1957).

(2) J. D. Roberts, *et al.*, *J. Amer. Chem. Soc.*, **91**, 7445 (1969); D. Grant and B. Cheny, *ibid.*, **89**, 5315–19 (1967); P. C. Lauterbur, in "Determination of Organic Structures By Physical Methods," F. C. Nachod and W. D. Phillips, Eds., Vol. 2, Academic Press, New York and London, 1962, Chap. 4. Other helpful references are given in these three papers.

(3) R. R. Ernst and W. A. Anderson, *Rev. Sci. Instr.*, **37**, 93 (1965).

(4) A. Abragam, "The Principles of Nuclear Magnetism," Clarendon Press, Oxford, 1961.

(5) E. D. Becker, J. A. Ferretti, and T. C. Farrar, *J. Amer. Chem. Soc.*, **91**, 7784 (1969).

(6) J. Jonas and H. S. Gutowsky, "Annual Reviews of Physical Chemistry," Vol. 18 (Eyring, H., Ed., Annual Reviews, Inc., 1968).

CORRELATION SPECTROSCOPY

J. H. DAVIES
Barringer Research Ltd., 304 Carlingview Dr.,
Rexdale, Ontario, Canada

The potential applications of correlation spectroscopy techniques to new generations of laboratory analytical instruments are great. Benefits should result in certain fields of absorption, emission, and fluorescent spectrophotometry. Correlation techniques can also be extended to interferometry

THE DISTINCTIVE spectral signature of liquids, gases, and vapors, either in absorption or emission, can be used in electro-optical techniques to perform qualitative and quantitative measurements. One particularly powerful technique is that of correlation spectroscopy. In this technique, an incoming spectral signature is continuously cross-correlated in real-time against a spectral replica of the spectrum sought. The stored replica is contained within the spectrometer and may be optical or magnetic. The technique is applicable from the far ultraviolet through the visible to the infrared spectral regions. If suitable background radiation sources are available, the technique can become a complete remote sensing system for both qualitative and quantitative measurements.

Correlation spectroscopy has a significant role to play in both laboratory analytical procedures and in real-time on-line process control. Initial development has concentrated on gas analysis, particularly those gases associated with combustion processes and earth resource applications. The advantages of the techniques—high sensitivity, good selectivity, real-time readout—ensure a dynamic future for its applications to situations where optical inputs are available.

Basic Principle of Correlation Spectroscopy

The basic principle of correlation spectroscopy is the cross correlation of an incoming spectral signature against a replica spectrum stored within the instrument. The basic techniques have been well described in the literature (*1–5*) and only a brief outline is given here. Figure 1 shows the regular absorption spectrum of iodine vapor and typically represents the type of

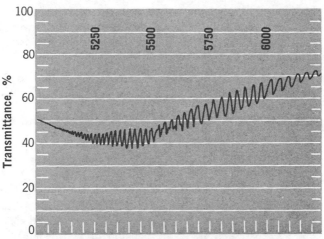

Figure 1. Absorption spectrum of iodine

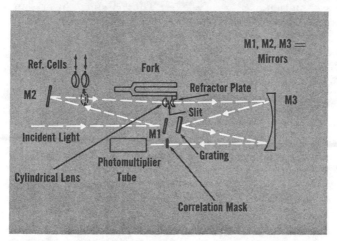

Figure 2. Dispersive system for vapor detection using spectrum correlation filter

spectrum that has been stored within the instrument. Figure 2 shows the type of correlation spectrometer which has been developed; this is a dispersive Ebert spectrophotometer.

The incoming radiation to be analyzed is passed *via* reflection from the Ebert mirror to a grating which is the dispersive element. The radiation emerging from the grating is now dispersed into its various spectral components to be reflected a second time by the Ebert mirror and to pass through the exit aperture of the spectrometer to a photodetector. The angular position of the grating determines the region of the spectrum that falls on the exit aperture. If the incoming radiation is now made to vibrate periodically through a limited wavelength excursion, and a mask replica of the sought spectrum is positioned in the exit aperture, the incoming spectrum will be moved periodically in and out of correlation against the mask. A beat signal will result and will be detected by the photodetector. Phase-sensitive synchronous detection techniques can be used for the signal detection using, as the reference, the drive signal that causes the spectrum to vibrate.

Signal Origin

The basic function of the correlation spectrometer is the Beer-Lambert law of absorption which describes the attenuation of a beam of monochromatic light transmitted through a medium which absorbs but does not scatter.

Then $\qquad H = H_0 e^{-acL} \qquad$ (1)

where $H_0 =$ incident irradiance at a fixed wavelength

$H =$ emerging irradiance at same wavelength

$a =$ absorption coefficient at the same wavelength

$c =$ concentration of medium

$L =$ pathlength through medium

If the absorption coefficients are different at two known wavelengths, while H_0 is the same, we have:

$$H_1 = H_0 \cdot e^{-a_1 cL} \text{ and } H_2 = H_0 e^{-a_2 cL}$$

where a_1 and a_2 are the absorption coefficients at wavelengths 1 and 2, respectively.

Thus, $\qquad \dfrac{H_1}{H_2} = e^{-(a_1 - a_2)cL} \qquad$ (2)

In general, the exit plane slits have a finite width and H_0 is not con-

stant with wavelength; similarly a_1 and a_2 vary with wavelengths. Thus, a more general power expression is:

$$\frac{P_1}{P_2} = e^{-(a_1-a_2)cL} \cdot \frac{\int_{\lambda_1}^{\lambda_1+\Delta\lambda} N_\lambda \cdot d\lambda}{\int_{\lambda_2}^{\lambda_2+\Delta\lambda} N_\lambda \cdot d\lambda}$$

(3)

In this expression a_1 and a_2 are the average cross sections per molecule in the bands $(\lambda_1, \lambda_1 + \Delta\lambda_1)$ and $(\lambda_2, \lambda_2 + \Delta\lambda_2)$. P_1 and P_2 have the dimension of watts, that is, power passing through the slits.

The passing power can be analyzed by means of a mask or array of slits for several exit slits at two different positions rather than for one slit in two positions. One would expect that if the slits are properly chosen, the effect of the spectral distribution of the light source and influences due to other absorbing gases, could be minimized. In general, not only the quotient of P_1 and P_2 can be used to determine cL, but also any other suitable relations between them. In this instrument, the physical effect selected to measure the product cL is the difference in power received by the phototube at two different wavelength positions on the mask. The power difference naturally results in a phototube current difference. The actual procedure used in the remote sensing spectrometer is as follows:

(1) At one position of the spectrum with respect to the mask, the incident power, P_1, on the phototube derives a current, J, which is then held fixed. The voltage, V, across the phototube dynode chain is variable and adjusted by an AGC

loop so that the response to P_1 watts is always J amps.

(2) For the second position of the spectrum with respect to the mask, the incident power will be P_2 and will consequently generate J' amps.

The correlation spectrometer measures the difference D between J and J'—i.e.,

$$D = J - J' = J \left(1 - \frac{J'}{J}\right)$$

When the two spectral positions are within the linear response of the phototube $J'/J = P_2/P_1$

Then
$$D = J \left(1 - \frac{P_2}{P_1}\right) \quad (4)$$

If this current is passed through a high impedance load, the instrument's response in volts is

$$R = \alpha J \left(1 - \frac{P_2}{P_1}\right) \text{ volts} \quad (5)$$

where α is the load impedance (ohms) and J is the phototube response (amps).

In general, the response of the spectrometer to a target of spectral radiance N_λ is:

$$R = \alpha J \left\{ 1 - \frac{\sum\limits_{i=1}^{n} \int_{\lambda_{2i}}^{\lambda_{2i}'} \beta(\lambda) \, N_\lambda e^{-a(\lambda) \int_0^l c(l)dl} \, d\lambda}{\sum\limits_{i=1}^{n} \int_{\lambda_{1i}}^{\lambda_{1i}'} \beta(\lambda) \, N_\lambda e^{-a(\lambda) \int_0^l c(l)dl} \, d\lambda} \right\} \quad (6)$$

If we assume the product cL is constant over the acceptance solid angle of the spectrometer:

$$R = \alpha J \left\{ 1 - \frac{A\Omega \sum\limits_{i=1}^{n} \int_{\lambda_{2i}}^{\lambda_{2i}'} \beta(\lambda) \, N_\lambda e^{-a(\lambda)cL} \, d\lambda}{A\Omega \sum\limits_{i=1}^{n} \int_{\lambda_{1i}}^{\lambda_{1i}'} \beta(\lambda) \, N_\lambda e^{-a(\lambda)cL} \, d\lambda} \right\} \quad (7)$$

where

$a(\lambda)$ is the absorption cross section per molecule, and is a function of wavelength, temperature, and pressure

$\lambda_{1i}, \lambda_{1i}'$ are the beginning and end wavelengths of the mask corresponding to slit (i) for the first position (1) of the spectrum with respect to the mask

$\lambda_{2i}, \lambda_{2i}'$ is the same for position 2

$\beta(\lambda)$ is the spectrometer's transmission function

n is the number of slits in the mask

A is the spectrometer's aperture and Ω is the solid angle of acceptance

The simplest response occurs when N_λ is constant throughout the wavelength of interest and is N_0; that $a(\lambda)$ is a periodic function of λ over the same waveband; that C is constant along the path L which is also constant, and that the instrument's response is lineal constant over the same waveband such that $\beta(\lambda) = \beta_0$.

Then $R =$

$$\alpha J \left\{ 1 - \frac{n \, \beta_0 \, \Delta\lambda \, N_0 e^{-a_2 cL} \, A\Omega}{n \, \beta_0 \, \Delta\lambda \, N_0 e^{-a_1 cL} \, A\Omega} \right\}$$

i.e., $R = \alpha J \left\{ 1 - e^{-(a_2 - a_1)cL} \right\} \quad (8)$

79

Instrument Techniques

The basic instrument has already been given in the section on basic principle of correlation spectroscopy. In the section on signal origins, a review of the physical processes which produce the instrument's voltage response was given. Equations 6 and 7 indicate the response obtained when scattered sky light was used as the target illuminating background; this is the true remote sensing role. One can naturally see that the problems of quantitative measurements rely upon a solution of these equations. This can be obtained once the response of the spectrometer is known

—*i.e.*, its output voltage (R) against gas concentration product (cL). This is achieved by inserting known (cL) calibration cells into the optical path of the spectrometer. This system has been mounted in an Aero Commander aircraft and has successfully undertaken many SO_2/NO_2 airborne surveys over such cities as Toronto, Washington, Chattanooga, Los Angeles, and San Francisco. A typical record is shown in Figure 3.

When control over the light source can be exercised, the problem becomes much simpler. This is the approach adopted by the ambient monitor technique (see Figure 4). Here the gas under analysis is

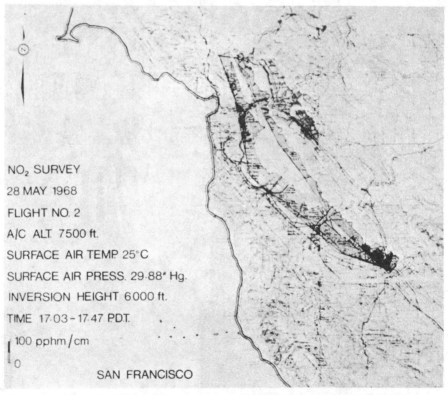

NO$_2$ SURVEY

28 MAY 1968

FLIGHT NO. 2

A/C ALT 7500 ft.

SURFACE AIR TEMP 25°C

SURFACE AIR PRESS. 29.88" Hg.

INVERSION HEIGHT 6000 ft.

TIME 17.03 - 17.47 PDT.

100 pphm/cm

0

SAN FRANCISCO

Figure 3.   Record of NO$_2$ airborne survey of San Francisco, May 28, 1968

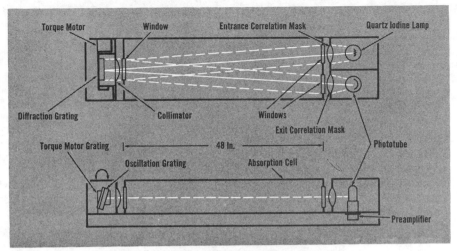

Figure 4. SO$_2$ ambient monitor schematic

passed through a multipass cell of fixed optical pathlength (L). Full control over the spectral radiance source (N_λ) is exercised. The ambient monitor uses a quartz-iodine lamp as the source and multi- entrance and -exit slits. Movement of the spectrum with respect to the mask is achieved by vibrating the grating with the torque motor driven under closed loop control. This technique essentially produces a response function of the type shown in Equation 8. The optical length is 2.5 m and, by use of a 100-sec. integration time, noise levels of 15 ppb for SO$_2$ and NO$_2$ have been obtained.

Another instrument technique is that of long-path monitoring (see Figure 5). This approach uses an actual quartz-iodide lamp positioned some distance away from the correlation spectrometer (up to 1000 m). The remote sensing spectrometer is accurately boresighted into the lamp, and the gas concentration between the lamp and sensor is

thereby obtained. This is obviously a more representative measure of the average ambient gas concentration throughout the long path than can be obtained from single-point ambient monitors. The long-path system is further modified by allowing the incoming spectrum to be slowly moved past the exit mask. This is achieved by slowly scanning the grating through a small angular movement. The net result is to cause the incoming spectrum to drift over the mask replica. There will be one grating position at which maximum cross correlation between mask and input spectrum occurs; at this position all lines match, and is in effect the optical counterpart to an electronic high Q network. For all other grating positions, the spectral match is less than optimum. Figure 6 well illustrates this for NO$_2$ while Figure 7 shows the NO$_2$ calibrations achieved for various increasing levels of gas in the test cell of a long-path sensor. A similar setup

81

Figure 5. Prototype long-path ambient monitor

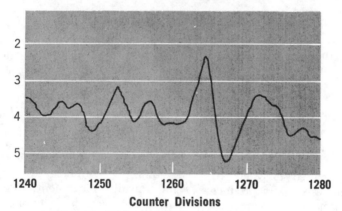

Figure 6. Wide-range NO_2 correlation scan

Run No. 15
3200 QI point source at 25 m
5-Slit optimized NO_2 mask (4300–4500 A)
86-Ppm meters reference cell

is used for SO_2. Table I shows the performance parameters for the long-path system.

The basic performance of the spectrometer, as mathematically expressed in equations similar to 6 and 7, has been transformed into a computer program. In this pro-gram the gas absorption coefficients filter and spectrometer transmission functions, spectral radiance of background light source, number of slits, wavelength jump and absorption coefficients of interfering gases are variable parameters. The computer output yields the cL product

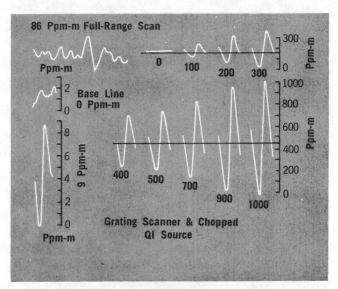

Figure 7. NO$_2$ calibrations achieved for various increasing levels of gas in test cell of long-path sensor

TABLE I. LONG-PATH SYSTEM PERFORMANCE

Gas Sensed	SO$_2$	NO$_2$
Spectral bandwidth	2800–3150 Å	4130–4500 Å
Threshold sensitivity (in 1000 m.)	2 ppb	1 ppb
Chart resolution (1000 m.)	2 ppb/ minor div.	1 ppb/ minor div.
Response linearity error (9–2500 ppb in 300 m.)	< 2%	< 3%
Speed of response	40 sec. to F.S.	20 sec. to F.S.
Grating oscillation:	Speed, Å/minute constant. Span, adjustable 5 Å to 500 Å usually set ~15 Å	
Sensor acceptance angle:	Azimuth 3 mR. Elevation adjustable 0 to 30 mR	
Light source beamwidth:	15 mR	
Permissible source to sensor misalignment:	6 mR max.	

of the sought gas against $R/\alpha J$, P_1, and P_2 as previously defined. By this means the operating wavelength position and correlation jump, correct number of slits and their widths can be optimized.

Correlation Interferometry

Because many gases and liquids have their fundamental absorption spectra in the near- and mid-infrared, where the luminous output of thermal sources is decreasing and

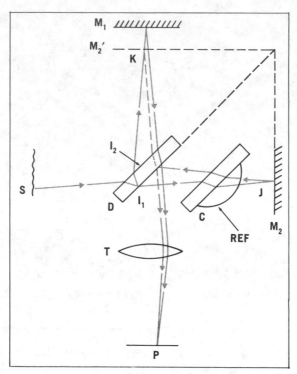

Figure 8. Correlation interferometer

the available detectors are generally less efficient and noisier than those used in the visible and uv, it is necessary to use the available radiation as efficiently as possible. Interferometers, by virtue of their circular symmetry, provide an increase in throughput of the order of 100 compared with an ordinary spectrometer.

The method by which the interferometer is made specific to certain gases is shown in Figure 8 which depicts a modified Michelson interferometer. Incoming radiation is optically filtered and then impinges on the beamsplitter D where it is amplitude divided into equal parts which are reflected, respectively, at mirrors M_1 and M_2 before

being recombined and focused on the detector, P. The compensating plate, C, which is of the same material and thickness as the beamsplitter, assures that both optical paths are equivalent. Now if mirror M_2 is displaced backward a distance of one-quarter optical wavelength from the location at which there is zero optical path difference between the two beams, the two components of radiation, which were constructively interfering to produce maximum radiation at P, will now destructively interfere to produce, under ideal conditions, no radiation. If the source radiation is monochromatic and M_2 moves with constant velocity, it is a simple matter to show that the detector output will be a cosine of constant electrical frequency given by $(\nu_0 v)$, where ν_0 is the optical frequency of the input radiation in cm^{-1} and v is the velocity of displacement of the mirror in cm sec^{-1}.

Thus it is possible to think of the interferometer as an optical analog to a superheterodyne receiver wherein each optical frequency is translated down to a much lower electrical frequency. In fact it can be shown that the detector output, as a function of optical path difference δ, is given by

$$D(\delta) = \int_0^{\nu_M} B(\nu)\,(1 + \cos 2\,\pi\nu\delta)d\nu$$

where the input radiation has power spectral density $B(\nu)$ and is band limited between 0 and ν_M.

Thus, apart from a constant term, the detector output as a function of δ is the Fourier cosine transform of the input power spectral distribution. In normal spectroscopy where it is desired to evaluate the power spectrum, $B(\nu)$, it is necessary to take the inverse transform of the interferogram $D(\delta)$ to achieve this. Such a calculation is usually performed on a computer, especially where high resolution spectra are required, and it is this computation which is the chief disadvantage.

Correlation is achieved in the following manner: mirror M_2 is actually fixed, and the optical-path difference is made to vary by rotating the compensating plate C about its center, thus changing the optical thickness of glass (or quartz) through which the radiation in this path must traverse. To each angular position of the plate

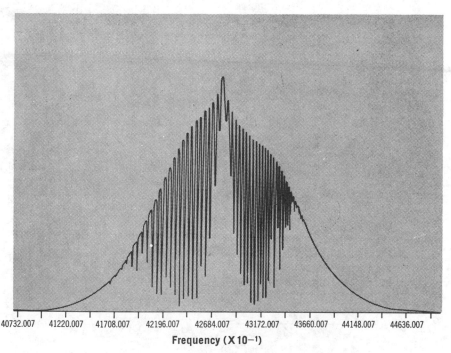

Frequency (X 10^{-1})

Figure 9. Carbon monoxide absorption spectrum 2.35 μ

there is thus a unique path difference. Rigidly fixed to the compensator plate, there is a magnetic recording drum, and one or more magnetic recording read-heads are located about the drum. To make the instrument specific to a certain gas, a sample of that gas is placed in an optical cell between the interferometer and a source of infrared radiation. The concentration is made large enough to achieve a good signal/noise ratio, and the output of the detector is connected to a magnetic head T_1. The compensation plate is rotated and a recording on the magnetic drum is made of the transform of that specific gas, as modified by the aberrations, etc. of the interferometer. This recording is the reference in the same sense as the transmission mask in the dispersive correlation spectrometer. The detector is now disconnected from the read-head. If the gas is present in the optical path, the detector will correlate with the reference signal in the electronic correlator, and the positive output voltage will be a measure of the gas concentration. The transform of an interfering gas will not correlate, assuming that an adequate path difference is taken, and the output voltage will be zero in their presence.

It is seen in Figure 9, for a specific gas, certain portions of the interferogram have a larger signal output than others, and may also have more information in the phase and amplitude variations. Thus, since only a portion of the interferogram is actually scanned, mirror M_2 is adjusted to optimize to the particular gas of interest. Saturation recording was used for the reference so that the correlation was against the phase of the reference rather than the amplitude. Contact recording and readout techniques were used after preliminary

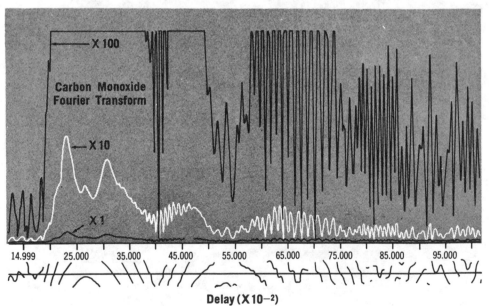

Figure 10. Carbon monoxide Fourier transform

tests had shown that over one million scans of the magnetic drum resulted in signal degradation of only 20%. The actual scan speed in operation was one scan per second. A floating head and flexible magnetic disc system has also been designed to reduce wear of the oxide coating to virtually zero.

The prototype interferometer was not designed for high sensitivity. It is quite feasible, however, to design a system that will yield sensitivities of the order of a few ppm—m when used in a remote sensing mode with the sun as a source of infrared radiation. It is also possible, of course, to use the device as an ambient or sampling sensor, whence if a cooled detector is used, the calculated sensitivities increase to the order of one ppb over a 10-cm pathlength for many gases of interest. However, just as the absorption spectrum of a given gas is unique to that gas, and can be used to quantitatively measure its concentration, so is the Fourier transform of the spectrum unique to the gas, and it too can be used for a quantitative measurement. That this is so is seen in Figures 9 and 10 which show the computed transmission spectra (modified by Gaussian limiting optical filters) of carbon monoxide gas in the region of 2.3 μ, and its computed transform as a function of δ, the optical path difference. As mentioned above, the detector output consists basically of an electrical cosine signal at frequency $\nu_0 v$, where ν_0 is now the center optical frequency of the entrance filter; but since the input radiation is no longer monochromatic, this cosine will be phase- and amplitude-modulated by the spectrum. The upper portion of the figure shows the mod-

ulation envelope of this cosine as a function of δ, and the lower portion illustrates the electrical angle of the cosine relative to the constantly increasing phase angle of $(\cos 2 \pi \nu_0 \delta)$. It is seen that both quantities are unique and quite different for the two spectra. Since, furthermore, the amplitude of the transform is directly proportional to the spectral transmission of the gas, correlations made on the transform will (through suitable calibrations) yield concentrations.

Conclusions

Correlation spectroscopy and interferometry still have considerable development potential ahead. The technique has presently only been used by gases and vapors in the uv, visible, and near-ir in absorption. However, the technique can be equally well applied to emission spectra, directly excited, flame-excited, or achieved by plasma. Other potential lies in the use of laser Raman spectra for correlation applications and also fluorescent spectra, particularly the quasi-linear fluorescent spectra obtained at low temperatures. Consequently the technique has a great future, not only in the airborne and satellite remote sensing roles, but also for new generations of laboratory analytical instruments.

The author acknowledges the continued support of the National Research Council of Canada for the development of correlation spectroscopy. Also the U.S. Department of Health, Education and Welfare, NASA Manned Spacecraft Centre, and AISI have contributed significantly to support this research at Barringer Research Ltd.

References

(1) A. R. Barringer and B. C. Newbury, "Remote Sensing Correlation Spectrometry for Pollution Measurement," Ninth Conference, Air Pollution & Industrial Hygiene Studies, Pasadena, Calif., February 1968.

(2) D. T. Williams and B. L. Kolitz, "Molecular Correlation Spectrometry," *Applied Optics*, **7**, 607 (1968).

(3) A. R. Barringer and A. J. Moffatt, "Recent Progress in the Remote Detection of Vapour and Gases," Sixth Symposium on Remote Sensing of Environment, Ann Arbor, Mich., Oct. 14, 1969.

(4) A. R. Barringer and J. D. McNeill, "Advances in Correlation Techniques Applied to Spectrometry," National Analysis Instrumentation Division of ISA Symposia, New Orleans, La., May 1969.

(5) M. Millan, S. Towsend, and J. H. Davies, "Study of Barringer Remote Sensor," University of Toronto Rept. 146, 1969.

THIS article is based on Dr. Davies' presentation at the Eastern Analytical Symposium, New York, N.Y., November 19, 1969.

A Miniature Mass Spectrometer

Achievements to date hold promise for the development of a relatively inexpensive, lightweight, rugged instrument with good sensitivity, moderate resolution, and capability of easy operation from a central multistation control

Peter H. Dawson

C.R.A.M., Universite Laval,
Quebec, Canada

J. W. Hedman

General Electric Co.,
Analytical Measurements
Business Section,
West Lynn, Mass. 01905

N. R. Whetten

General Electric Co.,
Research and Development
Center,
Schenectady, N. Y. 12305

A MINIATURE MASS SPECTROMETER with an unusually simple electrode structure, and with a unique method of operation is described in this paper. It is an ion storage instrument, with the ions being mass-selectively trapped in the volume enclosed by the electrodes. The mass-selected ions of a given charge-to-mass ratio are periodically pulsed out of the trap into an electron multiplier to determine the relative abundance of that species. By ramping the voltages applied to the electrodes, ions of each charge-to-mass ratio are trapped in sequence, and the mass spectrum is scanned. The performance, use, and potential of this device are discussed in the light of the development of mass spectroscopy as an analytical tool with special emphasis on recent efforts to develop simple, light-weight, low-cost instruments.

Ion analyzers can be divided into two broad categories—static and dynamic. Static analyzers depend on differences in ion trajectories under the influence of constant magnetic and/or electric fields. The familiar single and double focusing magnetic sector analyzers dominate this category. Another example is the cycloidal mass spectrometer in which uniform magnetic and electric fields are applied at right angles to each other. Dynamic spectrometers are of more recent origin (*1*). In these instruments, the ion separation is based on the time dependence of one of the system parameters. There are three subgroups: Energy balance spectrometers such as the omegatron; time-of-flight spectrometers; and path stability spectrometers.

Our particular interest (*2*) has been in the latter category, which includes the quadrupole mass filter, the monopole mass spectrometer, and the three-dimensional quadrupole or ion trap mass spectrometer. The mass filter and monopole type of instruments have been commercially available for about eight years. Our miniature spectrometer is a three-dimensional quadrupole ion trap.

The Evolution of Mass Spectrometry

The first crude mass spectrometers were Thomson's parabola instrument of 1910, and Aston's velocity focusing device of 1919. These were used to provide information on the existence of isotopes and therefore, on atomic structure. Dempster, at about the same time, introduced the 180° deflection magnetic spectrometer, the basic design of which is still in use. Large scale electromagnetic isotope separation has become technologically important since 1945. Early mass spectrometric work concentrated on improving the 180° design and achieving higher resolutions, particularly for determining the deviation of isotope masses from integral values. This is still an active specialized subdiscipline of mass spectrometry. A significant instrumental advance was the double focusing spectrograph using radial electric fields before the magnetic field to sort the ions according to their energies before mass analysis. One of the best known designs was by Mattauch and Herzog.

Mass spectrometers with 90° and 60° magnetic deflection became common after the work of Nier, beginning in 1940. They have the convenience, for many applications, of removing the

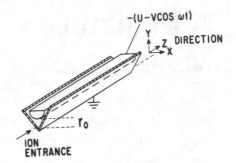

Figure 2. Schematic diagram of the monopole mass spectrometer

ion source and detector from inside the principal magnetic field and of reducing the size of the magnet. The use of mass spectrometry then began to extend into the field of chemical analysis of multicomponent mixtures, but before 1950 remained a specialized art rather than a standard technique. Progress was spurred by the interest of oil companies in analysis of complex hydrocarbon mixtures. For several years the emphasis was on deciphering mass spectra of complex mixtures and in improving stability and reproducibility. The resolution of laboratory size instruments was gradually improved. The development of the electron multiplier has been important, since the sensitivity obtained with the multiplier can be traded for additional resolution. The increasing demand for analysis of complex organic reaction products led to the construction of double focusing analytical mass spectrometers of high resolution. Spectrometers with resolutions of 10,000 can identify the exact composition of organic molecules by an accurate measurement of mass (3).

Meanwhile, in the sixties, developments in vacuum and surface physics and the increasing use of vacuum processing had brought a demand for small instruments of limited resolution but very high sensitivity, and able to withstand the baking that is part of high vacuum processing. These instruments are often known as "partial pressure

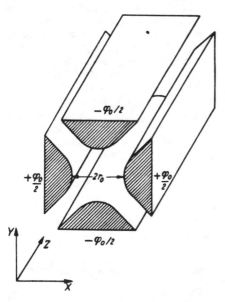

Figure 1. Schematic diagram of the quadrupole mass filter

analyzers." Space and upper atmosphere exploration also brought an interest in developing small spectrometers of limited resolution. The magnetic sector instruments were quickly adapted to this use, but the quadrupole field spectrometers soon became a more popular alternative.

Quadrupole Field Spectrometers

The focusing of ions by the use of alternating quadrupole fields was introduced to mass spectrometry by Paul and his colleagues at the University of Bonn in 1953. The first type of spectrometer to be developed was the quadrupole mass filter. A quadrupole field is formed by four rods located as shown in Figure 1. Rf and dc voltages are applied between opposite pairs of rods. Ions are sorted according to their path stability as they pass through the analyzing region in the z direction. By suitably adjusting the ratio of rf-to-dc voltage, the device can be made to pass only a given mass ion. Ions of all other charge-to-mass ratios are "filtered" out before reaching the detector, giving rise to the name "mass filter." The mass spectrum is swept by varying either the voltage or the frequency. The maximum resolution is limited by the number of cycles an ion spends in the field. The larger the number of cycles, the higher the maximum possible resolution. The resolution can be adjusted up to this maximum by adjusting the ratio of the rf-to-dc voltage. The quadrupole requires no magnet. If the rf frequency is increased, the length of the analyzer can be reduced. Thus the mass filter has the advantages of small size and light weight. Although the original applications were to problems requiring only low resolution, high sensitivity quadrupoles are now available with a mass range up to 800 and with a resolution of $2m$ at any mass. Scan speeds of faster than 1 msec/amu are available, so that a quadrupole type instrument can be coupled with a gas chromatograph.

Figure 3. Photograph of the three-dimensional quadrupole mass spectrometer, showing the stainless steel mesh electrodes

The monopole has characteristics and uses that are similar to those of the mass filter (4). A sketch is shown in Figure 2. A V-block is placed in the quadrupole ground plane so that ion trajectories are limited to one quarter of the quadrupole field. Thus the monopole utilizes both the path stability and focusing properties of the quadrupole field. One advantage is that for a given mass range, less rf power is required than for a mass filter of comparable size and frequency. The performance specifications for commercially available monopoles are very similar to those for quadrupoles.

The possibility of using a three-dimensional quadrupole field to actually trap ions with a small volume was recognized by Paul. In 1959, Fischer described an early model. It was difficult to use this model for mass analysis owing to problems in detecting trapped ions and interactions between the various species of ions which were trapped simultaneously.

The idea received little attention. We only accidentally became aware of Fischer's work in 1966, even though one of us was a self-confessed mass spectrometrist. We were intrigued by the principle and thought that the problems could be overcome. The principle of operation is so different from other mass spectrometers that it should give rise to a different set of properties. In particular, the trapping of ions can occur in a small volume, and there is no longer a relatively long z direction as in the mass filter and monopole (5).

The Three-Dimensional Quadrupole

A three year program of research and development, building devices, improving circuitry, studying the theory of operation, and computing the trajectories of ion motion culminated in the device shown in Figure 3. This is a three electrode structure, with a ring or donut-shaped electrode and two cap electrodes. The structure is rotationally symmetric about the principal axis. The electrodes are formed very approximately—in fact, by hand—to hyperbolic shapes from a coarse stainless steel mesh. Positioning of the electrodes is not very critical.

Ions are formed inside the trap (the ion source and analyzer regions are combined) by a beam of electrons from a simple electron gun. By applying an rf and dc voltage combination to the ring electrode (with the end cap electrodes at ground potential) only ions of a corresponding mass-to-charge ratio, m/e, are trapped inside the electrodes. All other ions are ejected. The stored ions have complicated trajectories but remain within the trap unless they are scattered out by multiple collisions with neutral gas molecules or other ions. Following a suitable storage or sorting time, ions of the particular m/e ratio are detected by applying a negative pulse to one of the end caps. The ions

are pulled through the mesh electrode into an electron multiplier.

At 10^{-6} torr a typical sequence is to have the electron beam on for 1 msec, switch it off for 50 μsec to complete the ion sorting process, and then to apply a 3-μsec pulse to one end cap electrode to empty the trap and draw the ions through the mesh electrode into the electron multiplier. The cycle is then repeated. The mass spectrum can be scanned by varying the rf and dc voltages, but with a fixed ratio between them. The resolution may be electronically controlled by varying this voltage ratio. The voltages needed are given approximately by

$$V_{rf} = 1.23 \; mz_0{}^2\omega^2/2e$$
$$U_{dc} = 0.67 \; mz_0{}^2\omega^2/4e$$

where ω is the angular frequency of the applied rf voltage and z_0 is half the distance between the two end caps. A 1.5-cm long device (end cap spacing) can operate up to about mass 200 with an rf supply of 1800 volts at 500 kHz. Higher masses could be scanned with a larger voltage, a smaller device, or a lower frequency. Each mass peak is observed at the electron multiplier output as a train of pulses. These are converted to a conventional display (Figure 4) by a peak-height reading circuit. The circuitry to operate the device has been constructed in a single module.

The effectiveness of the ion storage property of the ion trap is shown in Figure 5. Here, the relative number of ions remaining in the trap (after the ionizing filament was turned off) are given in a log-log plot as a function of time. These data were taken at a pressure of 3×10^{-10} torr and in a low resolution mode. It is shown by the shape of the decay curve that at these low pressures, ion loss results primarily from ion-ion scattering events. That is, ions are eventually scattered out of the trap by other trapped ions. At higher pressures, neutral gas scattering is important, although the ions may undergo many scattering events before being

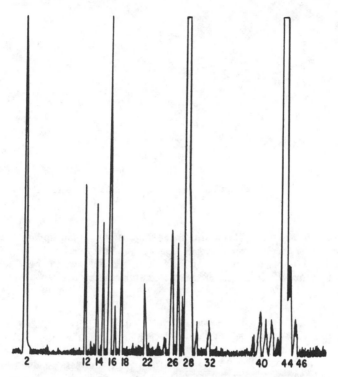

Figure 4. **Mass spectrum taken with a three-dimensional quadrupole**

lost from the trap.

There are, in principle, two unique features of the three-dimensional quadrupole. First, it is an ion storage device. The storage of the ions allows the ion current to be integrated with time. If the ion formation rate is low, one can store for longer times and still obtain an easily measurable signal. The second feature is the long path length in the field even though the field is small. A third feature, which was not predicted theoretically, but has been found experimentally, is that a simple electrode structure with a noncritical geometry works quite well for a moderate resolution up to about 150.

The achievements to date hold promise for the development of a relatively inexpensive, rugged, miniature mass spectrometer with good sensitivity and moderate resolution, and capability of easy operation from a central multistation control.

Some potentialities for this kind of device and the achievements to date are as follows:

Sensitivity. The ion storage feature should be valuable at low ion formation rates, that is, for very low pressures or for very low partial pressures or for trace analysis. This has still to be demonstrated at very low pressures. The lowest partial pressures measured to date have been 10^{-13} torr and the trace analysis capability has been about one part in 10^4. Experiments are in progress to extend these limits.

Resolution. The long path length of the ions in the field should permit a good resolution. Resolution can be traded for sensitivity, as with the quadrupole mass filter. Here the results have been good. Devices like that of

93

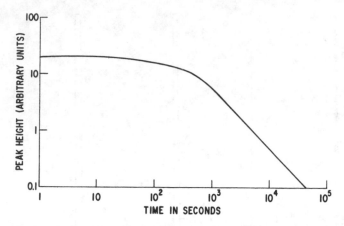

Figure 5. Ion loss from the three-dimensional quadrupole mass spectrometer operated in a low resolution mode. The relative number of ions remaining in the trap is given on the ordinate for the corresponding storage time on the abscissa

Figure 3 have had half-height resolutions up to 150. The resolution may be limited by the field perfection. In larger devices with carefully machined electrodes, we have observed resolutions up to 1000 over a limited mass range.

Size. Adequate performance in small simple devices has been demonstrated. Tube construction costs should be low.

Power. A limiting factor in rf quadrupole instruments is often the rf power. In all quadrupole instruments, the ions must stay in the field sufficient cycles of the rf field to be adequately sorted. For mass filters and monopoles of moderate length the frequency is high, typically 2–6 MHz. This limitation does not apply to the ion trap, and the long ion path length inside the trap allows reduction of the rf frequency to the hundreds of kilohertz range. The rf power depends on the fifth power of the rf frequency. A lower frequency also facilitates the remote use of the analyzer and multistation operation from a single control center.

Scan Speed. Due to its pulsed mode of operation, the quadrupole ion trap is not a rapid scanning device. The scan speed is determined by the number of ion pulses required in each mass peak. At high pressures where storage time can be reduced, the scan speed can be increased. Conversely, at low pressures where longer storage times are required to build up sufficient ions in the trap, the scan speed must be reduced.

Use of Miniature Mass Spectrometers

The demand for miniature spectrometers has led to several approaches by different workers. One solution has been a resurrection of the 180° magnetic deflection design with radius of only about 1 cm which has the disadvantages of low resolution and the inconvenience of a magnet (*6*). A novel approach has been the use of 90° magnetic deflection with multiple passages through the same field (*7*).

Some of the potential uses of miniature mass spectrometers are:

Standard Vacuum Applications. In research and production involving vacuum technology, a knowledge of the nature and quantity of the gases present is increasingly necessary. For many processes such as vacuum decarburization, outgassing studies, semiconductor processing, vacuum heat treating, and tube manufacture, the requirements on mass range and resolution are not severe. However, the spectrometer should be rugged, resistant to contamination, small and low in cost.

Process Control Applications. Mass spectrometers are a possible method of monitoring the composition of process lines in the petrochemical and chemical industries. Low cost, reliability, and ruggedness are essential. The speed of response of a mass spectrometer is a desirable feature.

Space Uses. Miniature mass spectrometers are well suited for space applications due to their size and weight and low power requirements. Ruggedness is another prime consideration.

Medical Uses. Mass spectrometers have medical application for blood gas, body fluid, and respiratory gas analysis (8). Continuous monitoring of patients is desirable during operations and in intensive care units. For blood and respiratory gas analysis, only low resolution is required. However, the instrument must be extremely reliable and easy to operate. Low cost and small size are other important needs.

Environmental Monitoring. Monitoring the environment to detect pollutants is another potential application. At present, miniature mass spectrometers do not have sufficient trace analysis capability for broad area monitoring. However, they can be used for monitoring emissions at the source such as at smoke stacks. Monitoring of the cabin atmosphere in airplanes and space stations is also possible. In many cases, the mass range need not be large, but the sensitivity should be high. In addition, the instrument should be extremely reliable.

Geological Uses. Two potential applications are the detection of helium anomalies near uranium deposits and the detection of mercury near certain mineral deposits. Ruggedness, reliability, and portability are essential.

Reaction Studies. Miniature mass spectrometers can be used to follow the rate of chemical reactions, the lifetime of metastable species, and the rate of ion-molecule reactions. For some of these applications, the trapping ability of the three-dimensional quadrupole should be useful.

These diversified applications suggest future interesting developments in instrumentation.

References

(1) E. W. Blauth, "Dynamic Mass Spectrometers," Elsevier, Amsterdam, 1966.
(2) P. H. Dawson and N. R. Whetten, "Mass Spectroscopy Using Radio Frequency Quadrupole Fields," in "Advances in Electronics and Electron Physics," L. Marton, Ed., Vol. 27, p. 58, Academic Press, N. Y., 1969.
(3) F. W. McLafferty, "Interpretation of Mass Spectra," W. A. Benjamin, Inc., N. Y., 1966.
(4) U. von Zahn, *Rev. Sci. Instrum.*, **34**, 1 (1963).
(5) P. H. Dawson and N. R. Whetten, *J. Vac. Sci. Technol.*, **5**, 1 (1968); **5**, 11 (1968). P. H. Dawson, J. W. Hedman, and N. R. Whetten, *Rev. Sci. Instrum.*, **40**, 1444 (1969).
(6) D. Allenden, R. D. Craig, and R. G. Johnson, 16th Annual Conf. on Mass Spectrometry and Allied Topics, unpublished, Pittsburgh, May, 1968.
(7) M. Baril, Int'l. Conf. on Mass Spectroscopy, unpublished, Kyoto, Japan, September, 1969.
(8) S. Woldring, D. C. Woolford, and G. Owens, *Science*, **153**, 885 (1966), also M. Mosharrafa, *Res./Develop.*, **21** (3), 24, (1970).

Total Internal Reflection Enhancement of Photodetector Performance

Photocathode quantum efficiencies may be increased by

optical modification of existing photomultipliers.

Total internal reflection cathodes offer the greatest promise for

enhancing the photodetector performance of modern

absorption and emission instruments

TOMAS HIRSCHFELD

Block Engineering, Inc.,
19 Blackstone St.,
Cambridge, Mass. 02139

R ECENT DEVELOPMENTS in electro-optical technology, involving break-throughs in light sources, optical components, electronics, and data handling, have gradually created a situation where photodetector quantum efficiencies place a ceiling on the performance of electrooptical instruments. In analytical chemistry this is particularly obvious for fields such as Raman spectroscopy, luminescence spectroscopy, or the now emerging techniques of remote optical analysis, originated by space and environmental technology. However, the limitation can also be felt in more established fields, such as emission and absorption spectroscopy.

Among photon detectors today, the photomultiplier tube dominates the field, its near perfect amplification further optimized by high gain dynode stages, cooling, and photon counting within a narrow pulse energy window. But, before this stage is reached, a photoemissive cathode must be used to transform as many photons as possible into emitted photoelectrons. And here the bottleneck lies, for even the best commercial photocathodes are easily the least efficient component of most optical systems.

The reason for this can be seen by observing the two types of photocathodes in use today (Figure 1). The transmission photocathode (superior from an electron optical point of view) must be simultaneously thick enough to absorb most of the incident light, and thin enough so that the generated photoelectrons can traverse it while retaining enough energy to overcome the work function barrier at the vacuum interface. Similarly, in the reflection photocathode, those photons that are

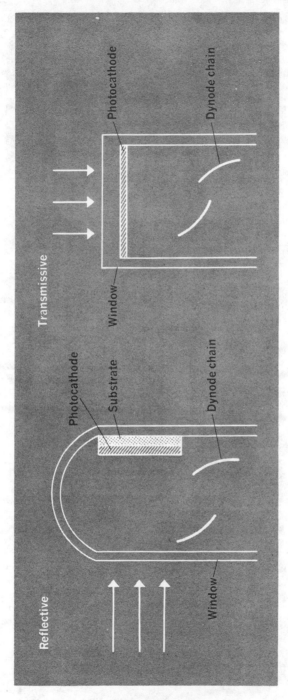

Figure 1. Photocathode structures

absorbed too deeply within the material will generate photoelectrons that can no longer get to its surface and be emitted.

For optimum efficiency a photocathode material must have the highest possible absorption coefficient for light and the lowest possible energy absorption coefficient for photoelectrons. The material must also have a low work function to extend its spectral coverage to longer wavelengths. Further demands concern the chemical and electrical properties of the cathode, as well as its suitability for production in a more or less controllable fashion.

It is, therefore, not surprising that research on improving photocathodes became centered on improving photocathode materials (1), leading to several decades of incremental improvements through very hard work. The final results of this effort were the extremely sophisticated, expensive, and temperamental semiconductor photocathodes of today, whose peak quantum efficiences of around 30% help us

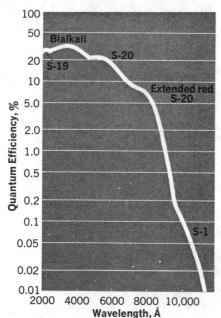

Figure 2. Envelope of commercial photocathode quantum efficiencies

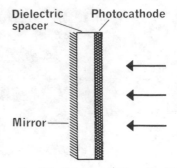

Figure 3. Interference photocathodes

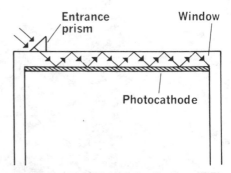

Figure 4. Total internal reflection (TIR) enhancement of photocathode quantum efficiency (5)

to live with their eccentricities and individualities.

However, these efficiencies are available only in a small fraction of the spectrum, below about 5000 Å, and decay rapidly towards the red, becoming exceedingly small in the infrared before failing altogether near 1.2 microns (Figure 2). This is owing to the seemingly universal tendency of photocathode materials to become more transparent at longer wavelengths, where, on the other hand, lower photoelectron energies mean reduced electron range within the cathode layer. Even these performances require using several different photocathodes, each of which is optimal only over a small fraction of the total range.

The situation clearly called for a new approach, and this was suggested in a 1957 paper by Deutscher (2), where he proposed improving the absorption of light in a photocathode by the techniques of physical optics, for a given material and layer thickness.

The method he chose for this purpose consisted of coating the photocathode on top of a dielectric spacer layer deposited on a mirror (Figure 3), the hole forming an optical cavity whose reflectivity could be made arbitrarily small by interference at preselected wavelengths. These are fixed for a given tube by the thickness of the spacer layer. This approach provided gains in quantum efficiency that increased toward longer wavelengths as the photocathode became more transparent, up to a maximum factor of three to five. Unfortunately, this gain was available only over a range of 300 to 500 Å on either side of the design wavelength, outside of which the quantum efficiency was actually lower than for a normal photocathode.

Along these same lines, Love and Sizelove (3, 4) studied photocathode behavior through a mathematical model in an effort to analyze the behavior of reflection, transmission, and intereference photocathodes. Their equations allow prediction of the relative performances of these photocathodes, as well as optimum photocathode and spacer layer thicknesses for them, with a fairly good accuracy. To the extent that the layer thicknesses had not been arrived at empirically, the results developed by this group allowed some further gains in quantum efficiency to be achieved.

But it was a third physical optics approach, introduced by Rambo in 1964 (5), that achieved the most significant results so far, and became the most widely used technique for physical optical enhancement of photocathode performance. As shown in Figure 4, it was based on illuminating a transmission photocathode at an incidence angle beyond the critical one for total reflection for the window-vacuum interface. Any fraction of the photons that was not absorbed in their first passage

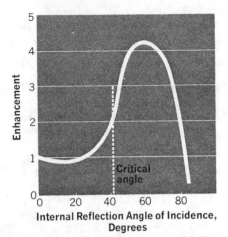

Figure 5. Angular dependence of TIR enhancement of photocathode quantum efficiency (S-11 photocathode) (6)

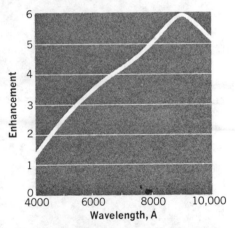

Figure 6. TIR enhancement of photocathode quantum efficiency as function of wavelength (S-20 photocathode) (16)

through the photocathode was then bounced back and forth between both sides of the window until eventually absorption was complete.

Independently rediscovered by Gunter and coworkers at NASA Ames, the technique aroused considerable interest, and research efforts on it were started at a number of institutions, notably the Electron Tube Division of ITT, the Electronic Components and

Devices Group of RCA, Optics Technology, and Block Engineering.

The technique in its initial form was applied in a more or less empirical fashion to photocathode layers of the S-11 (6), S-20 (7), extended red S-20 (8), and S-1 types (9). The typical behavior of the quantum efficiency enhancement as a function of angle and of wavelength can be seen in Figures 5 and 6. The largest enhancement was usually found in the vicinity of the critical angle, and grew from a factor <2 near the peak response wavelength to as much as 7 to 9 near the long wavelength threshold of the material. The improvement is quite polarization sensitive (10).

At about this time, another technique for enhancing photocathode quantum efficiencies was developed, based on using a strong external electric field to reduce the photocathode vacuum potential barrier and increase the fraction of photoelectrons crossing it (11). While the enhancement is modest, it is multiplicative with the one obtained by TIR, allowing even larger efficiency gains (12, 13).

An interesting variant of the technique was its inadvertent application by Samson and Cairns (14), who in-

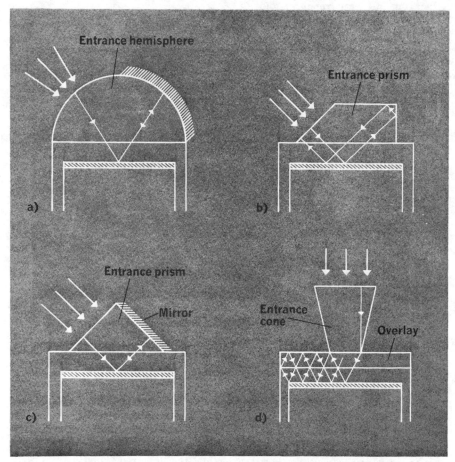

Figure 7. Various geometries used in TIR-enhanced photocathodes *(16, 18, 19)*

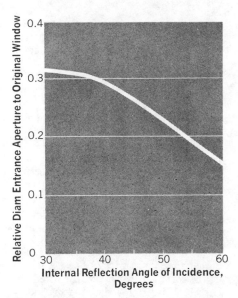

Figure 8. Relative aperture of cone entrance for TIR-enhanced photomultiplier for 4 reflections with f/5.5 beam as function of incidence angle (19)

creased the quantum efficiency of a far-uv metallic reflective cathode by using multiple reflections at grazing incidence inside a concave polygonal cathode. As the refractive index of metals is below unity in this spectral region, grazing incidence external reflection becomes identical with TIR for a high index material of unity index. As expected, their experimental results resemble those cited above. The phenomenon was later studied in detail by Pepper (15).

In absolute terms, very high sensitivities ($\sim 600 A/lm$) (7) and quantum efficiencies ($\sim 60\%$) (16) have been realized through the application of TIR to photocathodes. However, the geometry of these systems was quite inconvenient, as their angular aperture was limited to an f/6 cone and the entrance aperture was a small fraction of the area of the hypotenuse face of the entrance prism. Improved apertures were first described by Hirschfeld (17), who used high index materials for the

entrance prism. For a sapphire prism of index 1.76, an f/1.2 cone could be used. But even the optimum width of this entrance prism was of the order of the window thickness and too small for many applications. Optically contacting a glass disk to the window increased its effective thickness, and thus that of the aperture, but only by reducing the number of bounces across the tube diameter, and, therefore, lessening the enhancement. Other optical layouts giving higher apertures by reducing the number of reflections to two were proposed by Oke and Schild (18) (Figure 7a), Hirschfeld (19) (Figure 7b), and Gunter et al. (16) (Figure 7c). The largest apertures so far, however, resulted from a system (19) (Figure 7d) which combined the large aperture of a cone condenser-deflector with a thickened window, while maintaining a large number of bounces by double-passing the beam along the window radius. The aperture of such a system, in terms of the original window size, is shown in Figure 8.

The system of Figure 7c will reimage a line on itself, allowing simultaneous TIR-enhanced measurement of a whole spectrum in an image intensifier (20). With auxiliary optics (15, 21) even a two-dimensional image can be double-passed in this way for viewing at high quantum efficiencies with an intensifier. Still other geometries (16) have been developed to allow adding TIR enhancement to existing systems without displacing the detector.

With the primary goal of high quantum efficiencies within reach, efforts were made to analyze the enhancement numerically, in order to optimize it. The advantages expected from this were manyfold:

(1) Reducing the number of bounces required for a given enhancement improves the geometry, particularly for imaging applications.

(2) Higher enhancements still are needed in the red and infrared region for most photocathodes.

(3) For a large enough enhancement, a simultaneously thin and not very opaque cathode would be adequate. This removes both the optical and the electronic transmission behavior constraints on the photocathode material, and widens the range of compounds that may be used, allowing one to hope that a material may be found (18) permitting operation beyond the 1.2-micron long wavelength limit reached more than a generation ago. Another hope is to find somewhat more tractable materials than those in use today.

(4) Surface, low-level defect state, and phonon-aided photoemitters have never been practical because of low absorptions into these states, a problem that large enhancements might circumvent.

(5) A photocathode one or a few atomic layers thick often shows large reductions in work function (22), again enhancing IR performance.

Since initial calculations by Sizelove and Love (4, 23) were in error, as shown by Seachman (24), the first analysis published was one by Ramberg (25), who showed the possibility of a significant TIR enhancement in a single bounce.

Meanwhile, the theoretical understanding of the total reflection phenomenon had been enormously advanced by its use as a sampling technique for absorption spectroscopy through the efforts of Harrick (26). By analyzing the electromagnetic field in the vicinity of a totally reflecting surface, it was possible to predict to a good approximation the behavior of a thin absorbing film in contact with it. He furthermore showed that in some circumstances the effective thickness of such a film for absorption was much higher than its real one. These approximate formulae were then adapted by Hirschfeld (19) for the case of photocathodes, and used to generate optimization criteria, and predict the performances obtainable by using them.

Thus for perpendicularly polarized light, a low index window would allow a 15-fold enhancement in a single

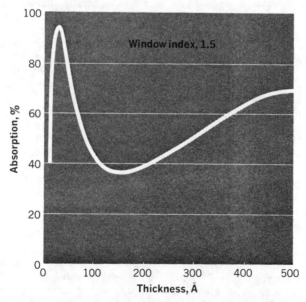

Figure 9. Absorption by very thin film of S-20 material in single total internal reflection at 60° incidence at 5000 Å as as function of thickness (21)

bounce near the critical angle and an 8-fold one for a cone having an $f/1$ aperture. For parallel polarization, the optimum window index of 1.9 allowed 3-fold enhancements at the optimum angle (about 10° beyond the critical) and

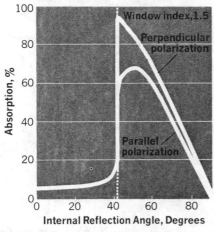

Figure 10. Absorption by 30 Å film of S-20 material in single total internal reflection at 60° incidence at 5000 Å as function of angle (21)

1.8-fold over even a 150° (!) field of view, all for a single bounce.

The development of an exact theory for absorption by thin films at a totally reflecting boundary by Hansen (27) allowed an exact analysis of TIR cathodes (20, 28), proving the validity of these earlier conclusions, and yielding a new and very surprising fact. The absorption by a very thin absorbing film in the total reflection region, after steadily decreasing with thickness as one would expect, suddenly changes behavior for extremely thin (<50 Å) layers, where nearly complete absorption occurs in a single bounce, even for low opacity layers! This behavior is shown in Figure 9, while the angular and polarization behavior of such an ultrathin cathode can be seen in Figure 10. These surprising results, first observed a decade ago for thin metallic films by Turbadar (29), arise from interference phenomena within the layer. These cancel the reflected beam, and since there is no transmitted one, all the radiation has to be absorbed. As the actual

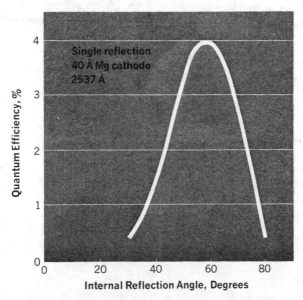

Figure 11. Experimental values of quantum efficiencies for single reflection ultrathin TIR-enhanced photocathode

thickness of the film is only a hundredth of a wavelength, the interference maximum covers a very wide spectral region (4000–12,000 Å for a layer optimized at 8000 Å), unlike the situation usually found in interference photocathodes.

These predictions were verified first by work with gold layers (*30, 31*) where 70% absorption (and 12-fold enhancement) was achieved in a single reflection at 2537 Å for a 25-Å thick layer. Later results (*21*) showed a 4% quantum efficiency for a single reflection on a 40-Å Mg layer at 2537 Å, a 13-fold enhancement over the performance of a thick reflective cathode of the same material at this wavelength (Figure 11). This work is now being extended to photocathode materials of practical interest.

One would therefore hope that the gains achieved so far by optical modification of preexisting photomultipliers will soon be exceeded by those provided by tubes designed from the beginning for this type of operation. The consequences of this in terms of optical instrumentation performance will no doubt be interesting to observe.

References

(1) A. H. Sommer and W. E. Spicer, in "Photoelectronic Materials and Devices," S. Larach, Ed., D. Van Nostrand Co., Inc., Princeton, N. J., 1965, p 175.

(2) K. Deutscher, *Naturwiss* 44, 486 (1957).

(3) J. A. Love and J. R. Sizelove, *IEEE Trans. Electron. Devices* 13, 921 (1966).

(4) J. A. Love and J. R. Sizelove, *Appl. Opt.* 7, 11 (1968).

(5) B. E. Rambo, Air Force Tech. Doc. Report **ALTDR** 64-19, April 1964.

(6) W. D. Gunter, E. F. Erickson, and G. R. Grant, *Appl. Opt.* 4, 512 (1965).

(7) G. R. Grant, W. D. Gunter, and E. F. Erickson, *Rev. Sci. Instrum.* 36, 1511 (1965).

(8) R. Laker and S. Poultney, *Appl. Opt.* 9, 2193 (1970).

(9) E. G. Burroughs, *Appl. Opt.* 7, 2429 (1968).

(10) H. Hora and R. Kantlehner, *Solid State Commun.* 4, 557 (1966).

(11) K. R. Crowe and J. L. Gumnick, *Appl. Phys. Lett.* 11, 249 (1967).

(12) W. D. Gunter, R. J. Jennings, and G. R. Grant, *Appl. Opt.* 7, 2143 (1968).

(13) C. D. Hollish and K. R. Crowe, *Appl. Opt.* 8, 1750 (1969).

(14) J. A. R. Samson and R. B. Cairns, *Rev. Sci. Instrum.* 37, 338 (1969).

(15) S. V. Pepper, *J. Opt. Soc. Amer.*, 160, 805 (1970).

(16) W. D. Gunter, G. R. Grant, and S. A. Shaw, *Appl. Opt.* 9, 251 (1970).

(17) T. Hirschfeld, *Appl. Opt.* 5, 1337 (1966).

(18) J. B. Oke and R. E. Schild, *Appl. Opt.* 7, 617 (1968).

(19) T. Hirschfeld, *Appl. Opt.* 7, 443 (1968).

(20) W. C. Livingston, *Appl. Opt.* 5, 1335 (1966).

(21) E. R. Schildkraut and T. Hirschfeld, Final Report, NASA Contract **NAS** 2-4661, 1969.

(22) F. J. Piepenbring, "Proceedings. International Colloquium on Optical Properties and Electronic Structure of Metals and Alloys," North Holland Publishing Co., Amsterdam, 1966, p 316.

(23) J. R. Sizelove and J. A. Love, *Appl. Opt.* 6, 443 (1967).

(24) N. J. Seachman, *Appl. Opt.* 6, 356 (1967).

(25) E. G. Ramberg, *Appl. Opt.* 6, 2163 (1967).

(26) N. J. Harrick, "Internal Reflection Spectroscopy," John Wiley & Sons, Inc., New York, 1967, p 41.

(27) W. N. Hansen, *J. Opt. Soc. Amer.* 58, 380 (1968).

(28) T. Hirschfeld, U.S. Patent 3,513,316, May 19, 1970.

(29) T. Turbadar, *Proc. Phys. Soc.* 73, 40 (1959).

(30) T. Hirschfeld and E. R. Schildkraut, paper presented at the 15th Annual Spectroscopy Symposium of Canada, Toronto, October, 1968.

(31) M. J. Block, *Proc. Infrared Information Symp.* 17, May 1969.

Radiation Sources for Optical Spectroscopy

Source technology is a vital and exciting field which

continues to add new dimensions to

analytical spectroscopy

AUGUST HELL[1]

Scientific Instruments Division, Beckman Instruments, Inc.
2500 Harbor Blvd., Fullerton, Calif. 92634

OPTICAL SPECTROSCOPY is one of the oldest and most useful techniques man has yet developed for probing and measuring his physical environment. Essentially, two factors are involved: a *property* characteristic of the object of study, and a *measuring system*. The property, electromagnetic radiation, either emanates from or acts upon the object. The measuring system can take many forms ranging from the human eye to complex photometric systems.

For convenience, spectroscopy is usually subdivided on the basis of the measuring system. Thus, the range of the human eye sets limits for the visible range. Other ranges extend beyond the visible, beginning with ultraviolet on the blue side and infrared on the red.

In this discussion, a different classification is utilized, one more closely related to the specific function of the radiation source than to spectral range. Three source types are considered:

[1] Present address, Agfa Gevaert Camera Werk, 8 Munchen 90, Tegernseer Landstr. 161, West Germany.

sample sources; sample-exciting sources; and sample-probing sources.

Sample Sources

In this category the object itself is the source of radiation measured. Various examples can be given, including celestial bodies and radiating objects in the terrestrial environment. Another type consists of systems in which the sample is brought into a radiating state by an auxiliary energy source such as a flame, arc, RF-plasma torch, or glow discharge. Even a second light source such as a laser can be considered here, providing it doesn't participate in the radiation measurement *per se* (Figure 1).

Spectroscopic studies on celestial sample sources represent an ancient field, but the past decade has brought dramatic changes. For the first time, these sources can be observed free from the attenuating filter of the earth's atmosphere. Although the significance of these spectral studies is mainly of interest to astronomers and other scientists, it can be foreseen that further improvements in orbiting telescopes and

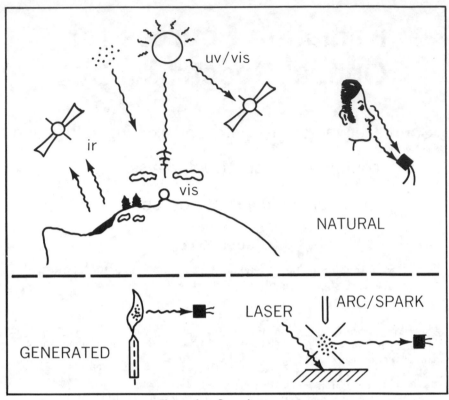

Figure 1.   Sample sources

spectrometers will bring a deeper and more accurate understanding of our universe including the sun and planets in the solar system. In the latter case, fly-by probes and actual landings will provide even better data.

Perhaps more exciting is the opportunity to examine the earth from outer space. As a large infrared source, the earth can be utilized to study weather, environmental changes, crop growth, and natural resources. The impact of these developments on our future life is so important that they certainly deserve mention.

Similar studies can be made on smaller objects including the human body. Here, infrared mapping can be used for detecting abnormal growths.

With the development of Fourier spectroscopy, such measurements may become even more precise and may be extended to objects having only weak emissions.

Of greater interest to analytical chemists are sample sources energized by an external means and used for atomic studies. In recent years, flames have received considerable attention. Mainly this has resulted from atomic absorption work, but much of the effort has brought benefits in emission as well (1). The premixed nitrous oxide–acetylene flame now commonly used for emission is even more suited with the advent of the separated flame (2). With it, a curtain of argon or nitrogen forms around a normal flame which lifts the secondary combustion zone and exposes the hotter inner flame zone. This reduces the effects of background emission by molecular species and improves detection limits for certain elements. The increased interest in flames also has led to the use of a premixed oxyacetylene flame which gives benefits in improved sensitivity.

Another nuance in flame-stimulated sources is the use of modulated sample feeding (3). Normally, the flame background radiation adds significantly to the noise level and causes zero offset in the signals. With modulated sample feeding, the background radiation remains constant while the sample signal is pulsed. Thus, an ac light signal, which can be processed and amplified with ac electronics, is generated.

Besides flames, various other energy sources are used to create sample sources. Arc and spark sources are widely used, particularly for simultaneous multielement analyses. Various methods are used to introduce the sample such that uniform and representative distribution is achieved. With solutions, this can be achieved by means of a rotating disk electrode.

With solids, analysis is more difficult because differences in melting temperatures and heats of evaporation complicate generation of a representative vapor sample. Improvements in solid sampling have been made by the use of lasers which produce highly localized energy, penetrating deep into the sample surface. The resulting vapor cloud is subsequently excited by the arc discharge. Undoubtedly, this will be an area of continued growth as more powerful lasers become available and rapid sequence firing is possible. Successive measurements can be visualized whereby readings can be averaged by photographs or evaluated by computer processing.

Work is also progressing with plasma torches (4, 5). These have a flame-like appearance but are RF-excited discharges in inert gases at atmospheric pressure. The sample is introduced in

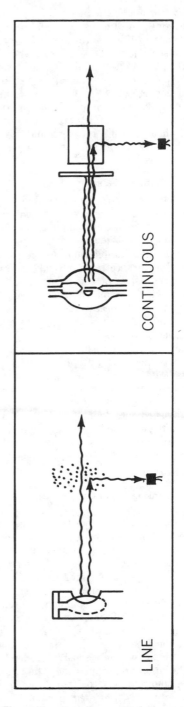

Figure 2.   Sample-exciting sources

107

mist form along with a carrier gas that sustains the discharge. Major benefits are obtained in improved sensitivity for some elements and reduced chemical interferences by elimination of oxides. To date, a major limitation has been quenching which occurs when sample quantities are increased as in trace analysis of complex organic samples. If this problem is overcome, plasma torches should attain widespread use in analytical work.

Sample-Exciting Sources

Sources of this type are used to excite the sample whereby characteristic radiation is re-emitted and measured. Included are various sources for molecular and atomic fluorescence as well as Raman spectroscopy. The spectral characteristics of the primary and re-emitted radiation may not coincide; nevertheless, spectral distribution is important and can be used advantageously.

In considering sources for molecular fluorescence, or phosphorescence, several factors are involved. Most analytical work is done in solution, so part of the energy of the exciting photon is converted into kinetic energy. Consequently, sharp lines are not required for excitation. However, the excitation efficiency varies with wavelength. In fact, if the wavelength of the primary radiation is varied, a characteristic emission spectrum often can be obtained.

Two general types of sources are used in molecular fluorescence: low-power line sources and powerful continua (Figure 2). Generally, these involve uv-visible excitation. With filter fluorometers, low-pressure mercury discharge lamps are common. Usually, the full excitation and fluorescence ranges are utilized, so sources of low to medium intensities are adequate. However, ultimate sensitivity depends, to a large extent, upon the ratio of sample fluorescence to that from other sources plus scattered primary radiation. Therefore, modifying the spec-

tral output of the source may be more beneficial than increasing the total source output. To this end, greater use is being made of fluorescent lamps in which the primary radiation is modified by a phosphor mounted either within or outside the lamp. In one type, a series of phosphors is arranged on a tabular sleeve around a mercury lamp. The desired spectral output is selected by simply rotating the sleeve.

With spectrofluorometers a new factor is introduced, namely that the available energy may be considerably less than with fluorometers. This results because: The f-number or aperture of a monochromator is usually smaller than that of a filter photometer; or, the spectral bandwidths of both excitation and fluorescence bands are kept small to increase selectivity and to provide high resolution.

Because of these factors, high-intensity sources are used almost exclusively. These are commonly known as high-pressure, compact arc lamps and include mercury arc lamps, xenon arc lamps, or Xe–Hg lamps. Respectively, these normally provide strongly broadened lines, broad continua, or continua with broadened lines.

During operation of high-pressure lamps, the arc is compressed in the narrow gap between the electrodes and becomes extremely bright. Lamps of this type are available from 75 watts to several kilowatts of power. Spectroscopic sources are usually operated below one kilowatt. Although intensity increases with power, benefits may not increase correspondingly because arc size becomes too large for effective use. Also, there is danger of sample degradation if source radiation becomes too intense.

A serious limitation of the compact arc lamp is instability which changes the illumination into the monochromator. This is caused mainly by arc wander rather than by actual change in brightness. Arc stability is better with dc arc lamps and these are generally preferred to ac lamps for spectroscopic work.

In a typical dc lamp, the cathode is much heavier than the anode and has cooling rings (Figure 3). Sturdy construction is required to withstand the intense heating caused by bombardment of positive ions. During operation, the electrode glows white and intense temperatures and pressures are generated. Special quartz envelopes are utilized to handle this harsh condition. With larger lamps, the intense uv radiation generates ozone which can create a hazardous condition if proper venting is not provided.

In recent years, extensive effort has been devoted to improve dc lamps, originally known as Osram types. Use of radioactive materials near the electrodes has improved both arc stability and firing characteristics. Furthermore, compact, better-regulated, solid-state power supplies have become available which make such lamps more attractive for analytical work.

Atomic fluorescence differs in several respects from molecular fluorescence and instrumental needs are somewhat unique. First of all, means must be available for generating atomic species. Flames can be used but considerable attention is being given to other techniques involving demountable hollow cathode lamps, laser-powered flash-vaporization cells, and other devices.

A second factor requiring attention is the critical wavelength match required between the exciting source and absorption line. Effectively, this requires a

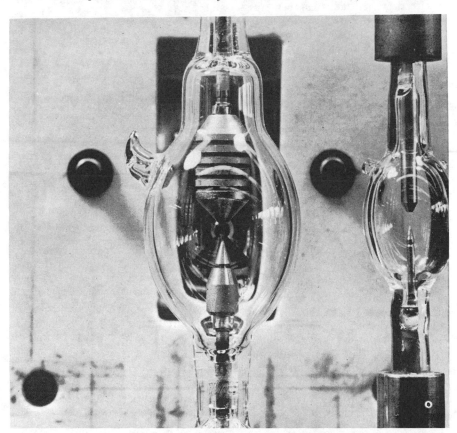

Figure 3.   Anode/cathode configurations of discharge lamps

correspondence within a fraction of an angstrom. Obviously, a line source such as a compact mercury lamp is useful only in the rare instances where atomic resonance lines coincide with the broadened source lines.

A high power continuous arc lamp—e.g., high-pressure Xenon lamp—would appear suited for this since it could be used for numerous elements of interest. However, two shortcomings limit its practical value: Only a small fraction of radiation is found at the excitation wavelength; and the ratio of scattered to fluorescent radiation is high.

These factors are related. As the percent of effective radiation decreases, the contribution of scattering increases. Although monochromators can be used to reject scattered radiation, their spectral bandwidth is large compared to the width of the atomic fluorescence line. Therefore, an appreciable quantity of nonfluorescence radiation reaches the detector limiting analytical sensitivity.

Obviously, the desirable source is one which gives intense radiation and is spectrally matched for the analyte of interest. Since the excitation spectrum of a ground state atom matches its absorption spectrum, hollow cathode lamps can be considered. Unfortunately, the typical low-pressure hollow cathode discharge is too weak for this, so attention has been given to the development of "high-intensity" lamps and vapor discharge lamps. These are useful but not totally satisfactory.

A more attractive development is the microwave-excited discharge lamp (6, 7). Although it has been available for many years, improvements were needed in terms of operating characteristics, useful life, and the number of elements covered. Significant progress is being made in this respect.

Sources of this type are relatively simple devices consisting of an argon-filled, sealed quartz tube about 1.0 to 1.5 in. long which contains a small quantity of an element or its volatile salt. In operation, the lamp is placed in a microwave cavity and subjected to microwave radiation. This generates a gas discharge which in turn vaporizes the enclosed element. When the vapor pressure is sufficiently high, a direct coupling is formed between the atomic vapor and the microwave. This creates an intense atomic line emission from the plasma. Consequently, the excitation radiation is strong relative to background and is selective for a given element.

It is conceivable that these sources will also find use in molecular fluorescence work for applications where intense and selective radiation is desired. They can also be used for wavelength calibration, but power input needs consideration to avoid excessive line broadening. The initial costs for power supply, cavity, and lamp are not excessive; in fact, they are comparable to those for high power compact arc lamps.

Another area where extremely high-power sources are required is Raman spectroscopy. This technique measures noncoherently scattered light which has been modulated by energy exchange with molecular oscillators. This provides valuable information similar to that afforded by infrared spectroscopy. In fact, the two techniques are complementary since Raman observations are possible on infrared inactive vibrations and vice versa.

Until recently, Raman spectroscopy was considered source-limited. Even with high-intensity mercury discharge lamps, measurements were made with difficulty. The advent of lasers dramatically altered this situation and significantly improved the growth potential of the technique. A preceding article in this series discusses lasers in detail (8).

Sample-Probing Sources

The third category of sources pertains to those used to probe samples either through absorption or reflection effects. Included are continuous-spectrum sources used in ultraviolet, visi-

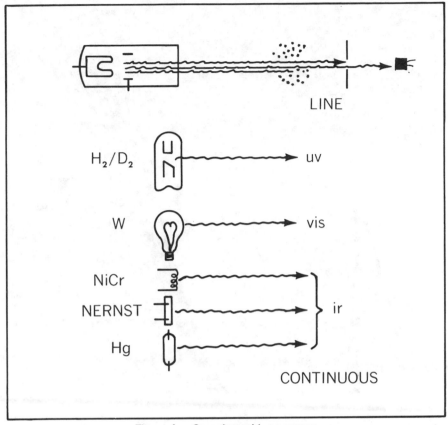

Figure 4.   Sample-probing sources

ble, and infrared spectroscopy, and the various line sources employed in atomic absorption spectroscopy (Figure 4).

For molecular measurements, the resolution of monochromators is normally adequate to permit continuous sources to be used. Generally, brightness is not a problem; in fact, extremely bright sources sometimes cause undue complications. Three general types of continuous sources are in common use: discharge lamps, hot-filament lamps, and ceramic or carbide types.

Work in the ultraviolet region is done mainly with hydrogen or deuterium discharge lamps (Figure 5) operated under low pressure and dc conditions. Heated cathodes provide the essential

function of maintaining the discharge. The discharge has negative characteristics, so a current-regulated supply is required.   A vital feature of these lamps is a mechanical aperture between the cathode and the anode which constricts the discharge to a narrow path. Normally, the anode is placed close to the aperture which creates an intensely radiating ball of light about 0.040 to 0.060 in. in diam on the cathode side of the opening.   The light is imaged at the slit of the spectrophotometer.

The unique emission of the hydrogen or deuterium lamp is generated by a power input of only 20 to 30 watts. Below 350 nm, a strong continuum is provided which fulfills most needs in

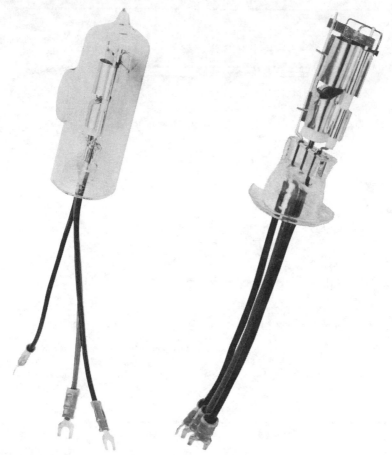

Figure 5.   Hydrogen discharge lamps

the ultraviolet. With quartz envelopes, work to about 160 nm is feasible. The use of deuterium in place of hydrogen slightly increases the size of the light ball but doesn't seem to enhance brightness.

Imaging of these sources at the lengthy spectrometer slit depends upon the aperture plate. In some lamps, the plate may include a small spheric or ellipsoid reflector around a circular hole. Others may have a reflector adjacent to a square opening. Lamps of this type can have either directly or indirectly heated cathodes and typically operate at a relatively low temperature.

At longer wavelengths, starting approximately at 360 nm, the hydrogen discharge has emission lines superimposed on the continuum. These can be used as a convenient means for calibration in much the same manner as the line spectra of low-pressure mercury lamps. However, in most work, the emission lines present a nuisance. Therefore, routine measurements above 360 nm and into the near-infrared are usually made with hot tungsten filament lamps which give continua over this range.

Filament lamps are rugged, low-cost units sufficiently bright for nearly all work. Special broad-band-type tungsten lamps are frequently used as standard sources. These can be calibrated and verified by the National Bureau of Standards. The shape of the radiating band on these lamps allows optimum imaging at the slit and thus uniform illumination. Spectral characteristics of the tungsten lamp are basically those of a blackbody. Therefore, measurements very far from the peak wavelength are susceptible to stray light effects.

In the region above 2 μm, blackbody-type sources without envelopes are commonly used. The same spectral characteristics cited for the tungsten lamp apply to these as well. The uneven spectral distribution is particularly serious for long-wavelength measurements.

A popular source of this type employs nichrome wire coils. These are simple and rugged, but less intense than other infrared sources. A black oxide forms on the wire which gives acceptable emissivity. Temperatures up to 1100°C can be reached. In one configuration, the hot coil is used to heat a ceramic sleeve which modifies the spectral output.

A still hotter, and therefore brighter, source is the Nernst glower which has an operating temperature as high as 1500°C. This source is generally preferred if long wavelength work is involved—e.g., to about 50 μm. Nernst glowers are constructed from yttria-stabilized zirconia in the form of narrow rods. Electrical connections are made with coated platinum wires.

Although more intense than wire filament sources, Nernst glowers have several disadvantages. The rod has a negative coefficient of resistance and must be preheated to be conductive. Therefore, auxiliary heaters must be provided as well as a ballast system to prevent overheating. Nernst glowers are fragile, particularly at the points of electrical connection. During use, platinum diffuses into the ceramic material particularly when local overheating or arcing occurs. This weakens the glower joints and eventually causes failure. Coating the joints with ceramic cement reduces the evaporation of platinum and increases life to some extent.

Until recently, another disadvantage of Nernst glowers has been their limitation to diameters of one mm or less. In the far infrared, where large slit widths are required to increase energy in spectrometers, such narrow rods are inadequate to illuminate the full slit. Fortunately, improved ceramic technology has led to purer, more uniform materials which can be used to construct rods with diameters up to two mm. Uniformity is essential to prevent current channeling which causes nonuniform surface temperatures. It is likely that these improvements will continue and even larger diameter sources will become available. These should have significant benefits in the far-infrared.

Another infrared source, with characteristics intermediate between heated wire coils and the Nernst glower, is constructed from silicon carbide. Commonly known as the globar, this source has an operating temperature near 1300°C. It has a positive temperature coefficient, and thus can be operated with a simpler power supply than that for the Nernst glower.

More recently, a source material which has aroused appreciable interest is molybdenum disilicate. Sources of this type can be operated to about 1450°C, thus approaching the temperature of Nernst glowers. The surface of the glower becomes coated with silica, providing favorable spectral characteristics. Unfortunately, resistivity is very low, so currents up to 80 amperes are required to heat a two-mm diam source. This necessitates heavy electrical leads which can get extremely hot. Formation of silica at the leads results in an increase in contact resistance.

In the very far-infrared, beyond 50 μm, blackbody-type sources lose effectiveness since their radiation decreases with the 4th power of wavelength. However, high-pressure mercury arcs give intense radiation in this region with maxima near 218 and 343 μm. Somewhat different from the compact mercury lamps described previously, these are high-vapor pressure discharge lamps which have an extra quartz jacket to reduce thermal loss. Output is similar to that from blackbody sources, but additional radiation emits from a plasma which enhances the long wavelength output.

In the field of atomic absorption spectroscopy, source requirements are quite different from those for molecular studies. Typically, atomic lines have widths of the order of 0.02Å; therefore, incident radiation must be of similar or smaller width to prevent serious loss of sensitivity and linearity. This requirement is beyond the capabilities of small and moderate size monochromators equipped with continuum-type sources. Therefore, work with continuum sources has been confined to research studies and has not entered the routine analytical phase. Development of AA

Figure 6. Electrodeless discharge lamp

spectroscopy as a major technique has occurred mainly because early workers selected a more suitable source and proceeded to develop practical units such as hollow cathode lamps (9).

The basic design of hollow cathode lamps is now well known. Essentially, they include a cathode containing the particular element to be tested and a noble gas at low pressure. In operation, the gas discharge sputters off cathode material to form an atomic cloud. This becomes excited and radiates the atomic line spectrum of the element in question. The lines are normally narrower than the corresponding absorption lines. Thus, both better linearity and sensitivity are achieved in the atomic analysis. A small monochromator is commonly used in connection with the hollow cathode lamps to reject undesired lines, yet, it has in most cases no effect upon the spectral width of the analytical line itself.

The simple one-element lamps first used have been followed by multielement lamps. These are constructed by including several elements within the single cathode. Generally, elements are selected on the basis of construction compatibility and freedom from overlapping lines. Use of intermetallic compounds helps provide uniform sputtering rates as desired to avoid preferential loss of one element over others. Also, consideration is given to group combinations commonly related to each other in analytical work.

Another trend in the development of hollow cathode lamps has been to increase brightness. This is particularly desirable in situations where signal-to-noise ratios are critical. Improvement in this regard has resulted from the use of auxiliary electrodes mounted in front of the cathode. These serve to improve atomic excitation without increasing the sputtering rate.

Still another modification has been the development of selective modulation lamps which in some instances eliminate the need for a monochromator (*10*). Actually, two lamps are used. The first generates a line spectrum in the conventional manner. This radiation enters a second hollow cathode lamp of the same element. The second lamp is periodically turned on and off. As a result, its atomic vapor acts as a selective and modulating absorber for the radiation from the first lamp. If the photometric system is tuned to the frequency of modulation, it responds only to the analytical line. The dc component may increase the noise level but this can often be reduced by using solar blind detectors.

The electrodeless discharge lamps (Figure 6) discussed in connection with atomic fluorescence are also useful in AA spectroscopy work. These are particularly attractive for elements where conventional hollow cathode lamps have proved ineffective—*i.e.*, volatile elements such as arsenic and selenium. However, great care must be taken with electrodeless discharge lamps to avoid line broadening with the resulting loss of sensitivity.

Summary

This discussion on sources has provided an opportunity to describe different types of sources and to point out various aspects in selecting sources for a particular analytical function. Success or failure often depends upon availability of suitable sources; in fact, many techniques became practical only after significant breakthroughs were made in this area. Atomic absorption spectroscopy is one example; Raman spectroscopy is likely to be another. Even in relatively successful fields such as infrared, new sources could still bring significant improvements. For this reason, source technology remains a vital and exciting field which will continue to add new dimensions to analytical spectroscopy.

References

(1) E. E. Pickett and S. R. Koirtyohann, *Spectrochim. Acta*, **24B**, 325 (1969).

(2) G. F. Kirkbright and T. S. West, *Appl. Opt.*, **7**, 1305 (1968).

(3) A. A. Javanovic, *Spectrochim. Acta*, **25B**, 405 (1970).

(4) C. D. West and D. N. Hume, ANAL. CHEM., **36**, 412 (1964).

(5) R. H. Wend and V. A. Fassel, *ibid.*, **38**, 337 (1966).

(6) J. M. Mansfield, Jr., M. P. Bratzel, Jr., H. O. Norgordon, D. O. Knapp, K. E. Zacha, and J. D. Winefordner, *Spectrochim. Acta*, **23B**, 389 (1968).

(7) R. M. Dagnall, K. C. Thompson, and T. S. West, *Talanta*, **14**, 551 (1967).

(8) R. G. Smith, ANAL. CHEM., **41** (10), 75A (1969).

(9) W. G. Jones and A. Walsh, *Spectrochim. Acta*, **19**, 249 (1960).

(10) J. V. Sullivan and A. Walsh, *Appl. Opt.*, **7**, 1271 (1968).

Fourier Transform Approaches to Spectroscopy

Gary Horlick

Department of Chemistry, University of Alberta,
Edmonton, Alta., Canada

Fourier transformation techniques have already led to significant advances in methods of spectral data handling. Increase in the use of the Fast Fourier Transform program should facilitate further developments in this area and enhance its value to analytical chemists

SPECTROSCOPISTS HAVE, in the past, dealt primarily with spectra. However, a consideration of the Fourier transformation of a spectrum can often result in a more complete understanding of several aspects of spectroscopy and spectroscopic measurements. The Fourier transform is basic to the very nature of a spectroscopic measurement since the dispersion step is, in effect, a Fourier transformation of the electromagnetic signal. The Fourier transform is intimately related to instrumental measurements through the convolution integral, and the important topic of spectral resolution falls in this area.

Certain types of mathematical operations, such as convolution, may be carried out on spectra using Fourier transformations. The calculations for this type of data handling are difficult to carry out. Recent advances in the machine calculation of Fourier transformations have removed this problem, and wider use of the Fourier transformation is sure to be seen in the data handling of spectra.

In some experiments the Fourier transformation of the spectrum may be the final desired result, rather than the spectrum. The work of Gordon on molecular correlation is an example of such a measurement (1).

Finally, two fairly new instrumental areas are presently being developed that necessitate an understanding of the Fourier transform operation in order to understand the measurement and subsequent data analysis. These techniques are Fourier transform spectroscopy in the optical region (2) and Fourier transform NMR spectroscopy (3). The measurement step in both these techniques results in the recording of a signal that is the Fourier transform of the conventional spectrum.

The above points indicate that analytical chemists need a basic understanding of Fourier transformations. A brief introduction to them is presented in this paper. Then several aspects of spectroscopy are interpreted on the basis of Fourier transformations. The coverage is not meant to be comprehensive but simply representative of this approach. Pictorial Fourier transforms are used to illustrate several points. It is often easier to get an intuitive feeling for the mathematical operation that is taking place by looking at a picture rather than an equation. The Fourier transform pairs illustrated pictorially in this paper are not schematic representations but are CALCOMP plots of the actual transformations as carried out on a digital computer.

Introduction to Fourier Transformations

This section provides a brief introduction to Fourier transformations. Terminology that will be used later in the paper is defined, and two of the most important properties of Fourier transformations, with respect to spectroscopic application, are illustrated. There are several comprehensive treatments of the theory of Fourier transforms. A particularly useful source is the book by Bracewell (4).

The Fourier integral is a mathematical means of relating two functions $f(x)$ and $F(v)$. It may be stated as:

$$f(x) = \int_{\infty-}^{\infty} F(v)e^{2\pi ixv}dv \quad (1)$$

An analogous integral exists such that

$$F(v) = \int_{-\infty}^{\infty} f(x)e^{-2\pi ixv}dx \quad (2)$$

These two equations indicate the reciprocal property of the Fourier integral. In the case of Equation 1, it may be stated that $f(x)$ is the Fourier transformation of $F(v)$ and for Equation 2, $F(v)$ is the Fourier transformation of $f(x)$. Thus, functions $f(x)$ and $F(v)$ constitute a Fourier transform pair.

The exponential of Equation 2 may be written as: $\cos(2\pi xv) - i\sin(2\pi xv)$. When $f(x)$ is an even function, Equation 2 reduces to:

$$F(v) = 2\int_0^{\infty} f(x)\cos 2\pi xv dx \quad (3)$$

This equation is often referred to as the cosine Fourier transformation or just the cosine transform. When $f(x)$ is an odd function, an analogous equation exists with the cosine term being replaced by the sine term. Both the sine and the cosine transforms have the reciprocal property indicated by Equations 1 and 2.

There are no rigorous constraints on the units of the functions $f(x)$ and $F(v)$. However, one of the functions is usually a function of frequency. For the purposes of this paper, $F(v)$ will designate the frequency dependent function and it will have units such as sec^{-1} or cm^{-1}. Often the function $F(v)$ is said to be in the frequency domain. The function $f(x)$ then becomes a function of time or distance and will have units of sec or cm. The function $f(x)$ is said to be in the time or space domain. Thus, very simply, $f(x)$ is a waveform and $F(v)$ is a spectrum, and the transform relationships provide a means of converting from one domain to the other. The integrals in Equations 1, 2, and 3 exist for any physically realizable functions $f(x)$ and $F(v)$—i.e., any waveform, in general, is composed of several frequencies. Fourier transformation is simply a technique for sorting out the intensities and frequencies present in any given waveform.

It is possible to solve this integral (Equation 3) for some simple functions. A common waveform is a cosine wave of finite length. This type of waveform occurs often in the physical world. Its transform is easy to calculate and illustrates several important characteristics of Fourier transform pairs. When we substitute $f(x) = \cos 2\pi xv'$ into Equation 3, the following simplified equation is obtained:

$$F(v) = \frac{\sin 2\pi x(v' - v)}{2\pi(v' - v)} \quad (4)$$

117

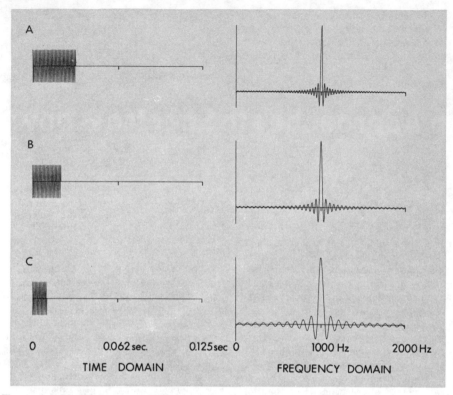

0   0.062 sec.   0.125 sec   0   1000 Hz   2000 Hz

TIME DOMAIN   FREQUENCY DOMAIN

Figure 1. Pictorial representations of the Fourier transformation of (A) 32 cycles, (B) 21 cycles, and (C) 10 cycles of a 1000-Hz cosine wave. Note the inverse dependence of the width of the frequency domain function on the length of the time domain function

where v' is the frequency of the cosine wave, and x is the length of the cosine wave. This equation represents the spectrum of the cosine wave. This transform is shown pictorially in Figure 1A. A finite length (32 cycles) of a 1000-Hz cosine wave was transformed, and the resulting spectrum is shown immediately to the right. Note that the function described by Equation 4 has a finite maximum at $v = v'$, and negative maxima on each side with intensities of about 20% of the central maximum. The width of this function depends inversely on x, the length of the original cosine wave. This is shown in Figures 1B and 1C where the Fourier transformations of 21 cycles and 10 cycles, respectively, of the same

1000-Hz cosine wave are illustrated pictorially. This inverse dependence of the width of the frequency domain function on the length of the time domain function is a very important characteristic of Fourier transform pairs.

A second important characteristic to note is that the functional dependence obtained for the spectrum of the cosine waveform is determined by the form of the truncation applied to the cosine wave. For the cases illustrated in Figure 1, the truncation was abrupt. The pictorial Fourier transform pairs for three other common truncations of a cosine wave are shown in Figure 2. A linear truncation of a cosine wave results in a $\sin^2 x / x^2$ functional depen-

118

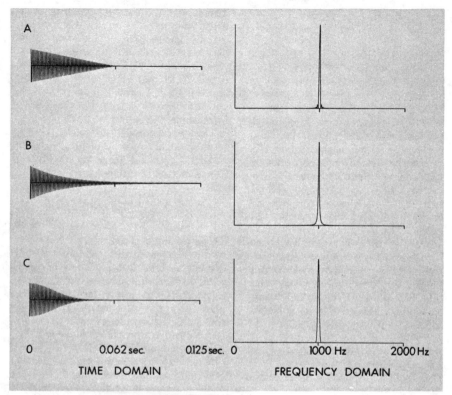

Figure 2. Pictorial representations of the Fourier transformation of a 1000-Hz cosine wave that has been damped in a linear (A), exponential (B), and Gaussian (C) manner. The respective functional dependencies of the frequency domain functions are $\sin^2 x/x^2$, Lorentzian, and Gaussian

dency for the spectrum (Figure 2A), an exponential truncation in a Lorentzian functional dependency (Figure 2B), and a Gaussian truncation in a Gaussian functional dependency (Figure 2C).

These dependencies can be shown mathematically by solution of Equation 3. However, the illustration of the Fourier transforms pictorially effectively indicates the main properties of the transforms. The book by Bracewell contains a dictionary of pictorial Fourier transforms for many additional functions (4).

Equations 1, 2, and 3 cannot easily be solved except for relatively few simple waveforms or spectra, such as those illustrated in Figures 1 and 2.

For general waveforms and spectra, the Fourier transformation is usually performed on a digital computer. The waveform or spectrum is sampled, and the evaluation of the Fourier transform takes the form of a summation. In 1965 Cooley and Tukey rediscovered and developed a technique for performing this summation efficiently. This is often referred to in the literature as the Fast Fourier Transform or the Cooley-Tukey Algorithm (5, 6). The development of this computer program has greatly facilitated the use of Fourier transformations in many data-handling situations. The Fast Fourier Transform was recently the topic of a special issue of the IEEE Transactions on Audio and Electroacoustics (7).

Utilizing the basic ideas covered in this section, we can discuss several aspects of spectroscopy by considering the time domain function in addition to the frequency domain function (i.e., the spectrum.) The Fourier transform provides the link between the two domains.

Fourier Transform Approaches to Spectroscopy

Frequency Decoding. The frequency-dependent nature of the interactions of electromagnetic radiation with matter provides the chemist with a vast amount of information. This information is most usefully interpreted in the form of a spectrum, a plot of the intensity of electromagnetic radiation as a function of frequency. However, this information is encoded in an electromagnetic waveform. To obtain the spectrum, this electromagnetic waveform must be analyzed for its frequency content—i.e., the frequency information must be decoded. This step amounts to taking the Fourier transformation of the electromagnetic waveform. In the electronic region this is relatively easy, as tunable components and systems are available that respond specifically to the actual frequencies of the electromagnetic waveform. A radio is an excellent example of such a system. Such components and systems are not yet available that respond in this fashion to the very high frequency electromagnetic waves that constitute the optical region.

In the optical region, somewhat indirect approaches must be used to carry out the Fourier transformation of an electromagnetic waveform. Prisms and gratings are, in a sense, powerful Fourier transformers. They decode the frequency information present in the electromagnetic waveform. The frequencies are spread out in space along the focal plane of an optical instrument (such as a spectrograph or monochromator) to form a spectrum. Another approach is to use a Michelson interferometer to generate a signal, called an interferogram, from the electromagnetic radiation. The Fourier transformation of this signal must be taken by the experimenter in order to obtain the spectrum of the original electromagnetic radiation. This step is usually performed on a digital computer. The more implicit presence of the Fourier transform step in this approach has resulted in the technique's being called Fourier transform spectroscopy. However, as can be seen from the above discussion, Fourier transformation is fundamental to all spectral determinations.

Fundamental Line Shapes. As was noted earlier, a damped or truncated cosine wave and the line shape of its corresponding spectrum are Fourier transform pairs. A real electromagnetic wave has a finite length and is damped or truncated in a specific fashion. Thus the fundamental width of a spectral line depends on how long the wave is and the shape of the line depends on the manner in which the wave is damped or truncated. Classical radiation theory leads to the conclusion that an emitted light wave, in the case of an unperturbed radiation lifetime, is exponentially damped (8). The Fourier transformation of an exponentially damped cosine wave is a line with Lorentzian functional dependency. This is the well-known line shape in the case of radiation damping.

This unperturbed line width and shape are seldom observed because of various line-broadening interactions, such as Doppler and collisional broadening. Doppler broadening leads to a Gaussian line shape, normally a couple of orders of magnitude wider than the natural line width. Collisional broadening leads to a Lorentzian line shape again—in general, significantly wider than the unperturbed line width. However, care must be exercised in concluding the type of electromagnetic signal that results in a specific line

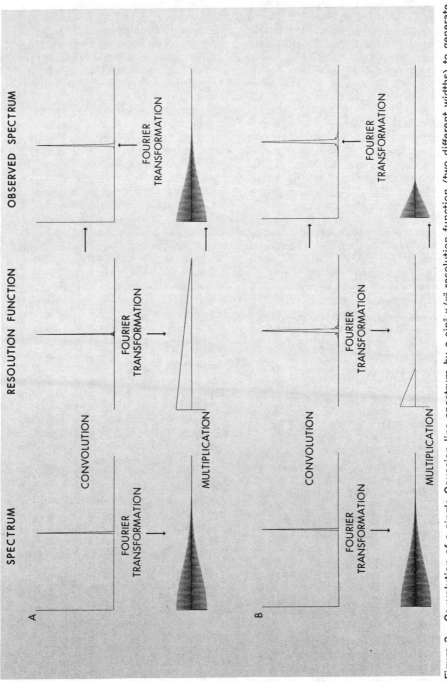

Figure 3. Convolution of a single Gaussian line spectrum by a $\sin^2 x/x^2$ resolution function (two different widths) to generate an observed spectral line. In each case the convolution is also depicted as a multiplication of the Fourier transforms of the respective functions

121

shape. In the case of collisional broadening, the Lorentzian line shape is not the result of an exponentially damped cosine wave but the result of the summation of several sinusoidal waves abruptly truncated by collisions.

Spectral Resolution. What shape does a spectroscopic instrument impose on an infinitely narrow spectral line? This is determined by the resolution or instrumental function of a spectroscopic measurement system. In prism and grating instruments, the resolution function may take two limiting forms. In the diffraction limited situation, the resolution function takes the form of a $\sin^2 x/x^2$ function (9), and in the slit-width limited situation, it may take the form of a triangular function.

The width of the resolution function imposes a limit on the resolution of the spectroscopic instrument, and the shape of the resolution function limits the ability of the instrument in accurately measuring spectral line shapes. If the resolution function is significantly wider than the width of the line being measured, the observed line width and shape will be that of the resolution function rather than that of the line itself; if they are approximately equivalent in width, the measured line shape will be a composite of the two, and only if the resolution function is significantly narrower than the observed line will the actual line shape and width of the source be measured.

The effects of the resolution function on the resulting spectrum are described by the convolution integral. The concept of convolution is generally useful in describing the effects of any particular instrument on an observed parameter (4, 10). However, it is often difficult to intuitively visualize the convolution operation. For the example stated above, the resolution function can be thought of as a scanning function that takes a "weighted average," or "running mean" of the spectrum to generate the observed spectrum. This type of terminology is frequently used to describe convolution.

The convolution operation may take on additional meaning and often ease of interpretation when it is realized that convolving two functions is equivalent to multiplying the Fourier transformations of the two functions. This is illustrated pictorially in Figure 3A.

The upper section of Figure 3A depicts the effects of convolving a Gaussian spectral line with a $\sin^2 x/x^2$ resolution function to generate an observed spectral line. The lower section of Figure 3A depicts the convolution operation as a multiplication of the Fourier transforms of the respective functions. In this case, the resolution function is narrower than the spectral line, and little broadening or distortion is observed. Figure 3B depicts the same convolution but with a wider $\sin^2 x/x^2$ resolution function. In this case the observed line is severely widened and distorted to the point of taking on the shape of the resolution function. That this should happen can be readily appreciated by noting the multiplication of the respective Fourier transforms (lower section of Figure 3B). The multiplication of the Gaussian damped cosine wave by the short linear truncation function results in a damped cosine wave with considerable linear character. This indicates that the observed spectral line will have a significant amount of $\sin^2 x/x^2$ functional dependence. Thus, the broadening and distortion of a spectral line when it is convolved by the resolution function of a spectroscopic instrument are readily understood on the basis of the two simple properties of Fourier transform pairs discussed in the first section—namely, the inverse dependence of the width of the frequency domain function on the length of the time domain function and the dependence of the shape of the frequency domain function on the form of the truncation applied to the time domain function.

The effects of convolving a spectrum by a resolution function are further illustrated in Figure 4. The format of this figure is analogous to that of Figure 3. In this case a spectrum consisting of wide and narrow Gaussian lines is convolved with a triangular resolution function to give the observed spectrum. This figure depicts the well-known situation where a narrow line in a spectrum may be severely distorted by a particular slit width, and a wide line in the same spectrum will not be significantly distorted (*11*). Again this can be readily appreciated by thinking in terms of the equivalent Fourier transform route of convolution. Note that in the lower section of Figure 4, the multiplication of the Fourier transform of the spectrum by the Fourier transform of the resolution function results in little, if any, truncation of the Gaussian damped cosine wave due to the wide line, while that due to the narrow line is truncated.

As mentioned by Savitzky and Golay (*10*), the observed spectrum is further convolved by time constants in the measurement electronics. The effects of this convolution could also be treated using Fourier transformations. However, the examples discussed in conjunction with the resolution function serve to indicate the approach. Thus, a number of instrumental effects on spectra can be appreciated, understood, and interpreted on the basis of a Fourier transform approach rather than by a direct application of the convolution integral. Also, an understanding of this approach leads to the development of data-handling techniques that can be performed on a spectrum to minimize or remove instrumental effects and also to the performance of operations on spectra not readily possible with hardware but easily implementable with software.

Data Handling Based on Fourier Transformations. Certain types of data-handling operations can be carried out on spectra by utilizing Fourier transformations (*12*). In this section a posterior convolution of observed spec-

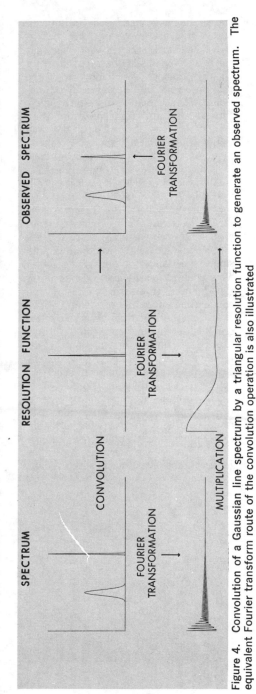

Figure 4. Convolution of a Gaussian line spectrum by a triangular resolution function to generate an observed spectrum. The equivalent Fourier transform route of the convolution operation is also illustrated

tra will be mentioned briefly to indicate one approach.

In the last section it was seen that the line shape in the observed spectrum was determined by convolution of the real spectrum by the resolution function. This convolution can be performed on a digitized spectrum using a digital computer. Thus, the line shape in an observed spectrum can be modified. For example, side lobes on a line shape function are undesirable if a small peak occurs close to a large peak, the side lobes easily being mistaken for real peaks. In this case the observed spectrum could be convolved on a computer with a mathematical resolution function that results in a line shape function of minimal side lobes. This could be easily carried out using the Fourier transform route of the convolution operation. The Fourier transformation of the observed spectrum is simply multiplied by the appropriate truncation function. This type of data handling is used extensively in Fourier transform spectroscopy and is called apodization.

This simple example illustrates a type of data-handling operation that is possible with spectra when utilizing Fourier transforms and convolutions. For the most part, extensive use of Fourier transformations in spectral data handling has not yet been made by analytical chemists. The Fast Fourier Transform program should facilitate further developments in this area.

References

(1) R. G. Gordon, *J. Chem. Phys.*, **43**, 1307 (1965).
(2) G. Horlick, *Appl. Spectros.*, **22**, 617 (1968).
(3) R. R. Ernst, "Advances in Magnetic Resonance, Vol. 2," J. S. Waugh, Ed., Academic Press, New York, N.Y., 1966, p 1.
(4) Ron Bracewell, "The Fourier Transform and Its Applications," McGraw-Hill, New York, N.Y., 1965.
(5) G-AE Subcommittee on Measurement Concepts, "What is the Fast Fourier Transform?", *IEEE Trans. Audio Electroacoustics*, **AU-15** (2), 45 (1967).
(6) L. Mertz, *Appl. Opt.*, **10**, 386 (1971).
(7) *IEEE Trans. Audio Electroacoustics*, **AU-17** (2), 65–186 (1969).
(8) W. Kauzmann, "Quantum Chemistry," Academic Press, New York, N.Y., 1957, p 556.
(9) R. A. Sawyer, "Experimental Spectroscopy," Dover Publications, New York, N.Y., 1963, p 33.
(10) A. Savitsky and M. J. E. Golay, ANAL. CHEM., **36**, 1627 (1964).
(11) W. E. Wentworth, *J. Chem. Educ.*, **43**, 262 (1966).
(12) G. Horlick, ANAL. CHEM., **44**, 943 (1972).

Ion Cyclotron Resonance Spectrometry: Recent Advances of Analytical Interest

MICHAEL L. GROSS and CHARLES L. WILKINS

Department of Chemistry, University of Nebraska
Lincoln, Neb. 68508·

Ion cyclotron resonance spectrometry is becoming established as an important analytical technique in its own right as well as a useful complement to conventional mass spectrometry. The introduction of digital computer techniques should further expand its usefulness

As MOST READERS ARE WELL AWARE, the instrumentation used in the broad field of mass spectrometry is diverse. There are many types of mass spectrometers, each with its own advantages and limitations. These instruments range from the simple "mass filters" and residual gas analyzers to very complex, high-resolution instruments only available in a few laboratories. However complex the instrument, the common objective is to determine the mass of ions produced in some source and to use their behavior as an analytical tool to obtain structural and thermodynamic information.

Since the design of these spectrometers dictates low ion residence times (ca. 10^{-6} sec), analytical techniques utilizing ion-molecule reactions have not been extensively employed. Naturally, when ions do not remain in the spectrometer for sufficient time to permit reaction with molecules, it is impossible to employ such reactions analytically. Of course, by suitable modification of the spectrometer and conditions, it may be possible to use higher sample pressures and thus to observe ion-moelcule reactions as a result of the increased reaction rates engendered. Now, even though the residence times are no longer, the probability of an ion colliding with a neutral molecule is much greater (since the concentration of neutrals is orders of magnitude higher), and reaction can occur before the residence time is "used up." Due to a number of complications which arise through the use of this method, it is often desirable to employ a different type of mass spectrometer which can allow ion-molecule reactions at low sample pressures. It is that instrument, the ion cyclotron resonance spectrometer, which will be considered in the present article.

Ion Cyclotron Resonance: Instrumentation and Principles

The basic phenomenon on which ion cyclotron resonance (icr) spectrometry depends is that observed when a charged particle is subjected to the influence of a magnetic field. Briefly, a magnetic field causes an ion to travel in a circle. As a result, the centrifugal force acting on the ion (mv^2/r is balanced by the force from the field (eHv/c), or $mv^2/r = eHv/c$. Rearrangement of this equation leads to an interesting result which is basic to ion cyclotron resonance

$$v/r = \frac{eH}{mc} = \omega_c \qquad (1)$$

where v is the velocity of the ion, r is the radius, e is the charge of the ion, H is the magnetic field strength, m is the mass of the ion, and c is the speed of light. Angular velocity, ω_c, is simply the number of radians per unit time, and the frequency of revolution in cycles per unit time is $\omega_c/2\pi$. Thus it can be seen that the cyclotron frequency, ω_c, at constant magnetic field strength depends only on the mass-to-charge ratio of an ion.

If, in addition to the magnetic field, the ion is further subjected to the simultaneous influence of an electric field applied at right angles to the magnetic, the particle departs from the circular motion it had previously pursued and proceeds in a cycloidal path at right angles to both fields. This motion is represented schematically in Figure 1. It still follows that the frequency (now the number of complete cycloids per unit time) is determined only by m/e at constant H.

This fact can be used to advantage to allow detection of the ions. If a rapidly alternating electric field is applied, as the frequency of alternation approaches the cyclotron frequency of an ion which is present, the ion will absorb energy. As this absorption takes place, the ion is accelerated, causing the radius of its cycloidal path to increase (since the frequency must remain constant). Figure 2 gives a pictorial representation of this process, and shows the result of an ion's resonance—i.e., the matching of frequencies discussed above. For ions of between 1 and 200 amu, with the magnetic field strength commonly employed, the frequencies range from 50 kHz to about 25 MHz. Therefore, a marginal oscillator detection system is practical. Using the cell as a capacitive element in the resonant circuit, a phase-sensitive detector can be used to observe the energy level changes of the oscillator. These changes take place only when ions absorb energy. Therefore, a mass spectrum can be obtained by plotting the output of the detector circuit vs. a time domain quantity (either the rate of scan of the magnetic field at constant frequency or, alternatively, the rate of change of the electric field frequency at constant magnetic field). As a result of the circuitous route which ions travel as they pass through a cell of approximately 10 cm in length, the residence times are on the order of milliseconds. Through the use of special techniques, residence times can be made as long as several seconds.

This method of detection has a striking advantage over conventional magnetic focusing mass spectrometers. We have mentioned that ω_c is dependent only on m/e and H, but it is not dependent on the velocity of the ion, v. An increase in velocity will increase the radius of the cycloidal trajectory, but will not change the frequency. On the other hand, the focusing properties of a conventional mass spectrometer depend strongly on

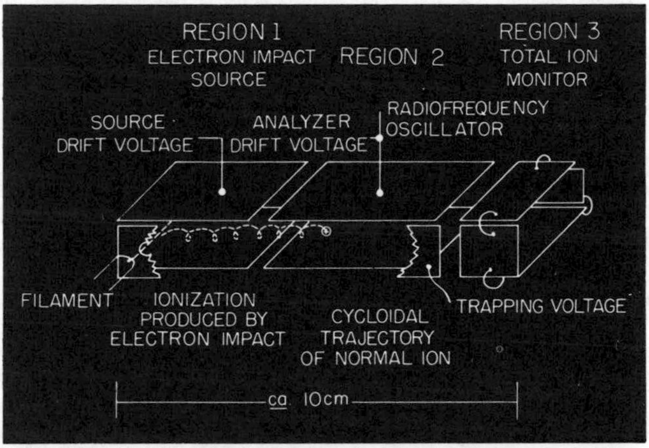

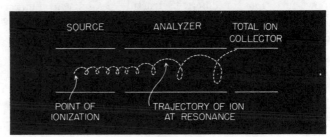

Figure 1. Schematic diagram of an ion cyclotron resonance cell showing the normal ion trajectory

REGION 1
ELECTRON IMPACT
SOURCE
REGION 2
REGION 3
TOTAL ION
MONITOR

SOURCE
DRIFT VOLTAGE
ANALYZER
DRIFT VOLTAGE
RADIOFREQUENCY
OSCILLATOR

FILAMENT   IONIZATION
PRODUCED BY
ELECTRON IMPACT
CYCLOIDAL
TRAJECTORY
OF NORMAL ION
TRAPPING VOLTAGE

ca 10cm

Figure 2. Trajectory of an ion in resonance

SOURCE   ANALYZER   TOTAL ION
COLLECTOR

POINT OF
IONIZATION
TRAJECTORY OF ION
AT RESONANCE

ion velocity. Thus, studies involving changes in ion kinetic energy can be made by icr without affecting the detection scheme.

Referring back to Figure 1, the operational procedure is to first form ions in the electron impact source (Region 1); they then travel down the cell through the analyzer section (Region 2) and eventually reach the total ion monitor (Region 3). To constrain the ions to a motion down the center of the cell, trapping voltages are applied to side plates. The reader is referred to any of the excellent review articles of Baldeschwieler (1, 2) or Henis (3) for more complete discussion of the theory. For illustrative purposes, we have included a photograph of the commercial icr spectrometer (Figure 3), a close-up of the vacuum can which encloses the cell being lowered between the poles of an electro-magnet (Figure 4), and a close-up of the icr cell removed from the vacuum can (Figure 5).

Analytical Advantages of Ion Cyclotron Resonance

As Henis points out (3), icr spectrometry is well suited for the study of ion-molecule reactions. Physical chemists have used the method for the investigation of collisional processes, gas-phase kinetic studies, and tests of theoretical predictions regarding fundamental molecular and atomic behavior. Although analytical chemists will find these topics interesting, it is our belief that the real value of icr spectrometry as an analytical tool for quantitative and qualitative analysis will lie in different areas. For example, physical chemists have elaborated elegant methods for quantitative kinetic studies of ions by icr. The analytical chemist can take advantage of these methods in attacking the difficult problems which sometimes face him as he attempts to interpret conventional mass spectrometry results. A complete understanding of a mass spectrum demands a thorough knowledge of not only ion mass-to-charge ratios, but also of ionic structures and fragmentation mechanisms. The new dimension added by icr makes possible the use of differences in ion-molecule reactivity to distinguish isomeric ions or neutrals. As a complement to conventional mass spectrometry (which is incapable of easily distinguishing isomeric ions), the icr method is uniquely powerful. Ions of different structure will rarely undergo the same bimolecular reactions. Therefore, structural identity of ions may be established by generating them from both known sources and compounds of interest and using ion-molecule reactions to demonstrate that identity. For instance, using this approach, we have been able to unequivocally demonstrate the nonidentity of the ions arising from the isomeric compounds, cyclooctatetraene and styrene (4), which were previously thought to be the same as a result of conventional mass spectrometric studies.

Similarily, neutral compounds which might otherwise be extremely difficult to analyze can be characterized by the use of suitable ion-molecule reactions. Bursey and co-workers (5) have recently demonstrated that a gas-phase acetylation ion-molecule reaction can be an important means of chemical ionization. The advantage of chemical ionization over ionization by electron impact is that the former is a "soft" ionization process—i.e., little internal energy is transferred to the ionized species. As a result, fewer fragmentations of this new ion will occur, and it is expected that these fragmentations will be more selective than those produced by electron impact ionization. Certainly, a significant limitation of

electron impact ionization is that molecules of similar structure produce nearly identical fragmentation patterns due to the large amount of internal energy imparted in the ionization process. Ion-molecule reactions, as studied by icr and other techniques, may be an important complement to conventional mass spectrometry in solving analytical problems concerning such similar molecules. The acetylation reaction mentioned above occurred with some selectivity depending on the nature of heteroatoms with unshared electron pairs present in an organic molecule. In our laboratory, we have used the 1,3-butadiene molecule ion as a chemical ionization agent and have found significant differences in the reactions of this ion with various isomeric pentenes (6). This series is typical of hydrocarbon isomers because nearly identical mass spectra are obtained by electron-impact ionization, and thus complete analysis is very difficult.

Implicit in the methods above is the use of kinetic data to distinguish isomeric species. It is also worthwhile to examine how kinetic data for reactions of interest can be obtained through use of a variety of techniques. For one thing, precursors of ionic products can be established by use of double resonance methods. This is extremely difficult to do by high-pressure mass spectrometric studies. Double resonance techniques depend on the principle that, if one irradiates at the cyclotron resonance frequency of a suspected precursor ion while observing the product ion, changes in the product abundances indicate their connection. The absence of such changes imply that the product does not arise from a reaction involving the suspected precursor. In the event that an ion product arises from two different precursors, one can be ejected, and thus the reaction of the other may be studied without interference. This "ion ejection" is either the result of irradiating in the source region at the interfering ion's cyclotron frequency

Figure 3. A commercial ion cyclotron resonance spectrometer

Magnet and cell are to the left of console

Figure 4. Cell container being lowered between the magnet pole faces

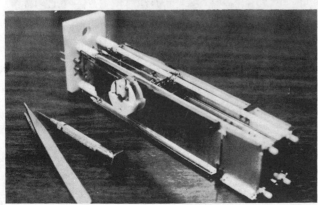

Figure 5. Typical ion cyclotron resonance cell

In use, this cell is contained in the holder shown in Figure 4; note the compact construction

129

and thus spiraling the ion into the cell wall before it can react, or of modulating the trapping voltage, which results in a similar occurrence. The technical details of the modulation method are discussed in detail in a recent paper by Beauchamp and coworkers (7). Another recent development is the use of a trapped ion analyzer cell to extend ion residence times to as long as several seconds (8). In this technique, ions are formed by a pulsed electron beam and then trapped in a completely enclosed cell, constructed so as to prevent ions from drifting out of the cell. This development will permit the study and use of even slower ion-molecule reactions than was hitherto possible. In addition, studies of the effect of long residence times on mass spectral fragmentations are now possible.

Of particular interest to analytical chemists are the potential applications of icr to such fundamental problems as the direct measurement of proton affinities and acidities in the gas phase. It is extremely important to be able to measure these if we are to fully understand the role of solvent in acid-base chemistry. Furthermore, the data which can be collected will be of inestimable aid in interpreting the large amount of solution data already available. Briefly, the method is to establish relative acidities by studying an ion-molecule reaction involving a proton transfer between the two compounds to be compared. In this way, it is relatively easy to measure data for a whole series of compounds and, ultimately, to relate that series to a similar series determined in solution. Icr allows the study of negative ions in exactly the same way. So, for example, the gas-phase basicities of alkoxide ions have been measured by icr and the order found to be different from that found in solution (9).

Other thermodynamic quantities including electron affinities and appearance potentials are readily measured by icr. Due to the open design of the cell, it is relatively easy to make provisions for studying interactions of ions with light. By open design, we mean the collector end of the cell is free, and it is therefore possible to use this region as an entry for a photon beam. Similar studies would be very difficult by conventional methods. So, by using negative ions and determin-

Table I. Appearance and Ionization Potentials for C_5H_{10} Isomers[a]

Compound	Ip	A(55+)	A(42+·)
Cyclopentane	10.91 ± 0.07 (10.53)[b]	11.36 ± 0.08 (12.1)[c]	11.74 ± 0.07 (11.4)[c]
1-Pentene	9.82 ± 0.06 (9.50)[b]	11.35 ± 0.07	11.61 ± 0.08
3-Methyl-1-butene	9.60 ± 0.03 (9.51)[b]	11.15 ± 0.12	11.54 ± 0.10
2-Methyl-1-butene	9.35 ± 0.08 (9.12)[b]	11.34 ± 0.07	11.66 ± 0.06
trans-2-Pentene	9.32 ± 0.03 (9.06)[d]	11.35 ± 0.03	11.73 ± 0.11
cis-2-Pentene	9.23 ± 0.02 (9.11)[d]	11.24 ± 0.02	11.54 ± 0.02
2-Methyl-2-butene	8.83 ± 0.11 (8.67)[b]	11.33 ± 0.12	11.70 ± 0.11

[a] The deviations are standard deviations of triplicate determinations, each measured on a different day.
[b] K. Watanabe, T. Nakayama, and J. Mottl, J. Quant. Spectrosc. Radiat. Transfer, 2, 369 (1962) as cited in J. L. Franklin, J. G. Dillard, H. M. Rosenstock, J. T. Herron, K. Draxl, and F. H. Field, Nat. Stand. Ref. Data Ser. Nat. Bur. Stand., 26, 55 (1969).
[c] J. Collin and F. P. Lossing, J. Amer. Chem. Soc., 81, 2064 (1959).
[d] J. Hissel, Bull. Soc. Roy. Sci. Liege, 21, 457 (1952).

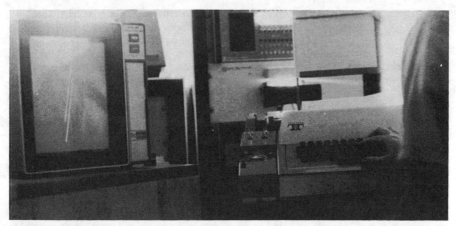

Figure 6. Computer system and real-time display used to obtain the appearance potential data contained in Table I

This system was directly interfaced to the icr spectrometer and experiments were carried out under computer control

ing the threshold for photodetachment of an electron, electron affinities can be determined with more accuracy than is possible by classical techniques (10). These data are quite important for understanding electrode processes, for example. The analogous threshold for detachment of an electron or an ion from a neutral species, the appearance potential, is of interest also. This fundamental quantity can also be readily measured by icr due to the ease of measuring ion abundances as a function of ionizing energy. Although this measurement can be made with a conventional mass spectrometer, the low electric field in the icr source and the longer residence times make it the superior tool.

Computer-Aided Ion Cyclotron Resonance

After considering all the apparent advantages which icr has to offer the analytical chemist, it is fair to ask why it has not been extensively used as a routine analytical method. At first glance, it might seem that the expense (ca. $50,000) of the commercial instrument might be a factor. How-

ever, many equally specialized instruments are widely used, so we must look more deeply for the causes. It is our suggestion that, although the data is rather easily collected, its sheer volume and the time required for its reduction has been one of the barriers to implementation. Until recently, icr spectrometry has been, as we mentioned previously, the province of physical chemists. Now it has passed the early development stage, and understanding of its analytical uses is possible by those who are not specialists in the field. As Henis noted (3) the initial development was in the direction of ion-molecule techniques and a different experimental orientation would be required for its development in the analytical area. With the application of modern computer techniques that different orientation is now possible. To cite an example from our laboratories, we have developed a computer-aided method for rapid and precise appearance potential measurements and applied it to the qualitative analysis of a group of very similar olefins which are difficult to distinguish otherwise. The precision can be judged from the data tabulated in Table I. Our estimate of the time required to gather and reduce the same data using previous approaches is approximately 100 times that required with computer assistance. Obviously, we would have been reluctant to undertake the analysis previously unless there were very strong incentives to do so. Now, we measure appearance potentials routinely, and typically, using our interactive system, can complete a measurement in 10 min. Figure 6 is a photograph of a storage oscilloscope with a plot of the curves used to obtain appearance potentials. This is simply representative of the sort of development to be expected in the near future. Our understanding has reached the point where chemists will soon be able to add icr to their repertoire of routine analytical tools.

Summary

Ion cyclotron resonance spectrometry is becoming established as an important analytical technique in its own right and a useful complement to conventional mass spectrometry. The introduction of digital computer techniques should further extend this usefulness. The body of knowledge to be gained through its use in studying ion-molecule reactions will be valuable in understanding and advancing the progress of other analytical methods such as plasma chromatography and chemical ionization spectrometry which depend, basically, on ion-molecule reactions. We think icr is making the transition from laboratory curiosity to analytical tool.

References

(1) J. D. Baldeschwieler, *Science,* 159, 263 (1968).
(2) J. D. Baldeschwieler and S. S. Woodgate, *Accounts Chem. Res.,* 4, 114, (1971).
(3) J. M. S. Henis, ANAL. CHEM., 41 (10), 22A (1969).
(4) C. L. Wilkins and M. L. Gross, *J. Amer. Chem. Soc.,* 93, 895 (1971).
(5) M. M. Bursey, T. A. Elwood, M. K. Hoffman, T. A. Lehman, and J. M. Tesarek, ANAL. CHEM., 42, 1371 (1970).
(6) M. L. Gross, P. H. Lin, and S. J. Franklin, ANAL. CHEM., 44, 974 (1972).
(7) J. L. Beauchamp and J. T. Armstrong, *Rev. Sci. Instrum.,* 40, 123 (1969).
(8) R. T. McIver, Jr., *ibid.,* 41, 555 (1970).
(9) J. I. Brauman and L. K. Blair, *J. Amer. Chem. Soc.,* 92, 5986 (1970).
(10) J. I. Brauman and K. C. Smyth, *ibid.,* 91, 7778 (1969).

Hadamard-Transform Spectrometry:
A New Analytical Technique

JOHN A. DECKER, JR.

Spectral Imaging, Inc., 572 Annursnac Hill Road,

Concord, Mass. 01742

Hadamard-transform multiplexing can allow the simultaneous measurement of several infrared absorption wavelengths, characteristic of several specific compounds of interest, with a comparatively economical and reliable instrument which needs only a single detector/preamplifier/data-analysis electronics chain

HADAMARD-TRANSFORM SPECTROMETRY (1–4) (or HTS) is an analytical technique probably completely new to most analytical chemists. This new class of infrared spectrometer can achieve the full "multiplex" performance of the very sophisticated (and very expensive!) Fourier-transform-interferometer-spectrometers (5) but employs the well-proven and comparatively simple technology of conventional dispersive (i.e., prism or grating) spectrometers (6). What this means to the analytical chemist, of course, is that he can now exploit the orders-of-magnitude decrease in required scanning time (or equivalently increase in signal-to-noise ratio and hence in sensitivity) possible with multiplex infrared spectrometers without incurring the comparative expense and complexity (and, thus, unreliability) that have been the unfortunate companions of multiplex instruments.

What, then, is a "Hadamard-Transform Spectrometer"? First, let us refer to the logic diagram (Figure 1) originally conceived in slightly different form by Fellgett (7) to make the point that "interferometric," "multiplex," and "Fourier-transform" are independent attributes of infrared spectrometric systems. First, a Hadamard-transform spectrometer is a multiplex instrument: It observes all the wavelengths in a spectrum at the same time and hence has the multiplex (or Fellgett) (8), advantage in signal-to-noise ratio over an equivalent conventional scanning-monochromator spectrometer. This advantage is approximately equal numerically to a factor of the square root of the total number of spectral resolution elements in the spec-

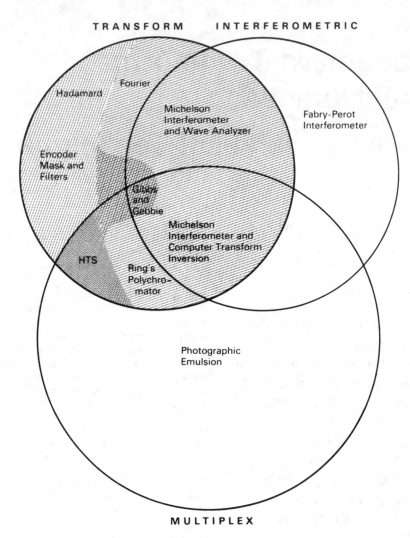

Figure 1. Multiplex/transform/interferometric logic diagram for infrared spectrometers (after Fellgett) (8)

trum; for a many-thousand-point high-resolution spectrum, this is obviously a substantial gain!

Second, it is a transform instrument in that multiplexing is accomplished by means of an optical coding process such that the measured data points are the mathematical transform of the input dispersed optical spectrum. Here, we use one of a class of binary (i.e., com-posed of "on" and "off" states only) "pseudo-noise" transforms based on the "Hadamard" matrices (9)—hence, the name of the technique.

Finally, it is not an interferometric instrument, as it uses conventional dis-persive optics (e.g., a prism or diffrac-tion grating) for spectral discrimina-tion. In addition, one can construct an HTS with the so-called "Multislit-Mul-

Figure 2. 2047-Slot cyclic mask code: "1" denotes transparent slot, whereas "0" denotes opaque slot

```
00000 00001 00000 00010 10000 00100 01000 01010 10100 10000
00011 01000 00111 00100 01101 11010 11101 01000 10100 00101
00010 01000 10101 10101 00001 10000 10011 11001 01110 01110
01011 11011 10010 01010 11101 10000 10101 11001 00001 01110
10010 01010 01101 10001 11101 11011 00101 01011 11000 00010
01100 00101 11110 01001 00011 10110 10110 10110 00110 00111
01111 01101 01001 01100 00110 01110 01111 11011 11000 01010
01100 10001 11111 01011 00001 00011 10010 10110 11100 00110
10110 01110 00111 11011 01100 01011 01110 10011 01010 01111
00001 11001 10011 01111 11111 01000 00001 00100 00010 11010
00100 11001 01011 11110 00010 00011 00101 00111 11000 11100
01101 10110 11101 10110 10101 10110 00001 10111 00011 10101
10110 10001 10110 01011 10111 10010 10100 11100 00011 10110
00110 10111 01110 00101 01011 01000 00011 00100 00111 11010
01100 01001 11110 10111 00010 00101 10101 01001 10000 00100
11000 01100 01100 11110 11111 10010 10000 11100 01001 10110
10111 10110 00100 10111 01011 00101 00011 11000 10110 01101
00111 11100 11100 10111 01011 01100 10111 01111 00100 00001
11010 00011 01001 00111 00110 11101 11110 10101 00010 00000
10101 00001 00000 10010 10001 01100 01010 01110 10001 11010
01011 01001 10011 00111 11111 11100 00000 00110 00000 01111
00000 11001 10001 11111 11011 00000 01011 10000 10010 11001
01100 11110 01111 10011 11000 11110 01101 10011 11101 11110
00101 00011 01000 10111 00101 00101 11000 11001 01011 11110
01101 00011 11100 10110 00111 00111 01101 11101 01101 00100
01100 11010 11111 11000 10000 01101 01000 11100 00101 10110
01001 10111 10111 10100 10100 10011 00011 01111 10111 01000
10101 00101 00000 11000 10001 11101 01011 00100 00011 11010
00110 01001 01111 10110 01000 10111 10101 00100 10000 11011
01001 11011 01000 10011 01011 00010 00100 10101 01011 00000
00011 10000 00110 11000 01110 11100 11010 10111 11000 00100
01100 01010 01110 10000 10010 01001 01101 10010 01101 10111
11101 10100 00101 10010 01001 11101 10111 00101 10101 11001
10001 01111 11010 01000 01001 10100 10111 10011 00100 11111
11011 10000 01010 11000 10000 11101 01001 10100 00111 10010
01100 11101 11111 10101 00000 10000 10001 01001 01010 00110
00001 01111 00010 01001 10101 10111 10001 10100 11011 10011
11010 11110 01000 00011 10101 01110 10000 01010 01000 10001
10101 01011 10000 00010 11000 00100 11100 01011 10110 10010
10110 01100 00111 11110 01100 00011 11110 00110 00011 01111
00111 01001 11101 00111 00100 11101 11011 10101 01010 10000
00000 01000 00000 10100 00001 00010 00010 10101 00100 00000
11010 00001 11001 00011 01110 10111 01010 00010 00001 01000
10010 00101 01101 01000 01100 00100 11110 10011 10011 10010
11110 11100 10010 10111 01100 00101 01110 01000 01011 10100
10010 10001 01100 01111 01110 11001 01010 11110 00000 10011
00001 01111 10010 01000 11101 10101 10101 10001 10001 11011
11011 01010 01011 00001 10011 10011 11110 11110 00010 10011
00100 01111 11010 11000 11000 11100 10010 10111 00001 10101
10011 10001 11110 11011 00010 11011 10100 11010 10011 11000
01110 01100 11011 11111 11010 00000 01001 00000 10110 10001
00110 01010 11111 11000 10000 11001 00101 11110 00011 00011
01101 10111 01101 10101 01101 10000 01101 11000 11101 01101
10100 01101 10010 11101 11100 10101 00111 00000 11101 10001
10101 11011 10001 10001 11100 10101 00111 00000 11110 10011
00010 01111 10101 11000 10001 01101 01010 01100 00001 11110
00011 00011 00111 11101 10011 00101 11111 11001 00000 01110
10000 11010 01001 11001 01111 01111 10101 01010 10000 00101
01000 01000 00100 10100 01011 00010 10011 10100 01110 10010
11010 01100 11001 11111 11111 00000 00001 10000 00011 11000
00110 01100 01111 11110 11000 00010 11100 10000 10110 01011
00111 10011 11100 11110 00111 10011 01100 11111 01111 10001
01000 11010 00101 11001 01001 01110 00110 01011 01111 10011
01000 11111 00101 10001 11001 11011 01011 01001 01001 00011
00110 10111 11110 00100 00011 01010 00111 00001 01101 10010
01101 11101 11101 00101 00100 11000 00101 11101 11010 00101
01001 01000 00110 00100 01111 01010 11001 00000 11110 10001
10010 01011 11101 10010 00101 11101 01001 00100 00110 11010
01110 11001 11010 11111 00001 10001 00101 01010 11000 00000
11100 00001 10110 00011 10111 00110 10101 11110 00001 00011
00010 10111 10100 00100 10010 01011 01101 10011 01101 11111
01101 00001 01100 10010 01011 01101 01101 01001 01011 01000
01011 11110 10010 00010 01101 00101 11100 11001 00111 11110
11100 00010 10110 00100 00111 01010 01101 00001 11100 10011
00110 01111 11101 11011 00100 00100 00100 01010 10001 10000
01011 11000 10010 01101 01101 11100 01101 00110 11100 11110
10111 10010 00100 11101 01011 10100 10000 00010 00100 01100
01001 11100 00000 10110 00001 00111 00010 11101 10100 10101
10011 00001 11111 10011 00000 11111 10001 10000 11011 11001
11010 01111 01001 11001 00111 01110 11101 01010 101
```

tiplex" mode of operation (*10*) (involving optical coding at both the entrance and exit of the spectrometer) which displays the full throughput (or Jacquinot) (*11*) advantage previously characteristic only of interferometric spectrometers; it then has an additional signal-to-noise advantage of several hundred or so owing to the increased light "grasp." Additionally, it has a unique spatial-imaging capability (*12*) lacking in all other multiplex spectrometric systems.

The Hadamard-transform spectrometer thus occupies the densely ruled area at center-left in Figure 1: a multiplex transform spectrometer which uses dispersive rather than interferometric optics for spectral discrimination and which uses the binary Hadamard transform rather than the usual analog Fourier transform (*13*) in its multiplexing logic.

The possibility of realizing the multiplex advantage with a dispersive spectrometer by means of optical coding based on binary orthogonal matrices was first pointed out by Fellgett (*8*), and an explicit technique for accomplishing this was first suggested by Ibbett et al. (*14*), and independently shortly afterwards by Decker and Harwit (*4*). These orthogonal binary codes were explicitly developed in terms of Hadamard matrices by Sloane et al. (*9*) who also developed the signal-to-noise analysis of their efficiency. We have described earlier the successful operation of a 19-slot feasibility-demonstration-prototype HTS (*3*) (that is, a spectrometer with 19 spectral resolution elements) and a 255-slot laboratory-prototype Hadamard-transform spectrometer (*1*), which was, to the best of our knowledge, the first dispersive multiplex spectrometer to quantitatively verify the multiplex advantage.

We will, therefore, limit ourselves here to the description of a recently completed 2047-slot HTS system completed under the support of the Air

Figure 3. Exit-focal-plane mask for 2047-slot HTS

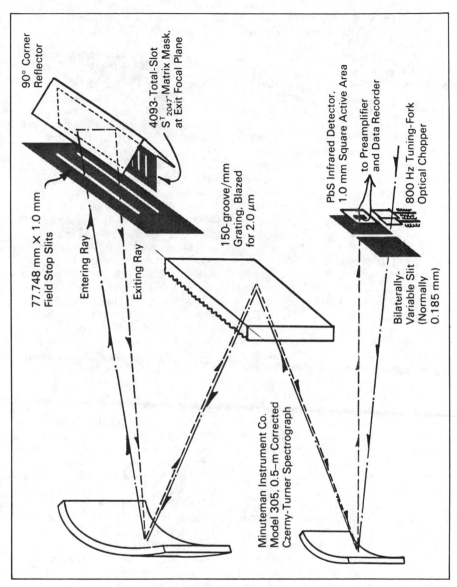

Figure 4. Optical schematic of 2047-slot HTS

Force Cambridge Research Laboratory, L. G. Hanscom Field, Bedford, Mass., while the author was with Comstock & Wescott, Cambridge, Mass. We will also include brief remarks on the application of this technique to the analysis of gas-phase molecular chemicals and, in particular, to the monitoring of such atmospheric pollutants as automobile exhaust.

Theory of Operation of HTS

Before we begin to discuss these instruments, however, further word is in order on the operation of the HTS. We have already mentioned that it multiplexes all the wavelengths in the spectrum onto a single infrared photodetector by means of an optical scheme based on orthogonal binary pseudonoise codes derived from Hadamard matrices. Basically, what is done is the insertion of a set of multislit masks consisting of transparent and opaque slots at the exit focal plane of an otherwise conventional dispersive spectrometer and the recording of the total intensity passing these masks for a number of different mask patterns (different combinations of opaque and transparent slots on the masks) (4, 14). This has the mathematical effect of creating a set of simultaneous equations for the spectral power densities: If one has N discrete spectral resolution elements in a given spectrum, then N different measurements of the total intensity passing N different mask combinations are sufficient to define the spectrum.

In practice these equations are rather easy to solve as the "coefficient matrix" elements are all either 1 or 0, denoting a transparent or opaque slot, respectively. There are, of course, a number of constraints on the coefficient matrix (that is, on the slot patterns on the masks) if the spectrum is to be reconstructed with maximum fidelity and efficiency. As a result, the family of "pseudo-noise" mask codes based on Hadamard matrices is optimum in the sense that the reconstructed spectrum has the maximum signal-to-noise ratio for any given experimental conditions (9).

For certain values of N (e.g., $N + 1 = 2^n$, where n is any integer) one can construct cyclic Hadamard-transform codes. These code matrices have the useful property that any given mask row of the matrix (i.e., any given mask pattern) is the immediately preceding row (i.e., mask pattern) shifted by one slot width (9). One can then construct a single mask with $2N - 1$ slots rather than N separate masks, each of which has N slots, for a total of N^2 slots to fabricate. This "cyclic code" has been used with all the previous operational HTS systems (1, 3, 10).

2047-Slot Hadamard-Transform Spectrometer

The C&W/AFCRL 2047-slot HTS used a cyclic binary code (Figure 2) derived from the S_{2047}-matrix (9) and computed with the shift-register computer logic given by Baumert (15). The resulting 2047-active-slot coding mask, containing a total of 4093 individual slot positions each 0.038-mm wide, is shown in Figure 3.

This mask was located at the exit focal plane of the spectrometer's dispersive optical system and mounted so that it bisected a 90° corner mirror which returned the encoded light beam back through the dispersive optics and shifted it from above to below the optical axis (see Figure 4). This "reversed pass" was necessary to "de-disperse" the encoded exit beam and allow it to be brought to a focus (at the entrance focal plane) on approximately the same size detector as would have been required by use of the optical system as a monochromator. This requirement was, in turn, caused by the unfortunate tendency of infrared-detector noise to increase in proportion to the square root of the detector area. The mask was moved sequentially by a stepping-motor-driven belt-drive controlled by an electronic indexing circuit which also provided the timing and gating pulses to the digital data system.

An optical schematic of the 2047-slot HTS system is given as Figure 4. The instrument was built around an existing AFCRL-owned Minuteman Instrument Co. Model 305 0.5-m fully corrected Czerny-Turner spectrograph specifically designed for wide-exit focal-plane applications. It was equipped with a 0.005–3.000-mm wide bilaterally variable entrance slit [normally used at a setting of 0.185 mm, which degraded the spectral resolution theoretically possible with the 0.038-mm wide coding slots but did not affect the "multiplex

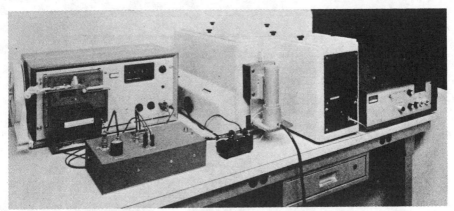

Figure 5.   C&W/AFCRL 2047-slot demonstration Hadamard-transform spectrometer

advantage" comparison with a single 0.038-mm wide slot *(1, 9)*] and a 150-groove/mm grating blazed for a wavelength of 2.0 μm (5000 cm^{-1}). In this configuration it had a first-order resolution of about 2.5 cm^{-1} at 5000 cm^{-1} (a resolving power of 2000) and an exit-focal-plane width of about 102 mm.

The existing encoded light rays emerged slightly above the optical axis and were detected by a room-temperature lead sulphide (PbS) detector with a 1.0-mm square active area. The signals from the detector were amplified by an Ithaco Model 144L low-noise pre-amplifier, passed through a synchronous demodulator triggered by signals from the American Time Products 800 Hz "tuning-fork chopper" at the entrance slit and fed to a Microwave Magnetics, Inc., Model ADA-183 digital data system which recorded the encoded data in standard ASCII format on punched paper tape. A photograph of the total 2047-slot HTS system is given as Figure 5.

The HTS's optical and mechanical systems were designed to operate with an AFCRL-owned digital-magnetic-tape recorder and hence can operate at data rates up to about 6000 data points per sec; the C&W-owned punched-paper-tape data system, used in the experiments described here, is limited to

a maximum data rate of about five readings per sec, and all data runs were therefore taken at this lower rate. The vast quantity of data involved in even a single 2047-slot spectrum, however, precluded the use of any sort of teletype unit for data input to the computer or output spectrum plotting.

The data were decoded on a Digital Equipment Corp. PDP-10 time-sharing computer controlled remotely from our laboratory. The data tapes were fed into the computer by means of the "onsite" (i.e., at the computer) high-speed punched-paper-tape reader, and the output spectra were plotted on an "onsite" high-speed Calcomp plotter. The computer decoding operation (reading in the data tapes, decoding the 2047 × 2047 coding matrix, and plotting the resulting output spectrum) took a total elapsed time of about 7–8 min, most of it spent in input/output. Since the decoding routine itself took only 1–2 min of running time with a simple FORTRAN IV matrix multiplication program, we have not yet used the considerably faster Fast Hadamard Transform (or FHT) algorithm *(16)*, although future systems will almost certainly use it. The FHT algorithm, incidentally, runs about 10 times faster than the equivalent Fast Fourier Transform (or FFT) algorithm *(17)*.

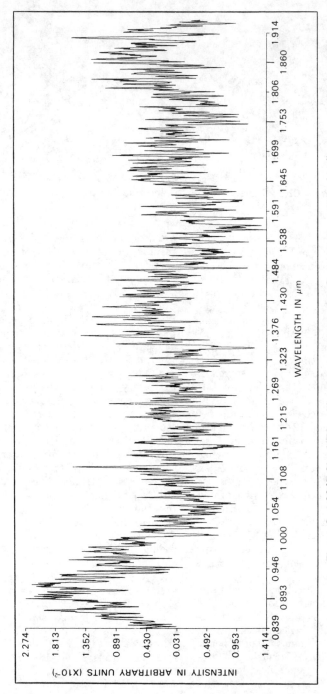

Figure 6. 2047-Slot HTS calibration spectrum of 1–2-μm mercury emission lines

The 2047-slot Hadamard-transform spectrometer had only become fully operational for about two weeks when some Air Force funding difficulties forced a halt to further experimentation. Therefore, no full signal-to-noise analysis similar to that published earlier for the 255-slot HTS (1) is available. Figure 6, however, shows a preliminary calibration spectrum of the 1–2-μm region of the mercury emission spectrum; the positions and relative intensities of the major lines match the predicted values, although the exact nature of some of the various absorption features is still under investigation.

As an example of the order of signal-to-noise improvement to be expected from this instrument, Figure 7 shows a typical spectrum recorded by the earlier 255-slot HTS whereas Figure 8 shows the same spectrum as recorded in the same total observation time under the same experimental conditions by the equivalent scanning-monochromator spectrometer (1). The signal-to-noise improvement of the 2047-slot instrument should theoretically be about a factor of 2.84 better than this, as the signal-to-noise-ratio gain is about a factor of 22.74 for the 2047-slot system vs. a factor of 8.01 for the 255-slot system. Since the 255-slot HTS displayed almost precisely the theoretically predicted signal-to-noise-ratio gain, we have no reason to doubt that the 2047-slot system will fully perform to specification. The primary reason for our lack of a quantitative noise analysis was the difficulty of measuring any noise in the HTS spectrum under experimental conditions when any spectral "signal" at all was detectable in the monochromator's output!

HTS Gas-Analysis Spectrometer

The Hadamard-transform spectrometer can be used in analytical chemistry in exactly the same way as any other infrared spectrometer, differing only in its increased performance and possibly in the use of its decoding computer for automatic data normalization, instrument calibration, and/or analytical interpretation.

If, however, one is only interested in a predetermined "list" of chemicals—the important constituents of automobile or power-plant exhausts which we will use as our example here—then a radical simplification is possible. The fundamental points of this simplifications are as follows.

First, a number of specific chemical constituents in the gas under analysis are measured quantitatively by a comparison of the intensities of transmitted infrared radiation at a number of specific adsorption ("signal") wavelengths characteristic of these chemicals with the transmitted radiation intensities at a number of selected nonabsorbing ("reference") wavelengths. Suitable "signal"/"reference" wavelength pairs for a number of exhaust constituents of recent interest (18) [all within the spectral sensitivity band of either room temperature Indium Antimonide (InSb) or thermoelectrically cooled Lead Selenide (PbSe) infrared detectors] are given in Ref. 2, which also describes the "signal"/"reference" wavelength-definition mask at the spectrometer exit focal plane, the HTS multiplexing channel allocations, the resulting HTS encoding mask, and the HTS code logic.

Second, this measurement is effected by means of optical chopping between the signal and reference wavelengths, such that the infrared detector only "sees" the difference between the "signal" absorption and the "reference" absorption.

Finally, all the various pairs of "signal"/"reference" fluxes are multiplexed onto a single infrared detector by means of Hadamard-transform encoding.

An artist's conception of a typical HTS automobile exhaust analysis system is shown in Figure 9. The system shown is quite compact and portable and utilizes a built-in "microcomputer" for data decoding and normalization, instrument calibration, and data-dis-

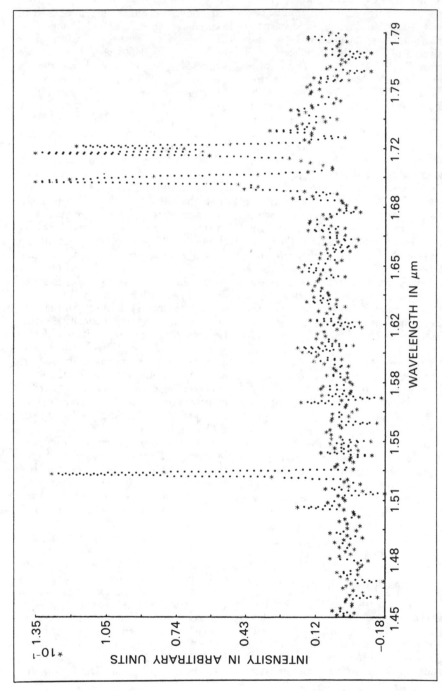

Figure 7. 255-Slot HTS spectrum of 1.5-μm mercury lines under identical experimental and instrumental conditions as Figure 8. * denotes computed spectral values, and dotted lines denote computer's linear interpolation between computed values

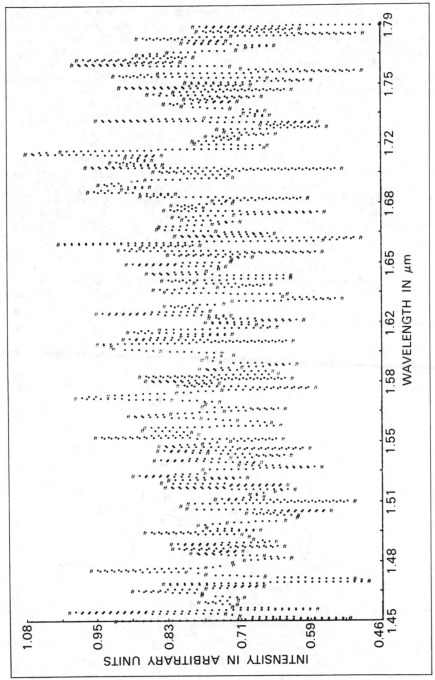

Figure 8. Conventional scanning-monochromator spectrum of 1.5-μm mercury emission lines under high-noise conditions

play formatting. Somewhat similar systems could be designed for powerplant or turbine-engine exhaust monitoring or for gas-phase process control application—for any special purpose analysis in which the analyst is interested only in one or more "lists" of specific constituents and does not mind being "blind" to any other (i.e., "nonlist") chemicals that may be present.

In summary, for the case of "listonly" analysis, the use of Hadamardtransform multiplexing can allow the simultaneous measurement of several absorption wavelengths, characteristic of several specific gases of interest, with a comparatively economical and reliable instrument which needs only a single detector/preamplifier/data-analysis electronics chain.

Conclusions

The Hadamard-transform spectrometer is an infrared analytical instrument which displays the classical multi-

plex sensitivity adventage (8) but uses the comparatively simple, reliable, and inexpensive technology of dispersive spectrometry. It shares with interferometric spectrometers, as with all multiplex spectrometers, the need for a digital computer to decode the transformed data from the spectrometer into a usable spectrum. However, since the FHT decoding algorithms (16) run up to an order of magnitude faster than the equivalent FFT algorithms used with interferometers, this decoding can be done "in real time" by a small integrated-circuit minicomputer integral with the spectrometer. In addition, HTS systems require only normal optical-shop tolerances in fabrication and alignment (not the "wavelength-of-light" tolerances typical of Fouriertransform systems, except for the grating ruling) and as noted above can easily be modified to record only those specific wavelengths of particular interest (2), difficult and/or inefficient to

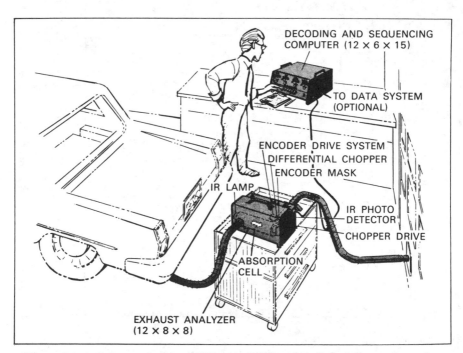

Figure 9. Artist's conception of HTS automobile exhaust-analysis spectrometer

accomplish with other forms of multiplex spectrometer.

On a more fundamental plane the data signal from a HTS system lacks the large zero-phase-difference "spike" characteristic of the more straightforward interferometric spectrometers (5) and hence has a substantially lower dynamic range. This is directly reflected in a lower required bit-rate for data transmission and a lower number of required bits in data recording and hence in a lower system cost. In general, we believe that the Hadamard-transform spectrometer will eventually be preferred over interferometric multiplex systems for those applications where the total system performance is limited by bandwidth, maximum data rate, computer time for data decoding, reliability, or cost (!); and that additionally, the simple exit-plane-encoded ("single-ended") HTS will be preferred for observations of faint point sources (e.g., stars and some laboratory sources) where the interferometer's "throughput" advantage does not come fully into play.

References

(1) J. A. Decker, Jr., *Appl. Opt.*, **10**, 510 (1971).

(2) J. A. Decker, Jr., *ibid.*, p 24.

(3) J. A. Decker, Jr., and M. Harwit, *ibid.*, **8**, 2552 (1969).

(4) J. A. Decker, Jr., and M. O. Harwit, *ibid.*, **7**, 2205 (1968).

(5) See, e.g., L. Mertz, "Transformations in Optics," Wiley, New York, N.Y., 1965, pp 1–79.

(6) See, e.g., F. Kneubuhl, *Appl Opt.*, **8**, 505 (1969).

(7) P. Fellgett, *J. Phys.*, **28**, Suppl. 3–4, p C2–165 (1967).

(8) P. Fellgett, *J. Phys. Radium,* **19**, 187 (1958).

(9) N. J. A. Sloane, T. Fine, P. G. Phillips, and M. Harwit, *Appl. Opt.*, **8**, 2103 (1969).

(10) M. Harwit, P. G. Phillips, T. Fine, and N. J. A. Sloane, *ibid.*, **9**, 1149 (1970).

(11) P. Jacquinot, *J. Opt. Soc. Amer.*, **44**, 761 (1954).

(12) M. Harwit, *Appl. Opt.*, **10**, 1415 (1971).

(13) J. F. Grainger, J. Ring, and J. H. Stell, *J. Phys.*, **28**, Suppl. 3–4, p C2–44 (1967).

(14) R. N. Ibbett, D. Aspinall, and J. F. Grainger, *Appl. Opt.*, **7**, 1089 (1968).

(15) L. D. Baumert, in "Digital Communications with Space Applications," S. W. Golomb, Ed., Prentice-Hall, Englewood Cliffs, N.J., 1964, pp 17–32, 47–64, 165–95.

(16) E. D. Nelson and M. L. Fredman, *J. Opt. Soc. Amer.*, **60**, 1664 (1970).

(17) W. K. Pratt, J. Kane, and H. C. Andrews, *Proc. IEEE*, **57**, 58 (1969).

(18) For individual spectra see, e.g., R. H. Pierson, A. N. Fletcher, and E. St. C. Gantz, *Anal. Chem.*, **28**, 1218 (1956): spectra for CO_2, CO, and n-butane (typical of hydrocarbons) are on p 1224, NO and NO_2 on p 1234, SO_2 on p 1237, and formaldehyde on p 1229.

The major part of this work was conducted while the author was with Comstock & Wescott, Inc., Cambridge, Mass., and was supported there by Cambridge Research Lab, USAF Systems Command, L. G. Hanscom Field, Bedford, Mass.

Ionization Sources in Mass Spectrometry

E. M. CHAIT, Du Pont Instruments
1500 South Shamrock Avenue, Monrovia, Calif. 91016

Traditional electron impact sources are giving way to alternative means of ionization for many fascinating special applications. Field ionization, chemical ionization, spark source ionization, surface ionization, and photo ionization are finding a place in mass spectrometry techniques in such diverse areas as forensics, toxicology, structure determination, polymer analysis, and plasma chromatography

THE MASS SPECTROMETER is an analytical laboratory for charged particles. Mass analyzers commonly in use (magnetic, time of flight, quadrupole, cycloidal, etc.) all derive their utility from their ability to separate ions which reflect the nature of the sample introduced into the spectrometer. The source of these ions is of primary importance and must be considered as the "heart of the mass spectrometer." Often the success of a mass spectrometric analysis depends upon the type of ion source chosen. The mode of production of ions often determines if they will indeed be representative of the sample in question. Analyzed ion signals, positive or negative ions, originating from a variety of ion sources may be interpreted by the chemist to yield such vital information as molecular structure, analysis of a gas mixture, identity of a gc effluent, trace dopants in a semiconductor, isotopic composition of a lunar rock—in fact, information on any form of matter that can be ionized in the mass spectrometer vacuum.

Ion Source as a Chemical Reactor

It is attractive for the chemist to think of the mass spectrometer's ion source as a chemical reactor, for in some way a chemical change occurs in the sample when it becomes an ionized plasma prior to injection into the mass analyzer. This analogy is easily understood in the case of the commonly used electron ionization source. Here the reagent is the stream of 70 eV electrons creating excited molecular ions of a complex molecule via a Franck-Condon process. The ensuing chemistry of unimolecular ion decompositions produces the fragment ions, rearrangement ions, and metastables we have learned to interpret to obtain molecular structure.

146

In the chemical ionization source, a different kind of chemistry may be exploited. Ions resulting from the electron bombardment of methane at high pressures (1 torr) may be used as a reagent for reaction with a neutral sample gas in bimolecular chemistry. The products of these ion-molecule reactions provide additional information on the nature of the sample.

Each ion source is a different kind of chemical reaction vessel and must be chosen as appropriate to the sample—be it a steroid or a semiconductor. Complex instrumentation technology has resulted from the development of multifarious ion sources (Figure 1), while at the same time ionization of materials for mass spectral analysis remains a challenge for future development of instrumentation. This discussion will elaborate on both the present technology (1, 2) and future challenge of ionization sources in mass spectrometry.

Requirements for Ion Sources

Monoenergetic Ion Beam. Ion sources not only have the function of producing ions but must also accelerate the ions into the mass analyzer for eventual detection. To insure maximum resolving power of the spectrometer, precautions are taken that the ions have a small spread of kinetic energies prior to acceleration. All ion sources are ion optical devices and contain lens elements which limit the energy spread of the ion beam. Ion sources should produce all masses without discrimination so that there is equal probability for both high and low masses being injected into the analyzer. Even so, some ion sources, as in the case of spark sources for analysis of solids, produce a large spread of energies in the ion beam and are usually used with double focusing mass analyzers which filter the ions according to energy prior to mass analysis so that high resolution may still be obtained.

Ionization Efficiency. The ionization efficiency of a source must be high so that a large proportion of the neutral sample particles presented will become ions to be analyzed and detected as a mass spectrum. The importance of high efficiency is accentuated in the everyday requirements of modern mass spectrometry for analysis of nanogram quantities of natural products and ppb impurities in semiconductors. An ion beam current of 10^{-10} A is a desirable source output, commonly used ion detectors having limits on the order of 10^{-15} A for electronic amplifiers and 10^{-18} A for electron multipliers.

Vacuum. Vacuum pumps of sufficient size must be provided to create a suitable low-pressure environment for ionization to take place, so that ions are not neutralized by collisions. The vacuum required is usually 10^{-5} to 10^{-7} torr. In cases of trace determination such as in the spark source mass spectrometer, much higher vacua are required, and this may require the use of ion pumps or cryogenic pumping methods. In electron impact mass spectrometers, the vacuum is also required to preserve the hot filament (tungsten at 2300°C).

In the case of sources in which ion molecule reactions occur in the ionization chamber, as for chemical ionization (3), very high-speed pumping systems on the order of 1000 l./sec are necessary to maintain the requisite low pressure surrounding the source. A good ion source vacuum also ensures that there is no cross contamination between successive samples. This is particularly important in combinations of gas chromatography and mass spectrometry where new samples are continually being introduced to the ion source on a second-to-second basis. To improve ion source vacuum conditions, such innovations as differential pumping have evolved, permitting a high pressure to exist inside the ionization chamber for maximum ionization efficiency and sensitivity while the surrounding source housing is at lower pressure. Pressure differentials on the

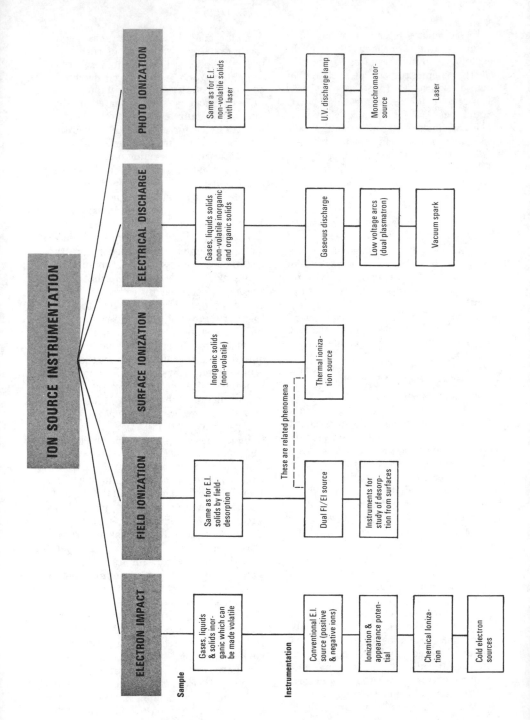

Figure 1.   Table of ion sources

order of factors of 10–100 are common.

Materials and Construction. The materials and construction for ion sources utilize the same considerations as in any good vacuum system practice, most commonly used metals for source construction being nichrome, monel, and stainless steel. Accurate machining and assembly are prerequisite to the ion optical performance required, and at the same time the metals must not be catalytic or reactive so as to change the sample prior to the ionization process.

Evolution of Modern Ion Source

In the pioneering work of Thomson and Aston at the Cavendish Laboratory in England, the requirements were met in a simple fashion by an electric discharge tube as the source of ions. The voltage of 10–50 kV used to sustain the gaseous discharge between the electrodes in an atmosphere of the sample gas also served to accelerate the ions generated through a collimating tube fashioned from a hypodermic needle. Not only did the source produce an intense beam of ions, but the useful resolution of the spectrometer was assured by the crude collimating device to compensate the large energy spread of the discharge. These simple beginnings had their limitations. The

discharge was unstable, samples were limited to gases, and the energy spread limited the spectrometer resolution. Further work by Demster, Bainbridge, and Nier produced the modern electron impact source which made mass spectrometry a versatile analytical technique. This source became the standard because of simplicity and ease of operation. Extensive libraries of electron impact spectra have been recorded.

Electron Impact Ionization

The electron impact ion source is the most commonly used and highly developed ionization method. Molecules in the gas phase are ionized by collision with energetic electrons. This results in a Franck-Condon transition producing a molecular ion, an odd electron ion usually in a high state of electronic and vibrational excitation. This excitation and its distribution over various modes of decomposition of the molecule determine the resultant fragmentation pattern.

Samples. The electron impact mass spectrometer is applicable to any volatile sample. Permanent gases, liquids, and solids all may be analyzed. In most cases a heated reservoir is sufficient to volatilize gases and liquids. Samples may be the gaseous effluent

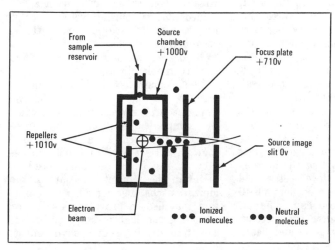

Figure 2. Electron impact ion source

of a gas chromatograph in a gc/ms system. Solid samples may be volatilized from a heated probe (-160 to $900°C$) for organic or inorganic solids or from a Knudsen cell (up to $2300°C$ with electron bombardment heating) for inorganic solids. Gas pressures in the ionizing region may be from 10^{-2} to 10^{-7} torr, or lower with high-sensitivity sources.

Description and Principle of Operation. Necessary elements for the construction of the source are the electron producing filament, a tungsten or rhenium wire, and an electron trap or anode so that a beam of electrons may be established through the ionizing region. In addition, there are high-voltage electrodes for accelerating the positive ions generated by electron impact, collimating slits, and other electrodes to focus the ion beam for optimum resolution. The details of a typical source are shown in Figure 2.

Electrons are produced by thermionic emission from the electrically heated filament and are accelerated at a selected energy (usually 70 eV) by a potential drop between the filament and anode. To produce molecular ions, the energy of the electrons must be greater than the highest ionization potential likely to be experienced for the sample gas. For maximum ionization efficiency it has been experimentally determined for most substances that an electron energy of 70 eV is practical. Not only are molecular ions produced for many compounds at 70 eV, but various fragment ions as well. Typical currents of electrons produced in the source are $100–200$ μA, which allows about one in a million molecules to become ions. Ion beam currents are on the order of 10^{-7} to 10^{-14} A.

A narrow energy spread of the ions is ensured by maintaining the ionizing region in a relatively low electric field established by the ion repellers ($1–10$ V/cm). Ions produced have different kinetic energies because the electron beam is not monoenergetic and has a finite width. When the ions are drawn out of the ionization chamber perpendicular to the electron beam into the accelerating region, these differences in position of ionization reflected as differences in kinetic energy will be small compared to the accelerating voltage of several kilovolts.

Negative as well as positive ions can be generated by electron impact. Electron capture processes are the usual mode of formation of negative ions. Many studies of negative ion mass spectrometry use conventional electron impact sources with the acceleration potentials reversed.

In addition to being a very general tool for obtaining mass spectra for a wide variety of samples, the electron impact source is extremely useful in obtaining fundamental physical chemical data. Since the energy of the bombarding electrons may be varied, the ionization potentials of molecules and the appearance potentials of fragments may be determined by producing the ionization efficiency curve. Special types of sources for accurate determination of ionization potentials have been developed in which the electrons are made monoenergetic by the use of an electrostatic analyzer or a retarding potential method. The experimental difficulties encountered in this method are great. Spectra obtained at low ionizing voltages (<20 eV) are often used in analytical determinations because less fragmentation occurs at lower energies and spectra are simplified.

The vacuum system is very important to the operation of the electron impact source, especially in cases when the sample comes from a high-pressure source such as a gas chromatograph. Here it is not only necessary to use extra pumping speed to maintain the requisite pressure of the ionization chamber for efficient ionization but also to provide a means for removing the high-pressure gc carrier gas from the ion source environment.

The use of a helium separator interface device between the gas chromatograph and the spectrometer is a necessity. Only in cases where extremely high-speed pumping is provided can the effluents capillary or SCOT column (about 5 ml/min) be introduced directly into the mass spectrometer while still maintaining the requisite low pressure in the ion source. Differential pumping is often employed to increase the sensitivity of the electron impact source by increasing the number of molecules present inside the ionization chamber while maintaining the low pressures in the source housing. This also serves to improve the general background of the spectrometer, eliminate memory effects, and keep the analyzer pressure low, preserving resolution.

Limitations. One of the most serious limitations of the electron impact source is that 70 V electrons used as an ionizing mechanism are an extremely drastic method of creating molecular ions. As a result, a large amount of vibrational and electronic excitation is experienced by the ion and in many cases dictates the destruction of the molecular ion so that it is not detectable in the spectrum. In such cases eventual structure determination is made very difficult since the molecular ion mass indicating the elemental composition of the molecule cannot be determined easily. In addition the electron impact process is inefficient and requires the sample to be in the vapor state. Since the electron impact source ionizes all gases presented to it, the background gases are ionized with the same efficiency as the sample and, therefore, in cases of high residual pressure in the ion source for previous samples, the interfering background spectra will be recorded. Vacuum requirements for the electron impact source also create engineering problems in source construction. Special sample introduction devices are required to maintain the low pressure in the ion source, and gc/ms interface devices are needed to keep high-pressure carrier gas out of the ion source.

Field Ionization

The main problem in the use of electron impact ion sources is that many organic molecules do not give molecular ions. Field ionization overcomes this difficulty for a great number of these molecules (*4*).

Samples. Field ionization may be used with the same types of samples and inlet systems as electron impact ionization. Nonvolatile solids are accessible to analysis by field desorption.

Description and Principle of Operation. Field ionization occurs when a molecule is brought near a metal surface in the presence of a high electric field _(10^8 V/cm). The phenomenon is observed because of the quantum mechanical tunneling of an electron from a molecule into the metal surface, producing a positive ion. The ion has very little internal energy in the form of electronic or vibrational excitation and is usually detected as a stable molecular ion. Apparatus to produce the high field in the source usually consists of a sharp metal point or edge for the anode and a cathode which is the exit slit of the ion chamber. A high voltage between 5 and 20 kV is applied between these closely spaced (1 mm) elements producing the requisite electric field for ionization. The design and materials of construction of the sharp point used as the ionization element in the source are very critical factors in the technology of field ionization. Since the inception of field ionization mass spectrometry with the work of Gomer and Ingram, where the electric field was generated by the sharp tungsten point of the field ion microscope, many other methods have been used. These include wires of micrometer diameter, arrays of tungsten points, and most recently the sharp edges of commercial razor blades. All of these field ionization sources have the disadvantage of being somewhat fragile in the high-voltage environment and subject to arcing. However, in field ionization sources presently in use, the ability to change the field ionization element via a vacuum lock permits greater ease of operation in spite of this disadvantage. The rest of the field ionization source consists of the lens type focus ele-

ments common to all ion sources. The field ionization capability for organic compounds is combined with a conventional electron impact source so that the methods of ionization are interchangeable and the spectra will be comparable. Figure 3 shows a typical dual field ionization electron impact source (5).

The ability of field ionization to

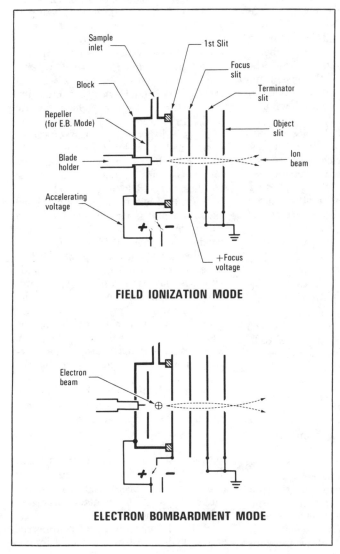

FIELD IONIZATION MODE

ELECTRON BOMBARDMENT MODE

Figure 3. Combination field ionization electron impact ion source

produce stable molecular ions for most of the compounds investigated has been exploited with great success. In organic mass spectrometry, especially in the study of natural products, and in combination with the high-resolution mass spectrometer, the dual field ionization electron impact source has allowed the measurement of the exact masses and, as a result, of the elemental compositions of those molecular ions which were previously unobtainable (5). Besides the obvious implications for structure determination, field ionization has proved to be a very valuable technique for studying surface phenomena. Adsorbed species and the results of chemical reactions on surfaces may be investigated. In a particularly brilliant adaptation of the field ionization phenomena,

Mueller has been able to demonstrate that selected molecules or atoms from the surface can be injected into a mass spectrometer for analyses (6). Because of the high field gradients between the field ionization anode and cathode, very fast reactions on the order of 10^{-14} to 10^{-11} sec may be observed so that the entire range of the kinetics of gaseous ion chemistry may be explored (7).

Limitations. Field ionization, in spite of its successes, is not without problems. In addition to the problem of fragile ionization elements, there is the added problem of low sensitivity. Sensitivity at best is an order of magnitude below that of electron impact ionization, usually producing ion currents with a maximum of 10^{-11} A. Sensitivity is also dependent on the

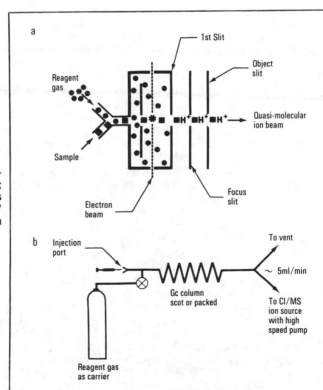

Figure 4. a) Chemical ionization source; b) combination gas chromatography / chemical ionization mass spectrometry

type of compounds, those with high dipole moments and high polarizability being the most sensitive.

Chemical Ionization

Recently the advantages of the versatility of the electron impact source have been combined with the ability of field ionization to obtain molecular species in a spectra of most compounds in the discovery of chemical ionization (3, 8). It is particularly useful in organic analyses where the spectra may be compared with those from electron impact and field ionization techniques.

Samples. The same kinds of samples accessible to electron impact and field ionization may be used with chemical ionization including the methods of sample introduction.

Description and Principle of Operation. In chemical ionization (Figure 4), a reaction gas such as methane is introduced into the ionization chamber of the mass spectrometer at high pressure (1 torr). The primary ionization of the reaction gas occurs by electron impact in the normal manner, and primary ions which are formed react with neutral methane molecules. The resulting ion molecule reactions produce products which are chemically reactive species and have such chemical properties as Lewis acid or Lewis base behavior. $C_2H_5^+$ ions, CH_5^+ ions, and $C_3H_5^+$ ions are produced from methane. These products of ion molecule reactions are the reagents which can react with the sample gas introduced as a smaller impurity. The spectra of the sample represents the chemical reaction products of the neutral sample molecules with the ions from the methane ion molecule reactions. The most common type of reactions are hydride transfers. Many times for compounds which do not normally show a molecular ion in electron impact mass spectrometry, a quasimolecular ion will be observed in which the molecular weight will be increased by the mass of one hydrogen atom. Hydrogen, on the other hand, seems to produce both the quasimolecular ion and increase fragmentation in the spectra as well. Chemical ionization produces very simplified mass spectra often containing only one peak, the quasimolecular ion. It is quite suitable in situations where rapid identification of compounds within a specific class needs to be made, especially if these compounds do not give molecular ions under normal electron impact conditions, as for barbiturates (9). Because chemical ionization requires both the presence of a reagent gas and the high pressure in the ionization chamber for its operation, it is possible to introduce gc effluents directly into the chemical ionization mass spectrometer if the reagent gas such as methane is used for the gas chromatographic carrier gas (Figure 4). Under these circumstances the gc chemical ionization mass spectrometer system is quite a simple one. It does not require an interfacing device to remove the carrier gas from the sample stream, as in the case of conventional electron impact ion sources. It has the dual advantage of high sensitivity, (because little sample is lost in an interface) and of obtaining a molecular species in the spectrum. To accomplish chemical ionization, certain modifications must be made in the conventional electron impact source. The ionization chamber must be gas tight so that high pressure may be sustained in the ionizing region. At the same time, the pumping speed of the vacuum pumps or source region of the mass spectrometer must be increased so that excess gas can be removed. This is accomplished by making all openings in the ionization chamber, including the exit slit for the ions, very small. Increased electron energies are necessary so that the electrons can penetrate the very dense gas plasma for efficient ionization of the methane reagent gas. Few such sources are available at present for commercial mass spectrometers, and most have been built in the laboratories where they are used. Since chemical ionization provides a unique opportunity to study gas phase ion-molecule chemistry as related to molecular structure, more of these sources should be available as commercial equipment interchangeable with conventional electron impact methods in the future.

Limitations. The high-pressure conditions under which chemical ionization must operate cause a problem with electrical discharges from the high accelerating voltages in the ion source. Various elaborate schemes to eliminate this discharge problem have been devised by people who have experimented with this technique in recent years. The hazard of discharge is of subsequent danger to personnel and poses a limitation on the present application of chemical ionization. Since the advent of chemical ionization spectra resulting from bimolecular gaseous ion chemistry, the interpretation of these spectra has become an interesting task, but not enough information has been obtained at the present time to make this interpretation as reliable as the interpretation of electron impact mass spectra.

Spark Source Ionization

Electron impact field ionization and chemical ionization are all limited to volatile samples. The spark source avoids this problem by heating a solid to high temperature in an electric discharge to produce ions of the elements in the sample.

Samples. Inorganic solids, metals, semiconductors, insulators, and organic materials.

Description and Principle of Operation. The vacuum spark produces ions by high voltage breakdown across two electrodes (Figure 5). One of these electrodes usually consists of or includes the sample material. In the spark ion source the spark is sustained between the two electrodes by a high voltage rf signal from 30–40 kV with a typical frequency of 800 kHz. Sources of this type can be used for analysis of a variety of samples: liquids, powders, conductors, semiconductors, and insulators. With a vacuum arc as the ionization element, ion optics are used to collimate the ion beam since the energy spread is broad. Generally these sources are used with double focusing mass spectrometers.

Spark sources are often used in the analysis of impurities in semiconductor materials. Such impurities may be analyzed into the ppb range. Obviously, much recent effort has been devoted to the improvement of the sensitivity and reproducibility of such a valuable analytical method. Spark source mass spectrometry is also applicable to non-volatile compounds, such as inorganic solids, and the method has been used with organic compounds producing spectra with a high degree of fragmentation.

Limitations. Spark sources produce ions with a large kinetic energy spread. To maintain the spectrometer resolution, expensive, complex double focusing analyzers must be used. Most often the integrating qualities of photographic plates are used for detection and require an expensive densitometer system to read the spark source data. Preparation of samples can be difficult since they must be incorporated into electrodes forming the spark.

Surface Ionization

Another approach to nonvolatile inorganic salts is surface ionization. It is not as destructive to the sample as spark source and produces stable ion beams of small energy spread.

Samples. Inorganic compounds, such as salts and minerals.

Description and Principle of Operation. The surface ionization source consists of conventional ion optics for the acceleration and focusing of ions into the mass analyzer. The ionization element itself is usually a hot ribbon of metal with a high work

155

function, such as tungsten coated with the sample (Figure 6).

At sufficiently elevated temperatures (2000°C), the emission of a neutral vapor from a heated surface will also be accompanied by positive ions—that is, some of the atoms or molecules will have lost an electron to the surface of high work function, particularly if the compound has a fairly low ionization potential, as for cesium chloride. The hot metal surface has a higher affinity for retaining an electron of an escaping atom than does the atom itself. The relationship for the ratio of ions ($n+$) to neutral particles ($n°$) is expressed in the following equation:

$$\frac{n°}{n^+} \propto \exp\frac{I - W}{kT} \qquad (1)$$

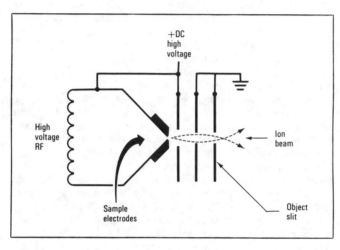

Figure 5. Spark source ionization

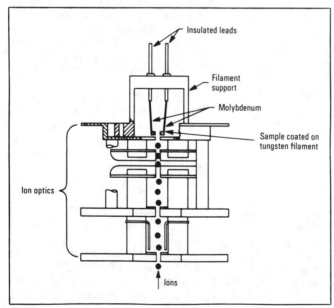

Figure 6. Surface ionization

where I is the ionization potential of the sample, W the work function of the filament, and T the temperature.

For efficient surface ionization it is important for the metal filament to have a high work function, the proper temperature, and the proper ionization potential for the element being evaporated. Surface ionization has a common bond with field ionization since in that case the tunneling of the electron into a metal surface promoted by the high electric field produces ions.

Since inorganic compounds generally have low ionization potentials (3–6 eV), this technique is very appropriate. For organic compounds with ionization potentials of 7–16 eV, surface ionization is difficult. Some examples of surface ionization of organic compounds have been shown in the literature by use of a tungsten oxide filament with an extremely high work function (10).

The surface ionization process is selective only for the material coated on the filament, and since no ionization of background gases in the mass spectrometer occurs, the spectrum produced is extremely clean. Very small quantities of samples (as low as 10^{-14} grams) may be detected under favorable conditions. Surface ionization is especially useful in determining isotope ratios in inorganic compounds for geochemical applications or in studies of elements involved in nuclear chemistry.

Limitations. The technique cannot be effectively used for nonvolatile organic compounds.

Photo Ionization

Molecules may be ionized by ultraviolet light. This method is related to electron impact ionization since the ionization occurs by a Franck-Condon mechanism. These sources are not in wide use in analytical mass spectrometry because of the complicated apparatus required.

Description and Principle of Operation. Ionization potentials of most molecules are between 7–16 eV so that ultraviolet light on the order of 1240–775 Å must be used. These wavelengths fall into the vacuum ultraviolet region, and in the lower end of this range below 1050 Å, there are no window materials which can transmit the radiation. Here it is necessary to use windowless ultraviolet sources. The simplest photo ionization sources make use of gas discharge lamps, particularly electrodeless discharge lamps with microwave power. A source of this type using the helium 584 Å line equivalent to 21.2 eV has been used by Brion (11) (Figure 7) in the analysis of organic compounds, preventing thermal decomposition caused by the hot electron emitting filament in the conventional electron impact source. Attempts to observe direct photo ionization of solids have been unsuccessful. Though such sources can produce spectra with the general appearance of conventional electron impact spectra, they have not found general use.

The most common application of photo ionization has been in sources equipped with a vacuum ultraviolet monochromator so that the ionization energy can be carefully selected. Such sources have been very useful in the precise determination of ionization and appearance potentials.

A most recent and futuristic application of photo ionization is laser ionization. A high powered pulsed laser is used both as a method of sampling and a source of ionization. Such samples as coal and other solids have been analyzed with the laser by microscopically focusing the laser beam on the area selected for study, perhaps an inclusion or an impurity of interest, and then firing the laser, producing a plume of vaporized material. This plume of vaporized material contains some ions produced by the high temperature of the laser beam and also contains neutral gases which may be ionized by conventional electron impact methods. Laser ionization seems to be a very promising method in working with

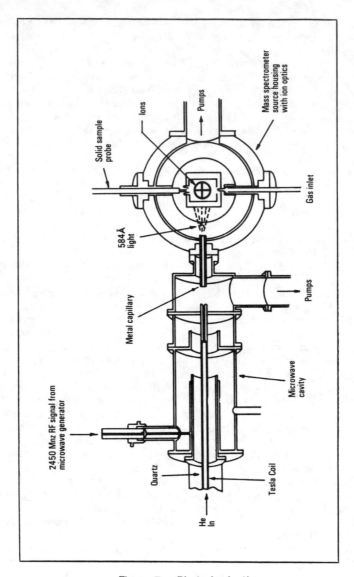

Figure 7.   Photo ionization

various solid materials, minerals, semi-conductors, and polymers.

Limitations.   Photo ionization sources give spectra similar to electron impact sources and are often more complex and difficult to operate. Attempts to analyze solids by direct photon bombardment have been unsuccessful.

Future Developments in Ion Sources

In spite of its limitations, the electron impact source has enjoyed great success in the history of mass spectrometry. Now mass spectroscopists are becoming interested in the possibilities of alternative methods of ionization. In organic mass spectrometry the trend toward more gentle methods of ionization has developed for very specific detection of limited classes of compounds, as in the forensic and toxicology fields. In addition, chemists realize the drastic fragmentation caused by electron impact ionization is a reason for loss of valuable structural data if the molecular ion is not observed. Low voltage ionization, although it reduces fragmentations, cannot increase the absolute abundance of a molecular ion. Both field ionization and chemical ionization will find increasing use in the area of increased molecular ion abundance and simplified spectra.

Chemical ionization offers a unique advantage for chemists to exploit the variety of reactions possible with different reagent gases to establish structure-reactivity relationships. In addition, by selecting reagent gases of different recombination energy for charge exchange ionization, the degree of fragmentation may be preselected in an exact fashion for the sample without sacrificing sensitivity. This type of research is just beginning to take shape in a few laboratories and will lead to very specific analytical techniques for particular structural types.

New ionization methods are being developed for previously intractable solid materials such as polymers. These cannot be successfully analyzed by electron impact techniques since they are nonvolatile. The spark source does not produce any structural information. Ion bombardment or sputtering will become the method of choice for such samples as synthetic and biopolymers. Already there has been some success in determining the basic structural units of Nylon 66 and Teflon by ion bombardment (*12*).

In spite of the success of alternative methods of ionization, vast improvements will be made in electron impact ion sources. Sensitivity improvement is the most important factor. High-speed pumping systems and increased electron currents will give better efficiency.

Cold electron sources, radioactive beta emitters such as Ni^{63} could be used, making ionization on high pressure possible without risk of filament burn out. This has already been successfully demonstrated in the plasma chromatograph, a very sensitive device for air pollution sampling (*13*). Cold electron sources would be a low-cost durable ionizer for very compact analytical mass spectrometers.

References

(1) H. D. Beckey, "Topics in Organic Mass Spectrometry," A. L. Burlingame, Ed., Wiley-Interscience, New York, N.Y., 1970, p 1.
(2) F. A. White, "Mass Spectrometry in Science and Technology," Wiley, New York, N.Y., 1968, p 57.
(3) M. S. B. Munson and F. H. Field, *J. Amer. Chem Soc.*, **88**, 2621 (1966).
(4) H. D. Beckey, *Angew. Chem., Int. Ed.*, **8**, 623 (1969).
(5) E. M. Chait, T. W. Shannon, J. W. Amy, and F. W. McLafferty, ANAL. CHEM., **40**, 835 (1968).
(6) D. F. Borofsky and E. W. Mueller, Sixteenth Annual Conference on Mass Spectrometry and Allied Topics, ASTM, Committee E-14, Pittsburgh, Pa., 1968.
(7) E. M. Chait and F. G. Kitson, *J. Org. Mass Spectrom.*, **3**, 533 (1970).
(8) M. S. B. Munson, ANAL. CHEM., **43** (13), 28A (1971).
(9) G. W. A. Milne, H. M. Fales, and Theodore Axenrod, *ibid.*, **43**, 1815 (1971).
(10) E. Ya. Zandberg, *Dokl. Phys. Chem.* (USSR), **172**, 91 (1967).
(11) C. E. Brion, Fourteenth Annual Conference on Mass Spectrometry and Allied Topics, ASTM, Committee E-14, Dallas, Tex., 1968.
(12) R. S. Lehrle, J. C. Robb, and D. W. Thomas, *J. Sci Instrum.*, **9**, 458 (1962).
(13) M. J. Cohen and F. W. Karesek, *J. Chromatogr. Sci.*, **8**, 330 (1970).

Double-Wavelength Spectroscopy

T. J. PORRO

Perkin-Elmer Corp., Norwalk, CT 06852

Photometric measurement of a material by passing radiation of two different wavelengths through the same sample before reaching the detector can extend the proven effectiveness of uv-vis absorption spectroscopic techniques by overcoming limitations in selectivity, thereby reducing interferences by one dimension

ELECTRONIC (uv-vis) absorption spectrophotometry has been an effective structurally diagnostic analytical tool for over 30 years, primarily because of its high signal-to-noise ratio and inherent sensitivity (high absorptivity of most absorbing species). This high sensitivity manifests itself in providing to users of the technique a high measurement precision and ultimate analytical accuracy.

It is not our intention to review uv-vis absorption spectroscopy since anyone acquainted with it can attest to its utility over the years, particularly as a quantitative tool. However, it might serve a useful purpose to indicate that electronic absorption spectroscopy holds a unique place in the history of analytical instrumentation as applied to the general solution of chemical problems.

Prior to the emergence of infrared spectroscopy as a structure elucidation aid, electronic absorption spectroscopy was employed most effectively in the late thirty's and forty's to unravel the identity of organic molecules, particularly the many steroids and alkaloids which the organic chemists were endeavoring to prepare synthetically. Biochemistry is deeply indebted to uv-vis absorption spectroscopy for its initial progress owing to all the reasons already stated, but particularly because the universal physiological solvent, water, had no appreciable absorption in the electronic region of the spectrum, thus allowing direct measurements. However, in spite of these useful advantages, two major limitations have served to restrict its use primarily to the quantitative analysis of clear solutions of low-molecular-weight solutes.

One limitation, which we shall label as one of poor *selectivity*, results from the inherently severe overlap of the vibrational bands within an electronic transition. Thus, for many absorbing species the vibrational bands show up as indistinct shoulders on the side of more prominent absorption peaks. This characteristic causes electronic spectra in their normal presentation of absorbance (or transmittance) vs. wavelength (or frequency) to be relatively poor compared to, say, infrared spectra for the identification of pure unknown materials or for the discrimination among components of a mixture. This is in spite of the fact that the uv-vis spectrum contains vibrational and rotational, as well as electronic, information.

The second major limitation, which we shall call one of *sensitivity*, results from the relatively poor transmission of radiation between 200 and 800 nm through solute and solvent combinations other than those of clear low-molecular-weight solutions. By this we mean that except for such solutions, radiation in the stated range tends to be scattered quite effectively by many important materials including high-molecular-weight polymers and proteins, colloidal suspensions, powders, paper, cloth, smoke, and the like, which comprise a sizable percentage of materials requiring absorption measurements. This effective scattering reduces the amount of light reaching the detector (photomultiplier), thereby causing the spectrophotometer system to be a relatively insensitive light collector, thus negating the fundamental sensitivity advantage previously mentioned.

The analytical chemist has handled these problems over the years either by getting his sample into an acceptable solution for uv-vis spectrophotometry or by searching for another analytical technique before deciding that there was no immediate answer to his problem.

Chance, in his brilliant work on the study of the mechanism of cell respiration by mitochondria, developed two separate instruments, a nonscanning double-wavelength instrument to overcome the selectivity limitation in the analysis of cytochrome mixtures (1) and a scanning double (split) beam spectrophotometer designed to measure small absorption differences between highly turbid mitochondria sus-

pensions (2). Several valuable methods based on Chance's dual-wavelength spectroscopy techniques have been reported (3–7), and Shibata and co-workers have applied the double-wavelength technique to the analysis of mixtures both organic and inorganic (8). A detailed description of the instruments and their use, mainly for turbid preparations, is available in the literature (8–13). In addition, the use of derivative spectroscopy instrumentation to overcome the selectivity problem has been described (14, 15).

In spite of the demonstrated power of the double-wavelength spectroscopy technique in the field of biochemical research, its exploitation as a general analytical tool remains below the threshold level, possibly because of the preoccupation many potential users have with their immediate analytical priorities.

The purpose of this article is to review the basic principles, instrumentation, and fundamental applications of double-wavelength spectroscopy, in the hope that these techniques may in the future be more rapidly applied to the many analytical problems extant.

Principle, Definition, and Types

The amount of fundamental information obtained in an instrumental measurement does not vary with the type of readout. However, the immediate usefulness of the information can and does depend upon the method of data presentation. For example, a uv transmittance spectrum provides the same information as a corresponding absorbance curve, but the latter is more directly useful in analytical situations since concentration is proportional to absorbance.

Double-wavelength spectroscopy provides information from two wavelengths per unit time, and all other factors being equal, the resultant data should be more useful than data from a double beam absorbance (single-wavelength) measurement. This is the fundamental principle underlying the application of double-wavelength spectroscopy. In fact, depending on the particular double-wavelength type employed, the measurement compensates for the presence of one parameter, be it an interfering impurity, scattering sample, or an indistinct shoulder on the side of a band. These benefits can be summarized as an improvement in the selectivity characteristic of the measurement. Thus, literally, two wavelengths are better than one.

The second major limitation of uv-vis absorption spectroscopy which we identified was one of sensitivity owing to the relatively poor light gathering power of conventional spectrophotometers when measuring any non-clear liquid on solids. Though this problem has no fundamental relationship to double-wavelength spectroscopy, the development of instrumentation to solve the absorption overlap or selectivity problem associated with these types of samples also includes the solution to the sensitivity problem. Several of the double-wavelength instruments available, including the one described in this paper, do provide this important capability, and one of the

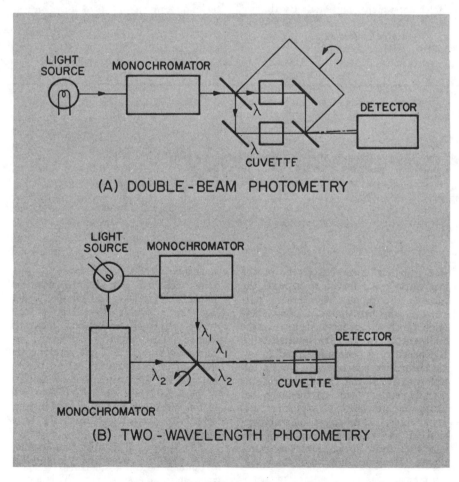

Figure 1. Sample arrangements

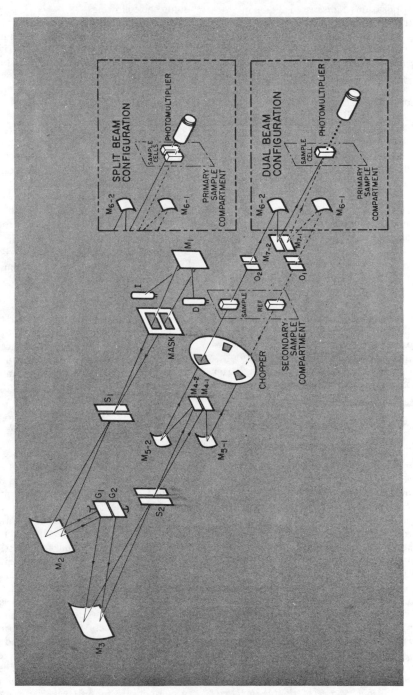

Figure 2.   Double-wavelength spectrophotometer optical schematic

applications discussed demonstrates this capability.

Let us define double-wavelength spectroscopy as we use it and identify several types used. In subsequent sections, double-wavelength instrumentation will be described, and selected examples illustrating a number of analytical benefits will be discussed.

Double-wavelength spectroscopy refers to the photometric measurement of a material by passing radiation of two different wavelengths through the same sample before reaching the detector. This technique, which has been referred to also as two-wavelength or dual-wavelength spectroscopy, can be compared with the more common *double (or split) beam absorption spectroscopy* which measures the ratio of light of the same wavelength transmitted (or absorbed) by a sample and reference material in separate cells. These two techniques are illustrated schematically in Figure 1. Note that for all types of two-wavelength spectrophotometry, the sample is positioned close to the detector to better compensate for any turbidity or scattering of the sample.

Derivative absorption spectroscopy refers to a measurement in which the first derivative of either transmittance (T) or absorbance (A) per interval of wavelength ($\Delta\lambda$) is displayed on a recorder. Several conventional spectrophotometers have been modified to perform this function by displaying the readout of the signal of a tachometer (speed measuring device) attached mechanically to the pen assembly of a chart recorder. A double-wavelength spectrophotometer provides the derivative function by scanning the two monochromators with a fixed small wavelength difference between them and is, therefore, a special example of double-wavelength spectroscopy. Several examples of this important type are given in the Applications Section.

Additional types of double- or two-wavelength measurements in use include the following:

Sample and reference beams are set at different wavelengths (nonscanning). This type is used in analytical situations where the interfering substance highly overlaps the analyte or when the effect of turbidity must be minimized. An example of the latter is given later.

Sample and reference beams are set at different wavelengths (nonscanning), and the output at each wavelength is measured independently. This mode is particularly suitable in reaction kinetics where absorbance changes of two species can be monitored simultaneously.

Sample beam scanning and reference beam fixed mode are used in determining the spectral characteristics of a highly turbid sample when no suitable reference can be prepared.

Instrumentation

There are a number of instruments available as of this writing that provide one or all of the double-wavelength functions indicated above in addition to the double beam scanning capability. It is not our intention to provide a comprehensive review of these here but rather to describe one such type to illustrate the utility and power of double-wavelength spectroscopy.

The optical layout of such an instrument, the Perkin-Elmer Model 356, is shown in Figure 2. Light from a tungsten-iodine (I) or deuterium (D) source passes into a Czerny-Turner monochromator through slit (S_1) to form an image of the mask on gratings G_1 and G_2. The dispersed radiation of both is focused by the collimator mirror (M_3), split, chopped, and sent separately through the secondary sample compartment used to minimize fluorescence effects of clear solutions. Bilateral optical attenuators O_1 and O_2, situated at the pupil image of each beam, are used to vary the intensity of radiation of each beam continuously to compensate for intensity differences, depending on the sampling situation.

Mirrors (M_{7-1} and M_{7-2}) either converge the two beams through a single cell in the primary sample compartment for all double-wavelength modes (dual beam) or diverge the beams so that they pass through two cells for double (split) beam spectrometry and then each image on the 2-in. diameter end-windowed detector. The time separated sample, reference, and zero signals are amplified, demodulated, and compared such that the sample signal is measured against a fixed voltage reference signal, controlled by a dynode feedback circuit, which makes the preamplifier output zero when no photo signal is received.

The compared signals are displayed on a recorder linearly in transmittance or linearly in absorbance through a log converter. The electronic processing of the photometric system is the same whether the two beams are similar in wavelength or not. Since each of the gratings can be independently driven linearly in wavelength separately or synchronously, double beam (same wavelength) and

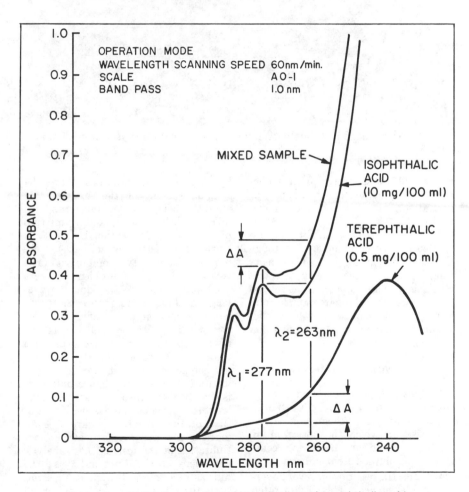

Figure 3.   Absorption spectra of isophthalic and terephthalic acid

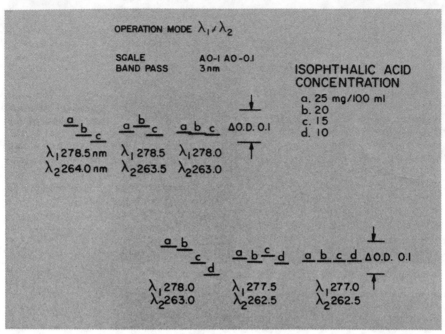

Figure 4. Optimization of parameters for two-wavelength analysis of terephthalic acid

the various double-wavelength modes previously mentioned, including the derivative, can be achieved conveniently.

A light gathering capability for increased sensitivity is achieved through the combination of a high intensity tungsten source, the ability to place a sample as close to detector as several millimeters in the primary sample compartment, and the use of a large end-windowed photomultiplier having low noise characteristics.

Analytical Applications

Analysis of Highly Overlapping Two-Component Mixtures by Two-Wavelength Technique. A common situation exists in which the spectrum of the analyte and an interfering component, possibly an isomer, have quite similar spectra resulting in severe overlap or interference. This is illustrated in Figure 3, as described in an article by Shibata et al. (8), which shows the solution spectra of the analyte, terephthalic acid, and an interfering isomer, isophthalic acid, and a mixture of the two. Conventional double or single beam (single-wavelength) techniques require a two-component analysis for this type of problem, whereas appropriate use of the double-wavelength technique can reduce the problem to a one-component analysis. The two conditions to be met are as follows. Choose two wavelengths such that the absorbance of the interfering component at both wavelengths is identical or their absorbance difference is exactly zero. Also, the sample wavelength (λ_2) should be a sensitive measure of the analyte. Figure 4 summarizes for the present sample, optimization of the instrumental analytic conditions; λ_1 is 277 nm and λ_2 is 262.5 nm, as determined by trial and error, such that A $(\lambda_2 - \lambda_1) = 0$ for all concentrations of isophthalic acid be-

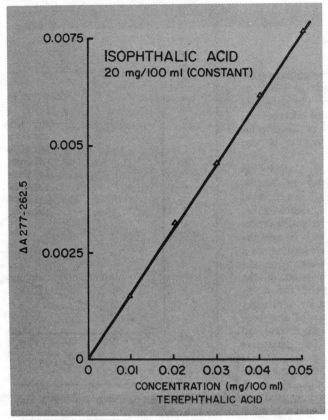

Figure 5.　Working curve of terephthalic acid

tween 0 and 25 mg/100 ml. Figure 5 is a linear working curve through zero absorbance resulting from the measurement of the absorbance difference at the two specific wavelengths for varying amounts of terephthalic acid. These data demonstrate that terephthalic acid can be analyzed directly and independently of isophthalic acid in any concentration of the latter to at least 25 mg/100 ml. The data also show that the absorbance difference of two independently linear absorbance systems at two wavelengths is also linear.

Analysis of Partially Overlapping Two-Component Mixtures by Derivative Spectrophotometry. We pointed out previously that in double-wave-

length spectroscopy, derivative spectroscopy is a special case. In analytical situations where the analyte · in a two-component system occurs on the side of the absorption band of the interfering component and the interference is approximated by a straight line, quantitative analysis of one component independent of the second can be accomplished with the derivative techniques.

This is illustrated in the next three figures, the first of which, Figure 6, shows absorption spectra of varying concentrations of an erbium salt solution in the presence of a constant but high 2000 ppm concentration of a cerium salt. Note the severe interference of the sloping cerium absorp-

167

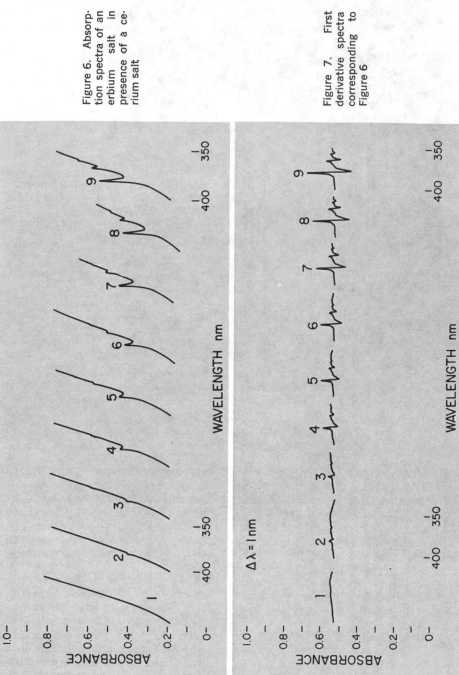

Figure 6. Absorption spectra of an erbium salt in presence of a cerium salt

Figure 7. First derivative spectra corresponding to Figure 6

168

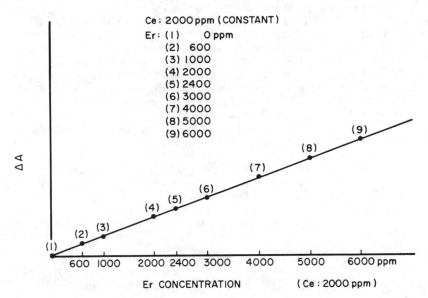

Figure 8. ΔA between maximum and minimum of derivative curve of absorption band at 379 nm

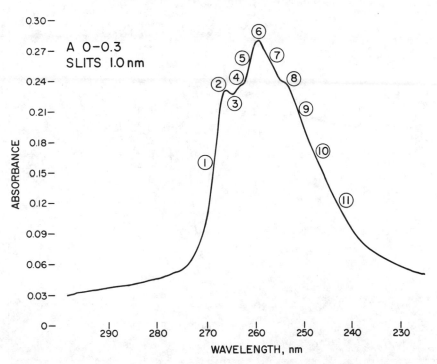

Figure 9. Absorbance spectrum of 1-methylpyridinium chloride adsorbed on Hectorite clay

tion throughout, making the baseline estimation in any single-wavelength analysis (single or double beam) most difficult at best. If the first derivative spectra of these solutions are scanned with a constant 1-nm interval through the same wavelength range, the data shown in Figure 7 are obtained. Note the absence of any interference from the cerium salt absorption. This is best seen by plotting the difference in absorbance (ΔA) between the maximum and minimum derivative absorption at 379 nm in each case against the erbium salt concentration as shown in Figure 8. The

excellent linearity and precision of the plot are a good indication that the quantitative measurement of a material obeys Beer's Law as well as the absorbance measurement. Also note that in this case the plot intersects zero absorbance at the 2000 ppm level. Though not shown, this analysis for erbium can be made independently of cerium interference up to at least 3000 ppm cerium.

Analysis of Turbid Solutions by Use of Derivative Mode. The final example to be shown demonstrates the combined capability of both selectivity in the derivative mode to differentiate

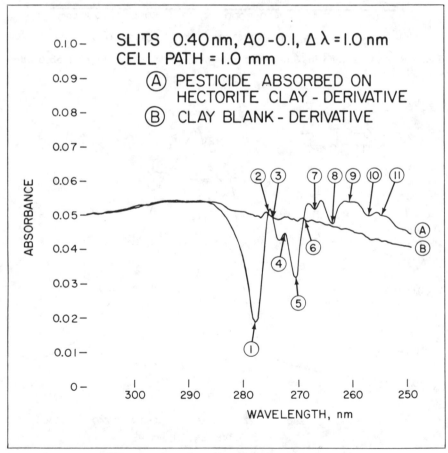

Figure 10. First derivative curve of 1-methylpyridinium chloride adsorbed on Hectorite clay

among overlapped absorptions and sensitivity required to overcome reasonably high turbidity. It also demonstrates the ability to measure extremely small absorbances with good signal-to-noise.

The problem was to determine the ability of uv absorption spectroscopy to detect small amounts of pesticides which have found their way into rivers and streams. The model system chosen was the pesticide 1-methylpyridinium chloride in films and emulsions adsorbed on clays. This example was part of a large experiment designed to develop accurate and reliable analytical procedures to determine pesticide transport by "runoff."

Figure 9 is the absorption spectrum of the pesticide adsorbed on clay, run as a dry film on a quartz plate. Every detail, including peaks, valleys, shoulders, and inflection points, has been labeled for comparison with the first derivative curve of the same sample shown in Figure 10, identically numbered. Except for the peaks or valleys (2, 3, and 6) which cross the blank clay line, all the points which were relatively indistinct in absorption show up as peaks or valleys in the derivative spectrum, particularly 7, 9, 10, and 11. Note that in spite of the high scattering of the clay powder, the signal-to-noise in both cases is quite high even at a 10X expansion of the ordinate in Figure 10.

Figure 11 shows a plot of the difference in absorbance between (1) (298 nm) and (2) (275 nm) in Figure 10 against the concentration of the pesticide adsorbed on clay in aqueous suspension measured in a 1-mm path cell in the derivative mode. Note the good linearity, high precision, and small absorbance difference measured in the concentration range of 10–30 ppm of the pesticide. Also, the equiva-

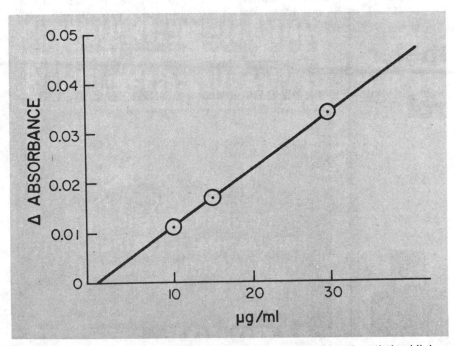

Figure 11. ΔA (peak 1 to valley 2 of derivative curves) vs. 1-methylpyridinium chloride concentration

171

lent absorbance owing to scattering of the clay at these wavelengths is in excess of two. This analysis would not be possible with conventional single-wavelength instrumentation. Such is the power of the double-wavelength spectroscopy.

Summary

The principles, instrumentation, and applicability of double-wavelength spectroscopy, including derivative spectroscopy, have been described. The ability of double-wavelength spectroscopy to extend the proven effectiveness of uv-vis absorption spectroscopic techniques by overcoming limitations in selectivity has been demonstrated certainly in the field of biochemistry but also in general analytical spectrophotometry.

Since the fundamental advantage of double-wavelength spectroscopy is its capability of reducing interferences by one dimension, many unsolved analytical problems, both qualitative and quantitative, can potentially be solved with presently available instrumentation.

Acknowledgment

All the applications examples used for illustrations in this article were obtained from Ref. *8* (Shibata et al.), as indicated, and the Spectroscopy Applications Laboratory at Hitachi Limited, Tokyo, Japan, and Perkin-Elmer, Main Avenue, Norwalk, CT.

References

(1) B. Chance, *Rev. Sci. Instrum.*, 22, 634 (1951).

(2) B. Chance, *Science*, 120, 767 (1954).

(3) B. Chance and G. Williams, *Advan. Enzymol.*, 17, 65 (1956).

(4) T. Ohnishi and S. Ebashi, *J. Biochem.* (Tokyo), 54, 506 (1965).

(5) T. Ohnishi and S. Ebashi, *ibid.*, 55, 599 (1964).

(6) B. Chance and L. Mela, *J. Biol. Chem.*, 241, 4588 (1966).

(7) B. Chance and L. Mela, *Proc. Nat. Acad. Sci. U.S.A.*, 55, 1243 (1966).

(8) S. Shibata, M. Furukawa, and K. Goto, *Anal. Chim. Acta*, 46, 271 (1969).

(9) R. Rikmenspoel, *Appl. Opt.*, 3, 351 (1964).

(10) R. Rikmenspoel, *Rev. Sci. Instrum.*, 36, 497 (1965).

(11) T. Ohnishi, B. Hagihara, and K. Okunuki, *J. Biochem.* (Tokyo), 54, 287 (1963).

(12) T. Ohnishi, *ibid.*, 55, 172 (1964).

(13) J. C. Cowles, *J. Opt. Soc. Amer.*, 55, 690 (1965).

(14) A. T. Giese and C. S. French, *Appl. Spectrosc.*, 9, 78 (1955).

(15) J. M. Goldstein, *Biophys. J.*, 10, 445 (1970).

Photon Counting for Spectrophotometry

HOWARD V. MALMSTADT, MICHAEL L. FRANKLIN,[1] and GARY HORLICK[2]
School of Chemical Sciences, University of Illinois, Urbana, IL 61801

Advantages over other light-measurement systems provided by the photon-counting method include direct digital processing of the inherently discrete spectral information, decrease of effective "dark current," improvement of signal-to-noise ratio, sensitivity to very low light levels, accurate long-term signal integration, improved precision of analytical results for a given measurement time, and less sensitivity to voltage and temperature changes

THE QUALITY of tens of thousands of research and routine spectrophotometers depends directly on the operating characteristics of photomultiplier (PM) tubes and their associated electronic readout and power supply circuits. The various types of instruments in which PM tubes are utilized indicate something about their importance. The PM tube is now found in essentially all of the commonly used molecular uv-visible absorption, fluorescence, phosphorescence, reflectance, laser-Raman, atomic absorption, atomic emission, and atomic fluorescence spectrometers and also in specialized rapid scan, submicrosecond time-resolved, T-jump, chromatogram scanning, multichannel spark-source direct readers, densitometers, and other types of instruments (1).

In nearly all instruments it has been standard practice to measure the output signal of the PM tube by using analog techniques. That is, the tube is operated under conditions and with measuring circuits so that an output current or voltage is obtained whose magnitude is directly proportional to the radiant power incident on the photocathode. This has proved to be a generally acceptable mode of operation. However, it has become increasingly apparent that in some spectrophotometers the output signal of the PM tube can be advantageously measured by using direct digital techniques. In the digital mode the PM tube and associated circuitry provide discrete electron pulses so that the number of counted pulses is directly proportional to the number of photons incident on the photocathode. This approach is commonly called "photon counting."

[1] Present address, Medical School, University of Colorado, Denver, CO 80210.
[2] Present address, Department of Chemistry, University of Alberta, Edmonton 7, ALT, Canada.

Compared to other light-measurement systems, the apparent advantages (2) provided directly or indirectly by the photon-counting method are direct digital processing of the inherently discrete spectral information, decrease of effective "dark current," improvement of signal-to-noise ratio, sensitivity to very low light levels, accurate long-term signal integration, improved precision of analytical results for a given measurement time, and less sensitivity to voltage and temperature changes.

The basic and practical characteristics are presented to illustrate that photon counting can be made widely applicable for many types of spectrophotometry including absorption, fluorescence, emission, and scattering methods. Also, some of the difficulties are considered. Even in those cases where the PM tube is used as an analog detector, the photon-counting concepts can be used for the analysis of its performance, and these can be very useful. Several applications of photon counting will be illustrated, including considerations for high-precision spectrophotometry.

PM TUBE AS DIGITAL OR ANALOG TRANSDUCER

The intensity of a light signal falling on the PM tube is directly dependent on the rate at which photons arrive at the photocathode (P_k). This rate fluctuates as a result of radiation noise (photon noise) inherent in the incoming light, and the fluctuation is, in general, random (3). When photons of a specific wavelength arrive at the photocathode, they eject photoelectrons with an efficiency that depends on the quantum efficiency, Q_λ, of the photocathode surface (4, 5).

The photoelectrons from the cathode are attracted to the first dynode by electrostatic focusing. However, not all of the photoelectrons reach the first dynode, and this collection efficiency, f, is typically about 75%. There is a high probability that each photoelec-

tron that reaches the first dynode will eject several secondary electrons (5).

The transfer efficiency, g, of electron bursts between dynodes is almost 100%, so that the number of anode pulses is nearly equal to the number of photoelectrons reaching the first dynode, assuming negligible dark pulses. The number of electrons in each anode pulse depends greatly on the PM voltage, but the number is typically 10^5 to 10^7 electrons per pulse. Even at constant PM voltage, the statistical nature of secondary emission from the dynodes introduces an amplitude fluctuation of the anode pulses.

If it is assumed that there is negligible pileup of anode pulses, i.e., nearly all anode pulses are resolved, then the number of anode pulses per second, N_a, can be written

$$N_a = gfQ_\lambda P_k \qquad (1)$$

The counting of the anode pulses, which are directly related to the number of photons incident on the cathode, provides the measurement technique appropriately called "photon counting."

Although the output at the anode of the PM tube is inherently a series of pulses, the output signal can be expressed as an average rate of flow of electrons per second or current, as summarized in Equation 2.

$$\underbrace{\left(\frac{\text{anode pulses}}{\text{sec}}\right)}_{N_a} \underbrace{\left(\frac{\text{electrons}}{\text{anode pulse}}\right)}_{\overline{G}} \times$$

$$\underbrace{e}_{\left(\frac{\text{coulombs}}{\text{electron}}\right)} = \underbrace{I_{(A)}}_{\frac{\text{coulombs}}{\text{sec}}} \qquad (2)$$

For example, if the number of anode pulses/sec, N_a, equals 10^6, and the average electrons/anode pulse, $\overline{G}$, equals 10^6, then

$$I_{(A)} = N_a \overline{G} e = 1.6 \times 10^{-7} \text{ A} \qquad (3)$$

However, this is an average current, and it is apparent from Equation 2 that the output current fluctuates as

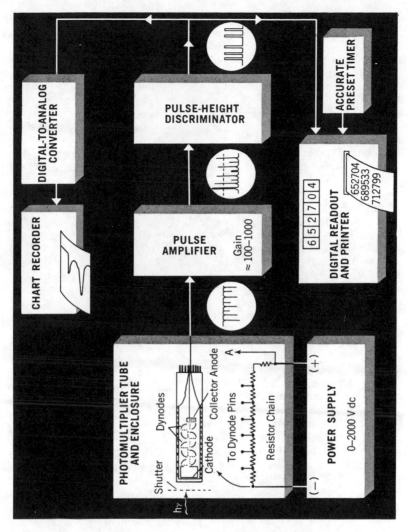

Figure 1. Block diagram of practical photon-counting system (2)

N_a and $\overline{G}$ fluctuate. That is, the random-time behavior of the anode pulses as a result of the incident radiation, and the amplitude distribution caused by secondary electron emission cause the so-called "shot noise" on the PM analog signal. By use of suitable electronic filters, the fluctuations are averaged or integrated so that the average current can be observed. Because an average current can be related to the number of pulses in unit time, it is frequently convenient to determine the noise component of an analog PM signal by utilizing count statistics.

PHOTON-COUNTING SYSTEMS

The required equipment used to implement photon counting in spectrophotometry is illustrated in Figure 1.

The photomultiplier tube is enclosed in a compartment that must be perfectly light tight, and it is powered by a regulated high-voltage (HV) power supply. The output signal current pulses are coupled through a resistive-capacitive load or pulse transformer into a pulse amplifier. All pulses within a preset discriminator voltage range are counted by the digital counter during an accurate preset time interval. The total count per integration period is read from the digital counter or printer. An analog readout is often made available to provide graphic monitoring. A digital-to-analog converter converts the photon counts to a dc voltage which is sent to a chart recorder. Each of the blocks of Figure 1 is briefly considered so as to summarize the important characteristics.

Photomultiplier Tube and Power Supply. Incoming photons eject photoelectrons with an efficiency that depends on the quantum efficiency of the photocathode surface. This efficiency is strongly dependent on wavelength; thus, a response characteristic is chosen that results in high efficiency for the wavelength of interest. When the PM is part of a spectrophotometer, the spectral characteristics of the source and monochromator must also be taken into account. The overall response characteristics are determined by the convolution of the spectral response of the detector with the spectral output of the source and optical system.

The gain, frequency response, and dark-count characteristics of the PM tube are very important and are discussed in the next major section.

The required operating voltages between PM tube electrodes are provided by connecting a chain of resistors across a stable HV power supply. The voltage drops across the resistors provide the voltage increments between successive electrodes (6).

The applied voltages influence the pulse-height distribution, but as the applied voltage is increased, a value is reached where changes in PM voltage cause relatively little change in pulse height.

Pulse Amplifier. A low-noise amplifier is required for photon counting because any noise induced by the amplifier increases the background count rate when the discriminator is set at a low level. The input of the amplifier should offer a minimum of shunt capacitance to reduce pulse degradation, and great care must be used in connecting from the anode of the PM to the pulse-amplifier input (2).

An ac-coupled amplifier is usually employed because it eliminates the dc drift from the PM. Also, reducing the low-frequency cutoff of the amplifier reduces a considerable amount of noise. Bandwidths from 10 KHz to 100 MHz are used in photon counting to reject the low-frequency noise but still provide adequate pulse fidelity.

The gain from the pulse amplifier must be stable so that the pulse-height shift does not change the number of photon counts getting through the discriminator. Sharing the gain requirement between the PM and the amplifier requires that the amplifier voltage gain be about 100–1000 for use with most PM tubes.

At high-count rates the amplifier duty cycle becomes large, and a baseline shift owing to ac coupling results. This changes the number of pulses getting through the pulse-height discriminator. It is often necessary to use a dc level restorer to provide a zero reference line. A simple diode clamp circuit can be employed with good results in some cases at count rates not exceeding 100 KHz. Other methods can be used to reduce the base-line shift with · increasing count rate, such as the Robinson clamp, bipolar pulses, base-line shift compensation at the discriminator, or a dc amplifier system. In many instances using a fast ac amplifier and keeping the duty cycle low result in acceptable base-line shift without dc restoration.

Pulse-Height Discriminator. A pulse-height discriminator useful for photon counting must be stable and sensitive to small voltages so that small signal pulses can be effectively counted; the frequency response must be high so that the upper count rate is not limited at this stage. Tunnel diode discriminators are available which operate at 50-mV sensitivity and at greater than 100 MHz.

Pulse-Pileup (Coincidence) Effects. Two effects are possible owing to pulse pileup at a fixed discriminator level. Two pulses which should each be counted can pile up so that only one count is recorded. A net loss of one count results. Two smaller pulses can pile up and sum to produce a pulse large enough to be counted when neither pulse should have been counted; thus, a net excess of one count is recorded. The discriminator level determines the relative importance of these two effects on the recorded count. At low discriminator levels when a large fraction of the theoretical count is recorded, the major effect is the loss of pulses owing to pileup.

Discriminator-Level Setting. Because of the counting statistics, the standard deviation of a measurement is determined by the square root of the number of counts. This means the more counts recorded for a given measurement time, the better the precision will be. The pulse-height discriminator level should, therefore, be set to accept the largest fraction of the theoretical signal counts possible without picking up a large amount of background pulses. By setting the discriminator level to accept the largest number of signal counts in a given observation time, both utilization of the frequency response of the system and less susceptibility to drift are realized. If the discriminator level is set so that only a small fraction of the theoretical count rate is admitted, a small drift in discrimination level causes a large change

in count rate because the signal count rate changes rapidly with discriminator voltage in this region.

Digital Counter. Fortunately, many new types of counters operate at rates of greater than 100 MHz so that this part of the system should not limit overall frequency response. Since the standard deviation is equal to the reciprocal of the square root of the number of events counted, the number of readout digits having significance can be estimated. Therefore, it is not necessary to provide readout of the total recorded events, but it is rather the scaled-down counts that are significant.

Accurate Preset Timer. Crystal clocks provide precision and accuracy better than 1 part in 10^6 for a wide range of time bases from 1 μsec to 10 sec or more. Integration times in photon-counting applications usually range from 0.1 to 10 sec or the duration of a specific event.

Digital-to-Analog Converter. The same pulses from the output of the pulse-height discriminator which are counted can be fed simultaneously to a digital-to-analog converter. This circuit shapes each input pulse so that each output pulse becomes equally weighted. The equally weighted pulses are then smoothed to provide a dc voltage output. A conventional diode pump circuit can be used after simple RC differentiation (6) to produce the analog output for locating trends in the counting data, such as peak-signal location during scanning.

Chart Recorder. After the pulses are converted to a dc voltage, the accuracy of the analog readout is determined by the reading accuracy and by the observation time. This integration time is determined by the time constant of the filter in the digital-to-analog converter. Chart recorders are usually accurate to 0.5% of full scale with a reading error of about 0.2%. The signal pulse-height fluctuation has been removed, but the random-time distribution is still present in the signal. In many analog photometric measurements the major portion of the unwanted fluctuation results from the reading error and not from the fundamental "shot noise limit" of the signal. Note that with digital readout of all the pulses, the reading error is eliminated, and wide dynamic range is available.

Commercial Systems. Several companies offer instrumentation packages or modules which can be used for photon counting at rates in excess of 1 MHz. Three of these systems are mentioned along with their specifications and special features.

Solid State Radiations, Inc. (SSRI) offers a pulse-amplifier-discriminator combination with gain of 2400 and bandwidth from 10 KHz to 100 MHz. The pulse-pair resolution specification is 10 nsec. A shielded PM housing is also available along with dc power supplies for the PM and amplifier. There is a separate module for each unit listed:

PM housing and dynode chain
PM and amplifier power supplies
Pulse amplifier
Digital computing counter

Approximate total cost, $4700

The digital counter has several modes of operation useful in photon counting. There are two separate registers so both the background and signal counts can be stored separately. Both background and signal counts can be integrated for equal preset periods in the synchronous mode. The background register can be subtracted from the signal register in still another mode of operation (7).

Elscint Instruments manufactures a system with PM and preamplifier power supplies, counter, and amplifier in one instrument package. This system has upper and lower level discriminators; thus, an energy window can be set to eliminate large as well as small pulses. In the synchronous mode

the background pulses are subtracted from the signal plus background by an up-down counter. However, it is often desirable to know both the background and the signal rate for error analysis, not simply the difference.

A novel feature is the utilization of the upper discriminator level to correct for first-order pileup effects. Pulses exceeding the lower discriminator level are counted normally, whereas pulses exceeding the upper discriminator level are counted twice. This corrects for the case where two pulses add together to give a single pulse of twice the amplitude. The total amplification of the preamp and the amplifier is 300 with a pulse-pair resolution of 12 nsec. The modules necessary for photon counting are the following (8):

Low-level counter spectrometer
PM housing
Fast preamplifier

Approximate total cost, $3300

Ortec offers a NIM bin system which can be used for photon counting. The amplifier has a gain of 200 with a 4-nsec rise time. The signal is fed to a 100-MHz discriminator and on to a dual counter timer. Both background and signal plus background counts can be stored. The NIM bin system offers some flexibility since it uses standardized plug-in modules. Additional modules can be plugged in to provide a teletype data logging system. The basic modules necessary for photon counting are as follows (9):

NIM bin and power supply
HV power supply
PM base
Amplifier
Dual counter timer with
  crystal time base

Approximate total cost, $3200

BASIC CONSIDERATIONS IN UTILIZING PHOTON COUNTING

The possibility of measuring low-intensity light with improved signal-to-noise ratio was one of the original reasons for interest in photon counting and remains a major basic consideration for any discussion of the topic (*10–16*). Other basic considerations which influence the applicability of photon counting and which are considered in this section are dynamic-range or frequency-response considerations, dark count, and pulse-height distribution.

Signal-to-Noise Ratio. The signal-to-noise ratio of a photon-counting measurement may be considered from the point of view of counting statistics. The fundamental noise in a photon-counting system is the fluctuation in the signal count as determined by the statistics of the photon arrival at the photocathode. In general, the arrival of photons at the photocathode is random; thus, the probability of fluctuations about the rate of arrival is determined by taking the square root of the number of counts (*17*). For the specific case of a Gaussian distribution, the reciprocal of the square root of the number of counts is equal to the standard deviation. On this basis the "signal-to noise ratio" for a single measurement can be estimated by dividing the count by the square root of the count.

Many photon-counting measurements are made in situations such that unwanted background pulses are present. In these cases the background count must be subtracted from the measured background plus signal count. This makes the signal-to-noise ratio expression more complicated (*10–13*). The fluctuation in the signal count (noise) is calculated by using the standard equation for the counting statistics of a difference. This equation can take the form:

$$F_s = (R_s + 2 R_B)^{1/2} T^{1/2} \qquad (4)$$

where F_s is the fluctuation in the signal count, R_s is the signal count rate, R_B is the background count rate, and T is the counting time. In this case it is assumed that the counting time is the same for the measurement of the background count and the signal plus background count. The signal is $R_s T$; thus, the signal-to-noise ratio can be expressed as:

$$\frac{S}{N} = \frac{R_s^{1/2} T^{1/2}}{(1 + 2\ R_B/R_s)^{1/2}} \qquad (5)$$

Equation 5 has two limiting forms. If the background rate is small with respect to the signal count rate, the equation reduces to $R_s^{1/2} T^{1/2}$, the square root of total signal count. Thus, for a total count of 10^6 the signal-to-noise ratio is 1000. If the background count rate is large with respect to the signal count rate, Equation 5 becomes the following:

$$\frac{S}{N} = \frac{R_s T^{1/2}}{(2\ R_B)^{1/2}} \qquad (6)$$

For a signal count rate of 2 cps and a background count rate of 100 cps, a signal-to-noise ratio of 4.5 can be achieved in 1000 sec. The stability of

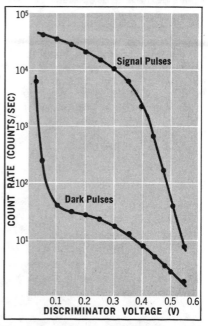

Figure 2. Integral pulse-height spectra of signal and dark pulses (*2*)

most photon-counting systems allows such long counting times.

The relative rates of the signal and background counts (and, hence, the signal-to-noise ratio) can, to some extent, be controlled by adjusting the pulse-height discriminator. Thus, it is important to consider briefly the pulse-height distributions of both background (dark) and signal pulses.

Pulse-Height Distribution of Pulses.
Current pulses which have been caused by incident photons and, hence, have undergone the full amplification of the PM tube have a pulse-height distribution closely approximated by a Poisson distribution (4, 18–21). The integral pulse-height spectra for signal and dark pulses for a 1P28 photomultiplier tube are shown in Figure 2. In this case all counts above a certain discriminator level were counted. Note that the pulse-height spectrum for the dark-current pulses is not the same as that for the signal pulses. It contains a larger number of smaller pulses than would be predicted on the basis of a Poisson distribution. This can be explained if the pulse components of the dark current are examined. Figure 2 should be considered specific to this tube. The specific shape can differ for different photomultiplier tubes and even for the same tube under different experimental conditions (different dynode voltages, magnetic defocusing).

Dark Current. There are a number of sources of dark currents in a photomultiplier. Some of the more common sources are discussed by Lallemand (22). Among these are thermionic emission, cold-field emission, radioactivity, and ohmic leakage. The first three are pulsed in nature. The pulses from thermionic emission and cold-field emission often originate down the dynode chain and, hence, do not undergo full amplification in the tube. These pulses give rise to the larger number of smaller pulses in the pulse-height spectrum of the dark current. In addition, a dramatic increase in the count rate at low discriminator levels is evident in the dark-current pulse-height spectrum (Figure 2). These pulses arise from stray electrical noise and amplifier-induced noise. Pulses originating from cosmic rays and radioactive potassium (β particle emission) in the glass envelope of the tube should result in a slightly elevated level of higher pulses in the dark current than would be expected on the basis of a Poisson distribution, but these are generally negligible for practical spectrophotometry. Therefore, an upper discriminator level is not required. Pulses induced from the 60-Hz power line also contribute to the pulsed dark-current component.

The different characteristics in the pulse-height spectra of signal and dark

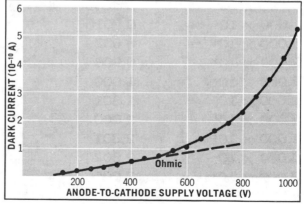

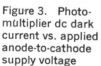

Figure 3. Photomultiplier dc dark current vs. applied anode-to-cathode supply voltage

current mean that pulse-height discrimination can be used to reject relatively low pulses and thus selectively discriminate against most dark-current pulses with respect to signal pulses (13). For a 1P28 PM tube used by the authors (at 900 V), a dc dark current of 4×10^{-10} A was reduced to an equivalent dark current of 3.9×10^{-12} A (41 counts/sec), as calculated by using Equation 2 with a $\overline{G}$ value of 6×10^5 electrons/count.

Another source of dark current is ohmic leakage. The photon-counting method is insensitive to the ohmic component of the dark current because of its dc nature. The rapid increase in dark count with applied voltage is shown in Figure 3.

Cooling a photomultiplier tube will reduce the dark current. Typically, the dark current will approach a minimum value by about $-30°C$ (23). However, note that the spectral sensitivity of a photomultiplier tube is also markedly affected by temperature, often in a complex manner. Several cases are summarized by Budde and Kelly (24). In addition, cooling of the photomultiplier tube does not necessarily improve the signal-to-noise ratio for photon-counting experiments (25), and in one case for signal count rates in excess of 8000 photons/sec, the signal-to-noise ratio was decreased by cooling (25).

Frequency-Response Considerations. The major problem in photon counting at the light levels often desirable for practical spectrophotometry is that the measurement system must be fast enough to preserve the high-frequency response of the photomultiplier detector. At present there is a crossover point where it is necessary to shift from photon counting to conventional dc current measurement which depends on the frequency response of the measurement system and incident light levels. In Figure 4 the outputs from two different photon-counting systems are shown. The frequency responses for

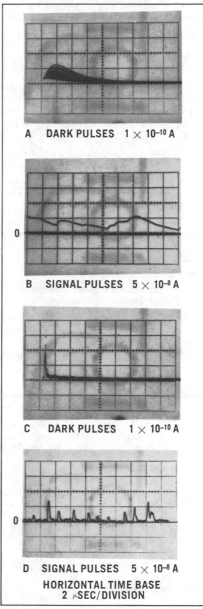

A   DARK PULSES 1×10^{-10} A

B   SIGNAL PULSES 5×10^{-8} A

C   DARK PULSES 1×10^{-10} A

D   SIGNAL PULSES 5×10^{-8} A
HORIZONTAL TIME BASE
2 μSEC/DIVISION

Figure 4. Comparison of pulse-resolution capabilities of two measurement systems which have different frequency responses. A, B: RC time constant is 4 μsec; C, D: RC time constant is 0.2 μsec

the two systems are measured by triggering the oscilloscope on individual dark pulses. For the first measurement system the RC decay time is 4 μsec, as shown by the envelope of current pulses in Figure 4A. The output signal shown in Figure 4B, which corresponds to a dc current of 5×10^{-8} A, indicates that individual pulses are not resolved. The second measurement system has a frequency response twenty times faster, as shown by Figure 4C. The output shown in Figure 4D, which also corresponds to a photocurrent of 5×10^{-8} A, indicates that nearly every signal pulse is resolved. These photographs illustrate the importance of frequency response of the entire system when using photon counting at light levels typically important in spectrometric methods.

Relationship Between Photocurrent and Count Rate. A plot of log pulse count rate vs. log photomultiplier current is shown in Figure 5 for a photon-counting system that we have used experimentally. The theoretical count-rate line shown on the figure was cal-

culated from the typical gain value of the 1P28 PM at an applied voltage of 900 V (RCA Phototubes and Photocells Technical Manual, pp 7–60). This curve was verified by extrapolating the signal pulses curve of Figure 2 to zero discriminator level at the value of dc photocurrent used to take the curve. The theoretical count-rate line represented the total number of signal pulses available for counting at a specific photomultiplier dc current level.

This dotted line will be translated horizontally depending on the amplification of the photomultiplier as the average charge content of the pulses changes with the voltage applied to the photomultiplier tube. However, the total number of signal pulses at a given light level remains constant.

In practice a pulse-height discriminator must be used to reject unwanted noise pulses in the measurement system (see Figure 2). This will also eliminate a certain fraction of the signal pulses. Two experimental curves taken at 70 and 27% of the theoretical count rate are also shown in Figure 5.

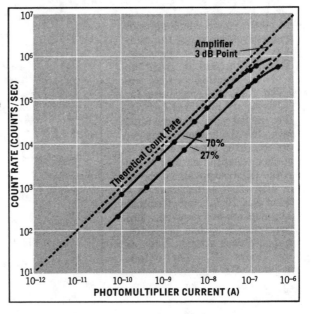

Figure 5. Count rate vs. photocurrent for theoretical and two experimental curves at different percent discriminator levels. Anode-to-cathode supply voltage is 900 V

They both have slopes of 45°, indicating that the photocurrent and the count rate are directly proportional. At higher count rates they both tail off at a point corresponding to the upper 3 dB response of the pulse amplifier. Note that the upper 3 dB response of a pulse amplifier may be misleading with respect to the maximum average counting frequency that can be measured without rolloff. The time distribution between photons from the light source is random; thus, at a certain instantaneous time the frequency of the pulse sequence may be greater or less than the average rate.

Experimentally, the measurement system frequency response should be at least 25 times greater than the maximum average pulse rate that is measured to keep the pileup error less than 1% at this maximum rate.

The crossover point between using photon counting or conventional dc current measurement for the experimental system used to obtain Figure 5 is about 10^{-8} A, which corresponds to a count rate of 100 KHz for the 1P28 PM operated at 900 V. Since the crossover point depends on the frequency response of the measurement system, it should be feasible to extend the system to equivalent dc currents approaching 10^{-6} A. This would correspond to a counting rate of about 10 MHz. Precision on the basis of counting statistics of about 0.03% would thus be obtained for a 1-sec observation time.

APPLICATIONS OF PHOTON COUNTING

Precision Absorption Spectrophotometry. The absorbance of a sample

Table I. Calculated Error in Absorbance for Assumed Values of Reference and Sample Counts
Reference counts = 2.000×10^6

Sample counts	Absorbance, A	Theoretical % error
1.990×10^6	0.0022	19.962
1.950×10^6	0.0110	3.973
1.800×10^6	0.0457	0.975
1.500×10^6	0.1249	0.375
1.000×10^6	0.3010	0.176
5.000×10^5	0.602	0.114
2.500×10^5	0.903	0.102
2.000×10^5	1.000	0.102
1.000×10^5	1.301	0.108
2.000×10^4	2.000	0.154
1.000×10^4	2.301	0.138
5.000×10^3	2.602	0.235
2.000×10^3	3.000	0.322

can be measured by using a photon-counting measurement system. The transmittance is calculated by dividing the total count obtained for the sample channel (N_s) by the total count obtained for the reference channel (N_r), provided the counting times for both measurements are identical. The absorbance is log (N_r/N_s). The analytical curve for acidic dichromate shown in Figure 6 was obtained by use of a photon-counting measurement system in conjunction with an absorption spectrophotometer.

Consideration of the measurement error from the viewpoint of photon counting results in an important clarification of the possible precision of an absorption measurement not readily apparent from analog measurements. The error in an absorbance value as a result of counting statistics can be calculated in two steps. First the error in one-over-transmittance $(1/T)$ is calculated by using the standard error equation for a ratio. The final equation is as follows:

$$E_{r/s} = \frac{N_r}{N_s} \left[\frac{(E_r)^2}{N_r} + \frac{(E_s)^2}{N_s} \right]^{1/2} \quad (7)$$

where $E_{r/s}$ is the error in N_r/N_s $(1/T)$. If background counts are neg-

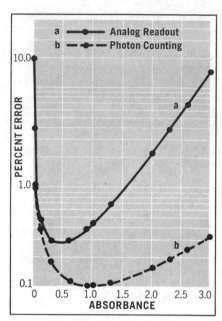

Figure 7. Precision of photon counting and analog systems as function of absorbance (26)

ligible, the errors in the reference count (E_r) and in the sample count (E_s) are equal to the square roots of the respective counts; if not, Equation 4 is used to calculate these errors. The final relative percent error in absorbance, % E_A, is calculated from the error in $1/T$ as:

$$\% E_A = \frac{\log (1/T + E_{r/s}) - \log(1/T)}{\log(1/T)} \times 100\% \quad (8)$$

Calculated theoretical errors in absorbance for assumed values of reference and sample counts are shown in Table I (26). The theoretical percent error has a broad minimum, changing only slightly between 0.3 and 2.3 absorbance units. This broad minimum is greatly different from the error curve determined by the classical reading error of an analog scale (27). This is illustrated in Figure 7. The photon-counting plot values were taken from

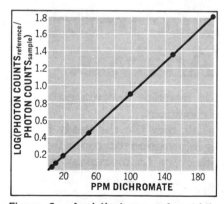

Figure 6. Analytical curve for acidic dichromate taken at 360 nm (2) by using photon counting where absorbance = log[(photon counts)$_{ref}$/(photon counts)$_{sample}$]

184

Table I, and the analog readout values are based on a photometric reading error of 0.1% in transmittance along a linear scale. The photon-counting measurement, with a reference count of 2.000×10^6, has an error of less than 0.2% from 0.3 to 2.4 absorbance units, whereas the error with the analog readout varies from 0.3 to 4.6%. The minimum error in the analog plot is 0.27% and occurs at an absorbance of 0.434, whereas the minimum error in the photon counting plot is 0.11% and occurs at an absorbance of 0.968. This position of the minimum in the photon-counting plot is specific to the case where equal times are spent measuring the reference and sample counts. The position of the optimum absorbance may vary considerably, depending on the specific measurement conditions. Several cases have been considered by Hughes (28).

At any specific absorbance value, the error may be decreased by simply increasing the total number of counts. If the numbers of sample and reference counts used to calculate the absorbance value of 0.0457 in Table I are both increased by a factor of 10, the percent error drops from 0.975 to 0.308. This emphasizes the importance of a high-frequency-response measuring system to minimize the counting time necessary to achieve a desired precision. To achieve the precisions indicated in Table I, counting times of 10 sec each for the sample and reference would be needed if the photon-counting system responds linearly to 200 KHz, and 100 sec would be needed if the maximum linear response was 20 KHz.

So far it has been assumed that background counts are negligible. If significant background counts are present and are not subtracted out, they cause

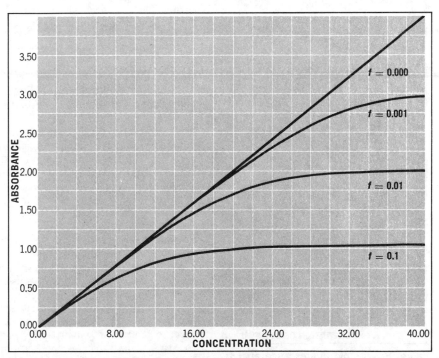

Figure 8. Effect of uncorrected background counts on analytical curve. Ratio of background counts to reference counts is f. Concentration axis is arbitrary

a negative deviation from Beer's law directly analogous to that caused by stray light, as shown in Figure 8. With a reference count of 2.000×10^6 and an ignored background of 2.000×10^3, there is a 2% negative deviation from Beer's law at an absorbance value of 2.0, and the analytical curve will be asymptotic to an absorbance value of 3.0. Of course, the background can be subtracted out to remove this deviation. The error terms in Equation 7 must be calculated by using Equation 4. For a reference count of 2.000×10^6 and a background count of 1.000×10^3, the % error at an absorbance of 2.0 increases from 0.154 to 0.161%.

Application of Small Computer. The comparator output of a photon-counting system in an absorption spectrophotometer can be sent directly to a low-cost (less than $5000) laboratory minicomputer. The computer can be programmed to subtract dark counts from both reference + dark and sample + dark counts and then provide an output of % transmittance (sample counts/reference counts) $\times$ 100, or absorbance [log(reference counts/sample counts)], or log absorptivity, etc.

The setup in our laboratory uses a programmed cell compartment (Model EU-721-11, Heath Co., Benton Harbor, MI) to alternately place reference and sample cells into a single light path. The desired time intervals for obtaining sample and reference counts are selected on the teletypewriter.

For recording spectra, the computer advances the wavelength by a selected increment as requested by teletype. At each wavelength the reference, sample, and background information are sent to the computer. To obtain high-precision photometric data at all wavelengths, the total reference count is maintained approximately at a preset level. This is accomplished by feedback to the monochromator, source, or the count integration timer so as to compensate for changes of overall relative sensitivity of the spectrophotometer.

Raman and Fluorescence Spectrophotometry. In both Raman and fluorescence spectrophotometry the intensities of Raman or fluorescence bands radiated from samples are generally very low. Therefore, the photon-counting technique has frequently been described and applied for such applications (29–31). An analytical curve obtained for the atomic fluorescence of cadmium (32) over a wide dynamic range of 10^5 is shown in Figure 9. For this example the photon counts were integrated over a 10-sec period for each sample. In Raman, atomic emission, absorption, and fluorescence techniques, the sought-for signal is often superimposed on a large background interference, so the use of synchronous photon counting is desirable for these applications (7, 12). This system is similar in principle to the computer system described in the previous section. It provides for two simultaneous channels of

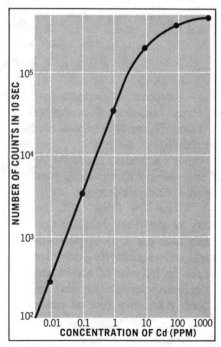

Figure 9. Atomic fluorescence analytical curve for cadmium obtained with photon counting (32)

count information to be accumulated and then manipulated.

Luminescence Decay-Time Measurements. Many molecules, such as aromatics, which are excited return to the ground state by photon emission. The exponential luminescence decay curves of the emitted radiation provide useful chemical information. The decay times might be in the nanosecond, microsecond, or millisecond time range. Although there are several techniques for making the longer decay-time measurements, most techniques do not work well for the very short decay times. However, a rather new technique referred to as the "single-photon technique" can be used even for decay times of a few nanoseconds. This unique procedure makes use of the probability of measuring a single photon pulse per excitation pulse. Details of the method have been presented (9), showing a specific experimental system that can be used for measuring decay times from 2 nsec to 250 μsec.

REFERENCES

(1) H. V. Malmstadt, M. L. Franklin, and G. Horlick, Progress in Nuclear Energy, Analytical Chemistry, Series IX, Pergamon, Elmsford, NY, 1972.
(2) M. L. Franklin, G. Horlick, and H. V. Malmstadt, *Anal. Chem.*, **41**, 2 (1969).
(3) R. C. Jones, in "Advances in Electronics," Vol V, p 18, L. Morton, Ed., Academic Press, New York, NY, 1953.
(4) G. A. Morton, *RCA Rev.*, **10**, 525 (1949).
(5) RCA Photomultiplier Manual, RCA Corp., Princeton, NJ, 1970.
(6) H. V. Malmstadt and C. G. Enke, "Digital Electronics for Scientists," Benjamin, New York, NY, 1969.
(7) SSR Instruments, Bulletin on Model 1110 Photon Counter, Santa Monica, CA, 1972.
(8) Elscint Scientific Instrumentation Catalog GMD-10, pp 71–73, Princeton, NJ, 1972.
(9) ORTEC, Application Note AN35, Oak Ridge, TN, 1971.
(10) R. R. Alfano and N. Ockman, *J. Opt. Soc. Amer.*, **58**, 90 (1968).
(11) J. J. Barrett and N. J. Adams, III, *ibid.*, 311 (1968).
(12) F. T. Areccki, E. Gatti, and A. Sona, *Rev. Sci. Instrum.*, **37**, 942 (1966).
(13) G. A. Morton, *Appl. Opt.*, **7**, 1 (1968).
(14) R. Jones, C. J. Oliver, and E. R. Pike, *ibid.*, **10**, 1673 (1971).
(15) M. Jonas and Y. Alan, *ibid.*, p 2437.
(16) F. Roblen, *ibid.*, p 777.
(17) S. Goldman, "Frequency Analysis Modulation and Noise," p 306, McGraw-Hill, New York, NY, 1948.
(18) M. Gadsden, *Appl. Opt.*, **4**, 1446 (1965).
(19) F. J. Lombard and F. Martin, *Rev. Sci. Instrum.*, **32**, 200 (1961).
(20) Z. Bay and G. Papp, *IEEE Trans. Nucl. Sci.*, **11**, 160 (1964).
(21) R. F. Tusting, Q. A. Kerns, and H. K. Knudsen, *ibid.*, **9**, 118 (1962).
(22) A. Lallemand in "Astronomical Techniques," Vol II, p 126, W. A. Hiltner, Ed., Univ. Chicago Press, Chicago, IL, 1962.
(23) R. Foord, R. Jones, C. J. Oliver, and E. R. Pike, *Appl. Opt.*, **8**, 1975 (1969).
(24) W. Budde and P. Kelly, *ibid.*, **10**, 2612 (1971).
(25) Y. D. Harker, J. D. Masso, and D. F. Edwards, *ibid.*, **8**, 2563 (1969).
(26) E. D. Jackson, PhD thesis, University of Illinois, Urbana, IL, 1970.
(27) W. A. Blaedel and V. W. Meloche, "Elementary Quantitative Analysis," Harper and Row, New York, NY, 1963.
(28) H. K. Hughes, *Appl. Opt.*, **2**, 937 (1963).
(29) M. R. Zatzick, *Res./Develop.*, **21**, 16 (1970).
(30) S. A. Miller, *Rev. Sci. Instrum.*, **39**, 1923 (1968).
(31) E. H. Eberhardt, Technical Application Notes of ITT, Fort Wayne, IN, 1968.
(32) K. P. Li, PhD thesis, University of Illinois, Urbana, IL, 1970.

Liquid Chromatography Detectors

By Ralph D. Conlon

THE TECHNIQUES grouped under the general term liquid chromatography are enjoying a comeback in the analytical arsenal. Liquid chromatography has long taken a place behind the more glamorous and sophisticated gas chromatography, and instrument manufacturers were willing to leave it there because of the limited market and because the area is so large that only a few specific techniques which gained prominence were defined well enough to instrument. Examples are amino acid and polymer separations. This very diversity combined with generally higher resolving power has finally forced interest and effort into liquid chromatography. The major reason for this Renaissance is that the compounds which are too high in molecular weight or too heat labile for gas chromatography now represent a formidable percentage of the chemical families of interest.

Whenever analytical effort is channeled into a new or renewed area, it soon follows that instrumentation must be developed and used to carry a part of the work load. Thus the rare separation, which required use of liquid chromatography and which could be developed and applied by taking numerous portions of the effluent and subjecting them to further chemical or physical tests, has now become a flood of work and the old methods cannot be economically tolerated.

The use of an appropriate detector to aid in developing and performing liquid chromatography separations is an obvious first thought in making the technique as useful and painless as possible. In methods development, the idea is to select the best column and solvent system based on the sample functional groups. It is generally necessary to evaluate several alternative systems until the optimum method is developed. If a suitable effluent stream monitor is available, the chemist need only study the recorded elution pattern to evaluate the system under test and make appropriate changes to column or solvent and try again until a satisfactory result is obtained.

Another valuable application for stream monitors is in checking the activity of commercial or prepared column material. The elution volumes and resolution of a simple test mixture are easily determined as a check on each batch of adsorbent. A similar use is the determination of elution volumes of individual components with a given column–solvent system. This is one of the simplest methods for determining if a given system will separate two or more materials where the materials are available in pure form. A stream monitor provides precise information as to time and milliliters required to elute and the degree of band broadening. If the elution volumes are sufficiently different to give a practical separation, the column and solvent are usable. Additionally this technique provides the order of removal from the column.

The advantages of efficient stream monitoring as an aid to methods development, and later application of the method, over collecting and testing or drying and weighing small portions of the effluent need not be belabored. As in gas chromatography, however, the detectors bring certain requirements of their own which must be accommodated to achieve satisfactory results.

188

Flow rate, stream temperature, and detector environment, for example, are important parameters and must be controlled. For this reason the best use is made of detectors in the context of a total system whether commercial or homemade.

The purpose and scope of this article is to suggest several approaches to liquid chromatography stream monitoring; briefly how they work, and what criteria to use in testing the chemical system to determine which are likely candidates. Since there is no true "universal" detector for liquid chromatography, it is necessary to select one on the basis of the problem at hand and in doing a variety of separations it is expected that additional detectors will be needed. The minimum practical sample sizes suggested are not the best values reported but should be obtainable without extreme precautions. Where possible the sample size should be increased to the practical limit of resolution to take advantage of signal-to-noise ratio. Five of the techniques discussed have been incorporated into instruments and are commercially available. A brief look at four systems which show promise as future monitors is also presented.

Refractive Index

Refractive index comes closest to being the universal detector for liquid chromatography. It has been compared to the thermal conductivity detector in gas chromatography because like thermal conductivity the refractive index changes whenever the carrier is contaminated by a solute. The trivial exception to this is the case where solvent and solute have the same refractive index. Refractive index has many features recommending it, level of sensitivity is good, measurement is nondestructive to the sample, and operation is simple.

Refractive index detectors for liquid chromatography are based primarily on two different measurement principles. The angle of deviation technique utilizes the actual bending of a ray of monochromatic light as it passes through the flowing effluent stream contained in a suitable optical cell. The bending angle changes as a function of sample concentration and is monitored by an electromechanical arrangement which moves either the source or a photodetector to maintain optical alignment. The movement is converted into an electrical signal by means of a retransmitting slide wire or balance potentiometer. Such devices are sensitive, stable, and expensive. The second approach makes use of the reflection principle first elucidated by Fresnel in which the intensity of the reflected component of a beam of light impinging on the surface of the effluent stream changes inversely with the refractive index.

The reflectance technique has certain advantages such as a smaller permissible flow, cell volume, and lower cost. Although there is no theoretical limit to the sensitivity obtainable with the reflectance technique, the commercial units available presently are less sensitive by a factor of about 50 to 100 based on manufacturers' specifications.

The application decision is based simply on the difference in refractive index between the solvent carrier and the sample. The refractive index of most of the common solvents can be found in handbooks of chemistry. If the sample material is not listed, this value can be measured with a laboratory refractometer of the Abbe type. Assuming that the sample will become diluted about 100 to 1 by the solvent carrier in the column (a general rule of thumb for all liquid chromatographic separations) and assuming that a 10^{-5} change in refractive index is readily detected, then the sample size in milligrams is roughly the reciprocal of the difference in refractive index between solvent and sample.

Temperature control of the column

189

effluent and detector is critical in refractive index monitoring. Many solvents used in liquid chromatography have a temperature coefficient of about 3×10^{-4} R.I. units per degree centigrade so control of better than a thousandth of a degree is necessary to achieve the higher sensitivity settings of commercial refractometers. Flow rate should be held constant for long term separations as an aid to maintaining thermal equilibrium. This mode of monitoring is usable with gradient elution but at a lower sensitivity since it is not possible to compensate the gradient completely even with differential type monitors. A reduction in sensitivity of 10 or more will be necessary which may be partially made up by increasing the sample size.

Electrolytic Conductivity

In electrolytic conductivity monitoring the specific conductivity of the effluent stream is continuously detected and entrained sample is indicated by a change in conductivity. This is one of the simplest and most dependable of the current monitoring techniques and its use is expanding rapidly as cell parameters and measuring circuitry are improved to permit use with both aqueous and nonaqueous effluents.

The use of conductivity measurement to determine water and reagent purity is a well known laboratory procedure. The standard apparatus consists of a pair of carefully spaced platinum electrodes contained in a probe assembly which is immersed in the sample. Alternating voltage in the range 60 to 1000 hertz at 14 volts is applied to the cell which forms one leg of a Wheatstone bridge. The bridge is brought to balance, as indicated by a phase detector, with variable resistors in opposing legs of the bridge. Conductivity or resistance parameters are then read out directly with proper correction made for the cell constant which is a geometric parameter based

on electrode size and spacing.

To utilize this technique for stream monitoring, the immersion cell is replaced by a low volume flow-through cell and variations in conductivity of the carrier liquid due to eluted sample are continuously recorded. Response is linear with concentration over a wide range so quantitation of the output signal is possible with suitable preliminary calibration. Best use is made of this monitor in single solvent chromatography since solvent gradients will cause a proportional shift in baseline. Detection sensitivity is a function of the difference in conductivity between eluting solvent and sample and so can be estimated from handbook values or determined experimentally by measuring a dilution series of sample in solvent with the monitoring cell. Sample sizes as low as 10 micrograms are detectable under conditions of constant flow rate and controlled temperature.

Gel filtration chromatography provides many applications for the electrolytic conductivity detector. The separation of proteins on gel columns using buffered aqueous eluting solvent has been successfully monitored in the lab as well as on a production scale in a pilot plant environment. The simplicity of operation, ease of cleaning, and freedom from maintenance and recalibration are important advantages of electrolytic conductivity monitoring.

Heat of Adsorption

The heat of adsorption detector is an excellent choice for the researcher doing a variety of liquid chromatographic separations which require different column media because it is also a "universal" detector within certain limitations.

The principle of this monitor involves detection of the minute temperature changes which accompany the interaction of sample and solvent with the separation medium in the column. As solvent molecules are displaced by sample molecules at the adsorbent surface a local temperature change reflects the exchange. A thermistor is used to detect the temperature change of the adsorbent and second thermistor imbedded in a nonadsorbing granular material is provided to aid in compensating for changes in steam or ambient temperature. Many approaches have been used in measuring these thermal reactions, including imbedding thermistors in the column, temperature monitoring at the point of emergence of the effluent from the column bed, and directing the effluent through a cell containing a small portion of the column bed material and the measuring thermistor. The latter approach is favored because of the flexibility offered. By repacking the cell or substituting a second pre-packed cell, the unit can be moved from column to column for a variety of studies.

All four major areas of separation: ion exchange, adsorption, partition, and molecular filtration provide a measurable change in temperature at the sample–column material interface. The real value of the heat of adsorption detector is in qualitatively visualizing the elution pattern and is therefore an excellent choice in methods development and in survey methods in which the presence or absence of a constituent is the important information, and not necessarily the amount.

The peak height is, in fact, a function of concentration, but frequent calibration is necessary in order to achieve what would be considered acceptable analytical accuracy. Unavoidable changes in the activity of the separation medium in the cell which occur from run to run and even during a run are primarily responsible for the lack of precision.

An apparent disadvantage could be considered to be the unconventional peak shape which has the general appearance of the first derivative of a Gaussian curve because the detector responds with a positive and negative going peak for each material eluted. This actually does not present any real problem because it is simply a matter of familiarity. Even the interpretation of two unresolved peaks or a peak with a shoulder is not difficult, and can in fact be facilitated by the first derivative presentation.

Two environmental considerations are essential to successful operation. The detector, and preferably the electronics should be maintained at an absolutely constant temperature with regard to short term variation. Since the detector is capable of measuring a change of about 10^{-5} degree Centigrade, it is not practical to control the temperature by adding increments of heat or cold. Most manufacturers instead provide a thermal ballast – thermal barrier environment usually consisting of a noncontrolled water bath which is well isolated from laboratory environment. This provides the best short term constancy of temperature but changes of 4 or 5 degrees in room temperature will result in excessive drift in long term studies.

The second requirement is that the flow rate be constant and surge-free. Flow rate rather than pressure must be constant, since only constant flow will guarantee that the stream velocity at the thermistor surface will be constant. Surging in the stream results in intolerable baseline noise. Positive displacement syringe pumps offer the best solution.

The heat of adsorption detector can be used in gradient elution chromatography but a shift in baseline will occur reflecting the change in solvent composition. As in the case of the refractive index detector the unit cannot be operated at maximum sensitivity for gradient elution applications but still can be useful.

The practical minimum sample size depends on the adsorption-desorption characteristics of the column material, solvent, and sample with ion exchange separations generally giving more signal. An average minimum sample injection is about one microgram per component for ion exchange separations and about 0.1 to 1.0 milligram for adsorption on silica gel.

The zero dead volume construction of the heat of adsorption detector makes it a valuable aid in studying band broadening in the column and extra-column peak effects. With proper calibration relative heats of adsorption can be measured for any liquid–solid system.

Ultraviolet and Visible Absorption

The application of ultraviolet or visible light absorption to stream monitoring is a natural extension of the use of spectrophotometers to examine individual portions of the effluent. In contrast to refractive index and heat of adsorption, UV or visible absorption is a specific parameter in that the sample material must exhibit a suitable extinction at some region of the UV spectrum or, in the case of visible absorption, combine with a complexing reagent to form a color derivative. This monitoring approach is among the simplest to evaluate since the UV absorption characteristics are either known from prior work on the system or can be easily obtained by running an absorption spectrum on a manual or recording spectrometer. The discovery of a suitable color developing reagent for visible absorption requires more chemistry but the evaluation procedure is the same.

After the proper wavelength has been established, it is desirable to substitute a single wavelength device for continuous monitoring. In this way an expensive versatile spectrometer is not committed for long periods. Optical filters with sufficient monochromaticity are available in any required wavelength from 2000 Angstroms in the ultraviolet up through the visible region.

The solvent system must be selected so that it is transparent at the ana-lytical wavelength. This then allows the UV or visible monitor to be used with gradient elution techniques. Other features of this monitoring technique are: no additional temperature control is normally required in the usual laboratory environment, the technique is sufficiently insensitive to flow rate that a simple pump or even gravity feed can be employed in less demanding situations, and the specificity allows a degree of qualitative information to be obtained.

Sample size requirement is dependent upon the absorption characteristic of the sample used but an acceptable minimum sample size under average conditions is 10 to 100 milligrams. An additional pump-to-meter color developing reagent into the column effluent stream must be supplied for continuous monitoring of colorless effluents in the visible region.

Flame Ionization Detector

The flame ionization detector combines high sensitivity, wide dynamic range, quantitative response, and the added advantage of being usable with gradient elution techniques. The singular limitation is that the sample must be relatively nonvolatile at normal laboratory temperatures. For this reason most applications to date have been made on lipids, proteinaceous materials, polymers, and other high boiling samples.

The flame ionization detector has long served gas chromatography, and the mechanics of thermal ionization and ion current monitoring are well known and demonstrated. The effluent stream sampling problem however is made more complex in liquid chromatography because the carrier must be separated from the sample material before combustion can take place, and this in turn creates difficulties in introducing the sample into the flame. The manufacturers of commercial flame ionization detectors for liquid chromatography solve these problems by applying all or a portion of the column effluent onto a moving flexible metal sample transport such as a small chain or wire. The carrier solvent is driven off by gentle heating in a stream of inert gas and the sample residue is then carried into the flame where volatilization and ionization occur. An increase in ion current is detected and a typical peak-shaped signal form is produced on a strip chart recorder. At least three mechanisms are presently being used to get the sample into the flame cone including:

a. burning residue directly on the wire,
b. using the flame heat to volatilize the sample before ignition, and
c. providing a heated chamber between the wire and the burner where the sample is removed from the wire by pyrolysis and the gaseous pyrolysis products directed into the flame.

Based on published data there appears to be an improvement in signal-to-noise ratio when the latter two techniques are employed.

As stated earlier, the selection of flame ionization is dictated primarily by the volatility of the sample. Since the metal carrier wire moves at a constant speed the loss of sample from the wire due to volatilization is also constant from run to run so sufficient precision is maintained so long as such losses are small. The technique of making a suitable dilution of sample in the solvent and applying it directly to the metal carrier with a syringe is very useful in evaluating this detector for both volatility losses and minimum detectable limit.

Commercial flame ionization detectors for liquid chromatography are equipped with temperature controls for the critical areas so additional protection from environmental variations is not required. Variations in flow rate will affect elution times as in gas chromatography but the detector noise level is not seriously affected by minor variations.

The flame ionization detector requires approximately 10 nanograms per second to shift the baseline a detectable amount above the noise level. A practical minimum sample size based on a 100 to 1 dilution in the column is about one milligram for a peak which elutes in ten minutes. Since a rate function is involved longer elutions require more sample.

New Detectors

The detectors described here are new only in that they are not now commercially available. The principles involved are familiar and promise to be very useful to liquid effluent monitoring.

Radioactivity. Separation and detection of radioactive compounds, although not widely used by analytical chemists is worthy of mention because of certain advantages accruing from this technique and because of recent major improvements in the scintillation medium.

The advantages of radioactivity in liquid chromatography where commercial tagged species, or suitable activation service is available, are good quantitation, wide dynamic range, usable with gradient elution, and low minimum detectable sample size.

The availability of calcium fluoride scintillators promises to extend this monitoring technique to many solvent systems which previously were impractical because of the solubility properties of the presently used anthracene scintillator. The solubility properties of calcium fluoride approach those of glass for most solvents commonly used in liquid chromatography.

Radioactivity monitors employing the scintillation technique normally consist of a tube or cell packed with a scintillator which has a suitable efficiency for C^{14} betas. The column effluent passes through the packed cell placing the isotope in direct contact with the scintillator. The light pulses

193

generated are detected with a photomultiplier and counted or converted to a voltage level with a ratemeter. Work to date with the calcium fluoride scintillator (primarily by the manufacturer, Harshaw Chemical) indicates a high efficiency for C^{14} betas and even some success with tritium-labeled compounds.

Fluorescence. The number of important chemical families which absorb ultraviolet radiation and re-radiate at longer, often visible, wavelengths is growing to such an extent that this highly specific measurement will probably be instrumented and packaged in a form intended for use in liquid chromatography. Workers are now using ultraviolet spectrophotometers and other general purpose instruments in utilizing fluorescence just as UV, visible, and infrared spectrometers can be used to monitor electromagnetic absorption at analytical wavelengths when equipped with flow-through type cells. In general however it is an inconvenience to commit expensive general purpose equipment to long-term duty as a monitor.

The major advantages of fluorescence monitoring are the very low minimum detectable sample size of about 10^{-9} gram and specificity. Used in conjunction with a second detector such as heat of adsorption the resulting chromatograms can yield valuable qualitative information.

It is essential that the solvents used in performing the separation be transparent to both the exciting ultraviolet energy and the fluorescence wavelength. With suitable solvents the fluorescence monitor is also usable with gradient elution chromatography. The minimum detectable limit is dependent upon the efficiency with which the wavelength conversion occurs with the individual sample but 10^{-8} to 10^{-9} gram is detectable with several fluorescent materials.

Polarography. One of the most efficient techniques for the determination of metal ion concentration is the simple two electrode polarograph in which a linearly increasing voltage ramp is applied to a suitable electrode pair which traditionally are a calomel reference and a dropping mercury electrode. The plot of current *versus* voltage indicates the amount of each metal ion present at the corresponding applied potential.

This technique has been applied to liquid chromatography monitoring by applying a fixed potential, chosen to be high enough to reduce the metal ions present, to a platinum electrode pair. Platinum electrodes are chosen because the calomel-DME pair introduce excessive dead volume to the system. Frequent regeneration of the platinum pair is necessary as the electrode surface becomes contaminated with the ions in solution. Sensitivities of 10^{-12} to 10^{-14} gram per second or 10^{-7} molar ion concentration are obtainable with polarographic detection. It is sometimes necessary to introduce a suitable electrolyte into the column effluent stream by means of a constant flow metering pump before the effluent passes into the measuring cell in order to reduce the voltage drop across the cell. With suitable cell design the three electrode polarographic cell could be employed for use with organic solvent systems in which the low concentration of electrolyte required may be included in the eluting solvent mixture as it is prepared before introduction on column.

Dielectric Constant. Physical property monitors are always attractive because they are nondestructive to the sample, provide instantaneous response, and, particularly in the case of electrical measurements, contribute very little to extra-column band broadening.

Dielectric constant or capacitance monitors respond to the polarity or, better, the polarizability of the material between the plates. The best use is made of this technique, therefore, when the polarity of sample and solvent are widely different.

The detector portion consists of a pair of parallel plates or concentric cylinders in close proximity through which the effluent is directed. An alternating potential in the range of 10 to 100 megahertz is applied to the cell connected in a standard capacitance bridge circuit.

Dielectric constant is the electrical analog of refractive index and so exhibits much the same advantages and disadvantages. As would be expected, use is not recommended with gradient elution techniques particularly if the gradient is a function of solvent polarity. The practical minimum sample size is in the one to ten milligram range. Response is linear with concentration over several decades so direct quantitation with area analysis is possible. To date the best use for this technique has been made with gel filtration chromatography.

Summary and Discussion

The use of a suitable detector to monitor liquid effluent is an important first step in reducing the time and effort required to develop and utilize liquid chromatographic procedures. In the absence of a true universal detector, however, certain tests must be performed on typical dilutions of sample in carrier so that a proper decision can be made on the choice of detector. Monitors such as refractive index, ultraviolet absorption, and electrolytic conductivity are readily evaluated for application on their laboratory counterparts. Heat of adsorption requires some experimentation to establish detection limits but is an excellent choice for general use where absolute quantitation is not essential. Flame ionization offers sensitivity, quantitative response, and freedom from baseline drift with gradient elution. This detector should be a first choice when working with low volatility samples. Special purpose detectors such as visible absorption, radioactivity, and fluorescence are used where the chemical system dictates.

Research in detectors is proceeding along predictable lines utilizing classical instrumental principles such as polarography, fluorescence, and dielectric constant. For the most part they are special purpose devices in that they utilize a specific chemical property or functional group. It is reasonable to assume that certain individual detectors will demonstrate a particular utility for a given area of liquid chromatography such as electrolytic conductivity and refractive index have shown in molecular weight separations. The need at present appears to be for a widely applicable detector which is not affected by solvent gradients. The ability to manipulate separation by varying the ionic strength or polarity of the eluting solvent during analysis is a powerful aid and adds a unique and indispensable dimension to liquid chromatography. More and better detectors will have to be built which can accommodate the gradient in a convenient and efficient manner without excessive loss in sensitivity.

Sensitivity will continue to be a paramount consideration as more classical methods and new procedures are adapted to narrow bore high speed columns which require as much as 50 to 100 times reduction in sample size.

Use of liquid chromatography as a preparative separation technique will continue to grow in importance in the preparation of pure fractions for use as reagents or for study and identification. The availability of equipment to carry out separations unattended will require that stream monitoring be employed. The wide dynamic range detectors such as refractive index and electrolytic conductivity are good choices.

The use of series multiple detectors to monitor the column effluent promises to yield a new dimension of utility to liquid chromatography by achieving quantitative and qualitative information from a single analysis.

Acknowledgment

The author thanks Dr. D. P. Raval for his helpful comments and for his suggestions regarding practical minimum sample sizes for several of the detectors discussed.

For information on manufacturers of detectors, see the "1968–69 ACS Laboratory Guide to Instruments, Equipment and Chemicals."

Ralph D. Conlon is Vice President and Director of Research at Perkin-Elmer Corporation, 2401 Ogletown Rd., Newark, Del. 19711

COMMENTARY
by Ralph H. Müller

Mr. Conlon has given cogent reasons for the revived interest in liquid chromatography and has indicated that new and improved detectors can contribute heavily to further improvements. We are inclined to believe that a breakthrough can be achieved in a variety of approaches, particularly if the immediate application to liquid chromatography is temporarily ignored. To be sure, one must keep in mind that very small samples must be used and all operating conditions must be compatible with the ultimate application. One impediment to progress has been the tendency to automate classical physical chemical techniques, largely because they are so well understood and generally accepted.

Let us consider the case of refractometry. The two main methods, that of angular deviation and the other involving the critical angle at a boundary, have been exploited in many ways and have been successfully automated. However, there are many other approaches which have not been accorded the attention which they deserve. It should be quite easy to automate the "schlieren," a striae effect, as a measure of refraction gradients. In this technique, a point source of light, after passing through the system, is brought to focus on a tiny stop which completely obscures the light. If any density gradient arises in the system, refraction will occur and some of the light will escape the stop. This method will detect refraction effects in gases of the order of ± 1 unit in the seventh decimal place. It is widely used in delineating shock waves in gases in ballistic and aerodynamic studies. The same can be done with much higher light economy by using a set of Ronchi plates. Another method which we have described involves the change in light scattering which occurs when a liquid or solution is in contact with a translucent matrix such as a small disc of filter paper or fiber glass or a ground glass surface [R. H. Müller, *Analyst*, **77**, 557 (1952)]. In general, as the refractive index of the liquid increases and approaches that of the matrix, more light is transmitted along the optic axis and less is scattered laterally. This is related to the ancient Bunsen grease spot photometer. A similar phenomenon is the Christiansen Filter in which a tube is filled with tiny glass beads. The latter are completely immersed in a liquid of almost the same refractive index, such as ethyl cinnamate. If the whole assembly is heated or cooled, the system becomes a monochromator because, with an initial beam of white light at a certain temperature, the liquid will have a refractive index exactly identical with that of the glass beads and will be transmitted freely with no scattering. This condition holds only for a given wavelength and we thus have a temperature controlled monochromator.

A great advance in instrumentation

was made when it was discovered that delicate galvanometers and bulky thermostats are difficult to operate in fighter planes or bombers. Unless one wishes to study a chemical system over a wide range of temperatures such as reaction velocities, we believe it is fair to say that temperature control in an instrument should be avoided if it is at all possible to provide automatic compensation for temperature changes. Thermistor networks are easy to design for electrolytic conductance measurements even though the temperature coefficient is quite high. The same is true for refractive index measurements. The cell first described by Harry Svenson for differential refractometry utilizes a tiny rectangular chamber with a diagonal glass partition. The unknown solution circulates through one portion of the chamber and the other holds a reference liquid or solution. The double refraction, in opposite directions, leads to a lateral shift in the light beam parallel to the optic axis. Because the cells are in intimate contact, they have essentially the same temperature and disturbance from this source as second or third order effects.

Some new thinking on conductance measurements is badly needed. Physical chemists still think of the Wheatstone bridge as it was used by Kohlrausch; two resistors in the ratio arms and a variable resistor in the arm adjacent to the conductance cell. At balance, the reciprocal of this resistor is calculated in order to get the desired conductance. They rarely think of using the resistor in the arm diagonally across from the conductance, the value of which at balance gives the conductance directly. Of course, the scale is not linear, but that is the nature of the beast. Under appropriate conditions of measurement, the use of d.c. can furnish conductance measurements in no wise inferior to a.c. measurements. It would be profitable to study brief pulses of d.c., possibly with periodic reversal of polarity, to minimize small polarization effects.

It is conceivable that direct automatic weighing of effluent droplets could be worked out. They would certainly be of general applicability. The Cahn Electrobalance could be used or a light cantilever beam with strain gauge or photoelectric readout. At least we are aware of the pitfalls to be encountered. For freely falling drops, the density as well as surface tension effects have to be considered. The latter are small for solutions of electrolytes but can be high for many organic systems. Most likely, it would be desirable to eject small, precisely defined volumes of effluent into a tiny, nonwettable weighing pan. After a second or two for recording the weight, the contents of the pan would be dumped into a receiver and become ready for the next sample.

There is also the question of whether it would be practical to receive effluent droplets in a light liquid immiscible with the sample droplet and measure its rate of fall through the liquid photoelectrically. In some systems, the sample could be recoverable, but there are obvious cases in which it would not. One may also ask if the famous Kelvin Electric Pail effect holds any promise. As he showed, droplets emerging from an orifice are electrostatically charged and by collecting them in a Faraday cup relatively enormous potentials can be built up. Presumably, such drops could be electrostatically deflected and preferentially so.

Mr. Conlon has done our readers a great service in outlining the reasons for renewed interest in liquid chromatography and in discussing the many possible means, many of them already commercially available, for detecting the various fractions. Our few observations and hunches may contribute little to the main problem but we think it must be realized that in a sense we have run out of phenomena awaiting automation. With present day instrumental techniques, some unusual and obscure effects may be worthy of study.

Gas Chromatography Detectors

C. HAROLD HARTMANN

Varian Aerograph/Varian Techtron,
2700 Mitchell Dr.,
Walnut Creek, Calif. 94598

A knowledge of the latest detector performance specifications and operating characteristics will aid the researcher in obtaining optimum results from a gas chromatography system

THIS DISCUSSION covers the most popular detectors that have been directly coupled to a gas chromatograph. Although several dozen different detectors have been suggested for use with gas chromatography, only 23 will be covered here. Six of the most important of these will be treated in detail (how they operate, performance specifications and operating characteristics, but no theory of operation). The significant performance features of the remaining 17 detectors will be summarized in a table.

This paper is directed not only to the analyst who is unfamiliar with the spectrum of fully developed detectors but also to the specialist who is interested in interpretations of the latest detector specifications. For the six detectors treated in detail, a number of operational suggestions will be given to assist the less experienced chromatographer in achieving optimum performance. However, the completeness of the information included here is by no means intended to approach that of an instruction manual.

One area of confusion surrounding gc detectors is that of terminology. The latest American Society for Testing and Materials (ASTM) proposals for definitions of detector characteristics will be given in simplified terms. Another area of confusion for those having less familiarity with gc detectors is that of performance expectations for a particular detector. Realistic, practical performance guidelines will be given for all 23 detectors. Detector specifications, of necessity, must relate to detector characteristics, so that meaningful comparisons can be made. However, specifications with units such as ampere-milliliters per gram do not directly assist the chemist who wants to know if a 1-ppm DDT residue in his alfalfa hay extract can be reliably quantitated.

Therefore, additional detectability guidelines will be given relating to the concentrations or amounts of sample that can be quantitated by the various detectors.

General Detector Characteristics

Performance Definitions. Before an adequate description can be given of any detector's performance, a brief discussion is needed to consider definitions of general detector characteristics. Although the definitions given here are by no means official or final, they do reflect the current, albeit incomplete, efforts of two task groups (of which the author has been a participating member) of the ASTM E-19 Committee on Chromatography. These definitions are simplified statements of much more definitive expressions being considered by the ASTM subcommittees.

Detector Type. Gc detectors can be classified by their mechanism of operation as concentration dependent, mass flow rate dependent, or a combination of both. These characteristics refer to the performance dependency of the detector on volumetric changes in the carrier gas flow rate. A concentration-dependent detector has an area response inversely proportional to the volume of carrier gas eluted with the sample. Theoretically, a mass flow rate detector will yield a peak area independent of the volume of carrier gas eluted with the sample. In practice, the carrier flow rate cannot be changed by more than about 25% without re-optimizing the detector. A thorough comprehension of the differences between concentration and mass flow rate detectors is required before optimum performance of any gc system can be assured. Halasz (*1*) has made a noteworthy contribution on this subject.

Sensitivity. This term has had broad and loose usage over the years, with a variety of definitions. "Sensitivity" is gradually being accepted as a measure of the effectiveness of the detector as a transducer in converting the sample into a measurable electrical signal. For concentration-dependent detectors, the sensitivity is determined by dividing the product of peak area and flow rate by the weight of the sample. For mass flow rate-dependent detectors, the sensitivity is found by dividing just peak area by the weight of the sample.

Noise. The so-called noise level of a detector is probably the most difficult parameter to define and quantitate. Detector noise most commonly refers to the random, relatively fast term peak-to-peak signal perturbation. In contrast to this noise, which has a period on the order of one second, other types of detector noise can have periods on the order of minutes. Noise is quantitated by the average peak-to-peak amplitude of the signal and expressed in electrical units. Characterization of noise is a science in itself and will not be expanded further here. The significance of knowing a detector's noise level is that it is the critical factor in determining the minimum detectable quantity of sample and the lower limit of the linear range.

Detectability. As the name implies, this term is related to the smallest detectable sample that can be read above the noise level. It has been almost universally, yet arbitrarily, decided that the smallest readable response above noise should be a factor of two. Although a gc peak of such minuscule proportions would hardly be considered a reliable response for good quantitative accuracy, it really does represent a reasonable standard for detector comparison. The detectability is calculated by dividing the sensitivity into twice the noise level. Although the detector's sensitivity and detectability specifications are meaningful to those interested in comparing detector performance, the analyst usually has difficulty relating these expressions to his practical problems of detection. Therefore, there will be given for each of the detectors discussed, a practical guideline of detectability for a component in the sample expressed in either of two ways:

weight or concentration. For a liquid sample these terms can be converted by using the following close approximations: 1 nanogram/microliter = 1 ppm, 1 picogram/microliter = 1 ppb. Under most circumstances the maximum liquid sample capacity for an analytical column (1–2-mm id) is about 10 μl. Similarly, gas samples can be converted by using the rough approximation of 1 ng/ml = 1 ppm and 1 pg/ml = 1 ppb. An advisable maximum sample volume for gases is about 5 ml. Another simplification used in these guidelines is that the retention of the component is on the order of 5 min and the efficiency of a column is not less than 250 plates/ft.

Specificity. Detectors can be classified as either universal or specific. A universal detector responds to all substances passing through it. A specific

detector responds primarily to a select group of substances or to groups of substances with a minimal response to all other interference substances. The specificity factor of a detector is the ratio of the detectability of a potentially interfering substance to the detectability of a desired substance. Specificity factors of 10,000 to 1 are considered good.

Linearity. Poor understanding of this familiar term has generated considerable confusion in the past. One of the best definitions of linearity is the range of sample concentration over which the detector maintains a constant sensitivity within a certain arbitrary deviation. A detector's linearity is commonly demonstrated on a log-log graph. After the peak height data are plotted over a wide range of sample concentrations, the curve of the best fit

Specifications. (For these TCD operating conditions: 250 mA; 220°C; 20 ml/ min He)

$$\text{Sensitivity} = S = \frac{\text{peak area (mV} \cdot \text{min)} \cdot \text{carrier flow (ml/min)}}{\text{sample wt, mg}} = 10{,}000 \text{ mV} \cdot \text{ml/mg}$$

Noise = N .... = 0.01 mV
Detectability = 2N/S .... = 2 × 10⁻⁶ mg/ml
Linear range .... = 10⁵
Background .... Not applicable

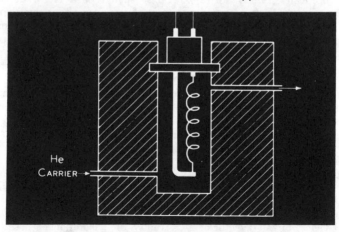

Figure 1.   Thermal conductivity detector

of the data points is obtained. Linearity is said to exist when the slope of this curve is unity (not necessarily 45°). This method of representing the linearity or calibration of a detector has one particular weakness. Because of the log scale, relatively large deviations from the curve appear insignificant. Another method which is being readily accepted as superior to the log-log plot is a semilog plot of sensitivity *vs.* log of concentration (plotting the same data as above will yield a slope equal to zero when linearity exists). The advantages here are that changes in slope are more easily detected with equal ease throughout the entire range of sample concentration.

There are two limits to the linearity curve. The lower concentration limit is set by the limit of detection and the upper limit is set by an arbitrary percentage deviation from the linearity curve, usually about 5%. The practical significance of a detector's linearity is that it alerts the analyst to concentrations for which he can be confident of a constant detector calibration factor. A detector with a small range of linearity can have a high quantitative reliability if precautions are taken by the analyst. A linearity expression which is beginning to fall out of favor is linear dynamic range (LDR). There is no more meaning to this term than to the less complex "linearity."

Response Time. As detector specifications become honed to a fine point, there will be more emphasis placed on suppressing the detector noise—*i.e.*, with electronic filters commensurate with high fidelity of response for rapidly eluting peaks. Little effort has been expended to date on measuring detector response time. It is a factor to be aware of even though few numbers are yet available. The definition now being considered for response time is the time required for the output signal to reach 63% of the new equilibrium value when the composition of the gas entering the detector is changed

in a stepwise manner. With typical response times on the order of fractions of a second, one problem with such measurement is that the standard strip chart recorder may be an inadequate measuring device.

Background Current. The background current or "dark" current is the constant output signal generated by the ionization detector when it is fully operational but has no sample passing through it. This parameter is a valuable diagnostic aid in recognizing gc system malfunctions. A highly recommended analytical technique is to monitor the background current on a periodic basis for shifts either up or down from expected levels.

Thermal Conductivity Detector (TCD)

Description. It is estimated that there are more TC detectors in service today than all others combined. This popularity is due in part to its high reliability and basic simplicity. Figure 1 shows a cross-section schematic of one TC cavity (of which there are usually four per TC detector). Although a two-filament detector is functional, four-filament detectors are more popular because of their factor of two increase in detectability. The filament usually consists of a 1-ml tungsten-rhenium helical wire mounted in a high mass block which helps provide a stable thermal reference for the filaments.

Physical Basis of Operation. The filaments are heated to up to 400°C with direct current. Two filaments are used as a reference and exposed to pure He or H_2 carrier gas (highest conductivities). Two filaments are used as sensors and exposed to the sample *via* the column effluent. These four resistances are connected electrically as elements of a Wheatstone bridge. Detection occurs when the bridge is imbalanced by an increase in resistance (increase in temperature) of the sample exposed filaments. The low thermal conductivity of the sample vapor as compared to the carrier gas causes the filament temperature increase. Heat loss of the filaments is due not only to conductivity to the gas but also to heat convection to the cavity walls. Minor heat losses arise from radiation and heat leakage through the electrical leads. Lawson and Miller (2) have published a very fine survey article on thermal conductivity detectors for those interested in studying the subject further.

Operating Characteristics.

• Although the sensitivity increases by the cube of the filament current, the minimum detectable quantity (MDQ) increases only by the square, since the noise also increases rapidly with increased filament current.

• The best linearity at concentrations above 10% is obtained with a constant current bridge supply.
• Oxidation of the tungsten filaments is the TC detector's most vulnerable characteristic. Oxygen at ppm levels in the carrier gas from diffusion-type leaks in the injector septum, column fittings, and filament seals can cause continuous unidirectional drift in the baseline as the filaments oxidize. Consequently, the practice of injecting air with the sample is not recommended.
• Recent advances in coating the filament with a protective gold layer have proved helpful in minimizing filament oxidation.
• The TCD is a concentration-dependent detector and, therefore, it is very important to have a constant

Specifications. (For these FID operating conditions: SE-30 @ 200°C; 20 ml/min H_2; 20 ml/min N_2; 300 ml/min air; 300 volts polarization voltage)

$$\text{Sensitivity} = S = \frac{\text{peak area (amps} \cdot \text{sec)}}{\text{sample wt, gm}} = 0.01 \ \text{coulombs/gm}$$

Noise $= N \ldots \ldots \ldots \ldots \ldots \ldots = 10^{-14}$ amps
Detectability $= D = 2N/S \ldots \ldots = 2 \times 10^{-12}$ gm/sec
Linear range $\ldots \ldots \ldots \ldots \ldots \ldots = 10^7$
Background $\ldots \ldots \ldots \ldots \ldots \ldots = 4 \times 10^{-11}$ amps

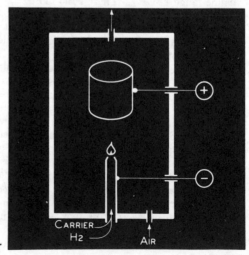

Figure 2. Flame ionization detector

flow rate when attempting quantitative analysis.

• It is a universal detector responding to all gases and vapors.

• One distinctive feature of the TCD is that it is close to ideal for preparative applications, since it is generally nondestructive to the sample and relatively unaffected by high concentrations passing through it.

Limitations.

• Every substance has a unique thermal conductivity; therefore, accurate quantitation usually requires individual calibration factors for all the substances of interest.

• The TCD has only moderate sample detectability under good gc conditions (stable column liquid phase, stable detector temperature, fast eluting peaks, oxygen-free helium) and can be expected to detect 10.0 ng of sample weight or about 5 ppm of sample gas concentration.

Flame Ionization Detector (FID)

Description. Although the FID lagged four years behind the TCD as an accepted gc detector, the number of FID's going into service at present is greater than the number of TCD's. The FID's popularity is primarily owing to its detection of small sample concentrations (about 1000 times lower than the TCD). Of all the ionization detectors, the FID has the best record of reliable performance. Figure 2 shows a schematic representation of the

Specifications. (For these ECD operating conditions: 35 ml/min N_2 flow; aldrin sample; 0.2 min peak width)

	³H	⁶³Ni
Sensitivity $= S = \dfrac{\text{peak area (amps} \cdot \text{min)} \cdot \text{flow rate (ml/min)}}{\text{sample wt, gm}}$ ...	$= 800$ amps ml/gm	40 amps ml/gm
Noise $= N$	$= 8 \times 10^{-12}$ amps	2×10^{-12} amps
Detectability $= 2N/S$	$= 2 \times 10^{-14}$ gm/ml	10^{-13} gm/ml
Linear range	$= 500 \ (10^4)^a$	$50 \ (10^4)^a$
Standing current	$= 3 \times 10^{-8}$ amps	2×10^{-9} amps
Specificity factor	$= 10^7$	10^7

^{a} Obtainable with electronic linearizing device (4)

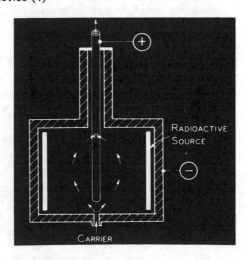

Figure 3.　Electron capture detector

essential components of an FID: burner jet; collector electrode (shown here as cylindrical); and appropriate plumbing for column effluent, hydrogen, and air.

Physical Basis of Operation. An oxidative hydrogen flame (excess of oxygen) burns organic molecules producing ionized molecular fragments. The resulting ions are collected by means of an electrical field. Usually it is the negative ions that are collected at the electrometer electrode; however, positive ions have also been collected in some designs.

Operating Characteristics.

• The FID sensitivity is nearly uniform to all pure organics composed only of carbon and hydrogen.
• With optimized operating parameters and good gc conditions, sample detection is about 20 pg or about 5 ppb of sample gas concentration.
• Although the FID is nearly a universal detector, there are a number of gases which give little to no signal —*i.e.*, all the fixed gases, oxides of nitrogen, H_2S, SO_2, COS, CS_2, CO, CO_2, H_2O, NH_3, HCOOH. The FID's insensitivity to CS_2 and H_2O makes these substances advantageous as solvents, minimizing the characteristic solvent "tail."
• The FID is a mass flow rate detector; however, if the column flow is changed more than about 25%, the hydrogen flow will require adjustment to maintain constant sensitivity.
• By increasing the orifice of the burner jet, air flow rate, and polarizing voltage, and by reoptimizing, the linearity can be extended to about 5 mg per peak for 5-min retention components. This is achieved, however, at a sacrifice of detectability of small sample concentrations.
• The following factors adversely affect the linearity of nonpolarized

flame jets: operation at temperatures above 250°C; contamination on insulator surfaces; concentrations greater than 0.1%. This is owing to partial grounding of the ion current through the flame jet. A more complete discussion of FID characteristics is reported by Sternberg *et al.* (3).

Limitations.

• It is obviously a destructive detector to the sample; therefore, a sample splitter must be used for preparative work.
• Atoms of oxygen, nitrogen, phosphorus, sulfur, or halogen in the structure of the organics cause significant decreases in sensitivity, depending on the degree of substitution.
• High detectability of organics requires special precautions—*i.e.*, the septum must be kept as cool as possible, particularly when the column is cooled to ambient temperatures and then programmed.

Electron Capture Detector (ECD)

Description. Although the ECD has many performance drawbacks which will be discussed below, it has gained broad acceptance in certain applications —principally because it can detect extremely small amounts of certain substances. Figure 3 is a schematic representation of one of several possible geometries for an electron capture detector. Essential components include a radioactive source (usually 250 mCi ^{3}H or 10 mCl ^{63}Ni), and an anode and a cathode.

Physical Basis of Operation. The carrier gas (either N_2 or Ar plus 10% methane quench gas) is ionized by the radioactive source to form an electron flow in the detector cavity on the order of 10^{-8} amps. Ionization background current is usually undesirable, but background current for the ECD is a primary necessity and is commonly referred to as standing current. The electrons forming the standing current are collected with a relatively weak field strength of either dc voltage or pulsed voltage. Under certain conditions pulsed voltage is somewhat more advantageous because of reduced space charge effects. Those substances which have an affinity for free electrons deplete the standing current as they pass through the detector cavity. One of the complexities of the ECD is that the magnitude of current depletion is not only a measure of the amount of capturing species in the detector, but also a measure of the electron affinity of the species. This means that every substance requiries individual calibration.

Operating Characteristics.

• The ECD is a specific detector sensitive primarily to halogenated hydrocarbons and certain other classes of organics such as conjugated carbonyls and nitrates which have the ability to accept a negative charge.

• 0.1 pg of high electron affinity substances such as lindane, aldrin, and SF_6 can be detected.
• It is a concentration-dependent detector.
• Since O_2 can be detected at 1 ppm, chromatographs unsusceptible to air diffusion are recommended. Typical trouble spots are injector septum and tightness of detector construction.
• Hot septa are notorious for eluting relatively large quantities of contaminants which interfere considerably with detector performance. A highly recommended analytical technique is to heat the injector no hotter than necessary (column temperature) and use on-column injection rather than flash vaporization.

Limitations.

• Because the ECD is sensitive to a variety of classes of compounds, it does not make an ideal specific detector. Although the ECD has had broad acceptance in pesticide resi-

Specifications. (For these HeD conditions: 60 ml/min flow; methane sample; 4000 volts/cm)

$$\text{Sensitivity} = S = \frac{\text{peak area (amps} \cdot \text{sec)}}{\text{sample wt, gm}} \dots = 100 \text{ coulombs/gm}$$

Noise $= N$ .... $= 2 \times 10^{-19}$ amps
Detectability $= 2N/S$ .... $= 4 \times 10^{-14}$ gm/sec
Linear range .... $= 5 \times 10^{3}$
Background current .... $= 3 \times 10^{-8}$ amps

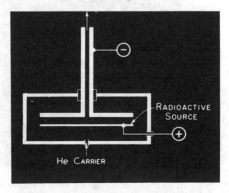

Figure 4. Helium ionization detector

due analysis, results are never completely certain—*i.e.*, a naturally occurring organic can be confused with the response of a pesticide residue.

• At best, the calibration curve of the ECD approaches linearity for small ranges of concentration. Much care is required to obtain accurate quantitation.

• Partial deactivation of the radioactive surface by condensation of high boiling compounds results in gradual deterioration of performance with time (particularly with the ^{3}H sources which are temperature-limited to 220°C by the U.S. Atomic Energy Commission).

• The ^{63}Ni ECD minimizes the contamination phenomenon but has about five times less detectability and a 10-fold narrower linear range than the ^{3}H ECD.

Helium Detector (HeD)

Description. Although the use of the HeD today is relatively limited, this detector does have a unique characteristic which makes it worthy of discussion here. It is the only gc detector that is capable of low ppb detection of all the fixed gases. Figure 4 is a schematic representation of the helium detector showing two electrodes as essential components of the detector. The anode is a thin square of radioactive

Specifications. (For these AFID conditions: 37 ml/min H_2; 220 ml/min air; 20 ml/min N_2; OV-17 @ 200°C)

$$\text{Sensitivity} = S = \frac{\text{peak area (amps} \cdot \text{sec)}}{\text{sample wt, gm}} = 20 \text{ coulombs/gm}$$
(Parathion)

Noise = N .... = 3×10^{-14} amps

Detectability = $D = 2N/S$
(as atomic phosphorus) ...... = 3×10^{-15} grams/sec

Linear range .... = 10^8

Background .... = 3×10^{-11} amps

Specificity factor .... = 15,000

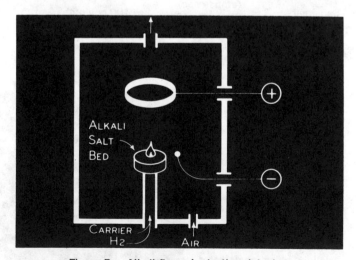

Figure 5. Alkali flame ionization detector

Specifications. (For these FPD conditions: 80 ml/min N_2; 170 ml/min H_2; 20 ml/min O_2; ml/min air)

Sensitivity

$$\text{Sulfur} = S = \frac{\text{peak area (amps} \cdot \text{sec)}}{\text{sample wt, gm}} \ldots = 400 \text{ coulombs/gm}$$

$$\text{Phosphorus} = S = \frac{\text{peak area (amps} \cdot \text{sec)}}{\text{sample wt, gm}} = 470 \text{ coulombs/gm}$$

Noise level = N . . . . = 4×10^{-10} amps
Detectability
    Sulfur = 2N/S . . . . = 2×10^{-12} gm/sec
    Phosphorus = 2N/S . . . . = 1.7×10^{-12} gm/sec
Linear range . . . . = None
Background . . . . = 10^{-7} amps
Specificity factors
    Sulfur mode—hydrocarbons . . . = 10,000
                          phosphorus . . . . = 100–1000
    Phosphorus mode—
        hydrocarbons . . . . = 10,000
        sulfur . . . . = 4

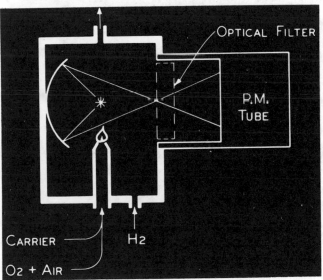

Figure 6. Flame photometric detector

foil (250 mCi ^{3}H) which is also used to collect the ion current for measurement. The cathode also serves two purposes: to carry the high polarizing voltage and to prevent air from diffusing into the ionization chamber with its capillary exit tube. Also important, but not shown, are the special gas passages which must be meticulously gas-tight.

Physical Basis of Operation. The combination of the ^{3}H decay and the high field gradient raise the He gas flowing between the closely spaced electrodes to the metastable state (He*). The excitation energy of He* is greater than the ionization potential of all other gases and vapors except neon; therefore, all samples are ionized. The field gradient between electrodes is typically 4000 volts/cm; however, greater detectability is achieved at even higher gradients. All fixed gases have different sensitivities (5). The limiting factor in detectability of the helium detector is not so much the detector capability itself but rather the purity of the He entering the ionization chamber (contamination-free gc system).

Operating Characteristics.

• Negative peaks are common when ppb determinations are attempted owing to the vacancy chromatography effect (sample component concentration less than the concentration of that same constituent in the carrier gas). Quantitation under these circumstances is dubious (6).
• The HeD has a combination characteristic of mass flow dependency and concentration dependency.
• The HeD has 50 times the detectability of methane of the FID (sub-ppb detectability).
• It is a universal detector.

Limitations.

• All sample gases except helium yield a "solvent" response which tends to interfere with peaks eluting later.
• Separations are limited to gas solid chromatography (molecular sieve, Porapak, silica gel, activated alumina) because any gas liquid column quenches the He* before it can be used to ionize sample molecules.
• The calibration system is most often the limiting factor in accurate quantitation of low ppb concentration. The exponential dilution flask is about the best method.

Alkali Flame Ionization Detector (AFID)

Description. The development of the AFID has been by far the most interesting from a phenomenological viewpoint. Although the AFID has a few noteworthy disadvantages, it has

Table I. Performance

Detector	Detectability, sample weight
Thermistor (11)	nanogram
Gas density balance (12)	high nanogram
Cross section (13)	high nanogram
Argon (14)	picogram
Microwave emission (15)	subnanogram
Photoionization (16)	picogram
Nonradioactive helium ionization (17)	subnanogram
Radio frequency (18)	nanogram
Radio gc (19)	subnano-Curie
Ultrasonic (20)	nanogram
Microcoulometric (21)	nanogram
Electrolytic conductivity (22)	nanogram
Reaction coulometric (23)	high nanogram
Infrared spectrometer (24)	high microgram
Mass spectrometer (25, 26)	high nanogram
Far ultraviolet spectometer (27)	high nanogram
Spectrophotofluorometric (28)	nanogram

received considerable attention from those analysts who are interested in a specific detector with picogram detectability for phosphorus-containing compounds. Figure 5 is a schematic representation of one of several possible AFID configurations. An amazing variety of geometrical arrangements have been used for this detector. They all incorporate three essential components: a collector electrode (shown here as a simple loop); an electrode to

Characteristics of Other Gas Chromatography Detectors

Universal (U), specfic (S)	Species detected	Carrier recommended	Complexity factor[a]	Linearity (some estimated)	Other comments
U		He, H$_2$	1	50 (est)	Best detectability at low temperature Very small detector volume
U	CO$_2$, Ar, He, H$_2$		3	10^5	Can be used for determining molecular weight of sample
U		H$_2$	2	10^5	Very dependable, can predict sensitivities
U		Ar	3	10^5	Uses unpopular Sr^{90} source No sensitivity to methane
S	halides, phosphorous	85% He, 15% Ar	9	10^3 (est)	Low detector pressures recommended Good specificity to all compounds
U		Ar, H$_2$, N$_2$	8	10^5	Application: fixed gas analysis
U		He	4	10^4 (est)	Application: fixed gas analysis
U		He, Ar	4	10^3	Application: fixed gas analysis
S	^{3}H, ^{14}C	N$_2$, He, Ar	8	10^4	Very high specificity factor to non-radioactive species
U		He, H$_2$, Ar, N$_2$, O$_2$, Air	6	10^6	Best for fixed gas analysis Use of any carrier eliminates solvent
S	halides, sulfur, nitrogen	N$_2$, He, Ar	6	10^4 (est)	High specificity factors to all interferences
S	halides, sulfur, nitrogen	N$_2$, He, Ar	5	10^3 (est)	High specificity factors to all interferences
U	organics	N$_2$, He, Ar	7	10^4	Can be used to detect O or H in sample molecule Self calibrating
S	ir absorber	He, N$_2$	10	none	Special interfacing to gc required
U		He, H$_2$	10	10^6	Special interfacing to gc required
S	uv absorber	He	9	None	Application: low molecular weight organics Special interfacing to gc required
S	fluorescein compounds	N$_2$	8	10^3	Special interfacing problems

[a] Author's arbitrary rating on scale from 0–10 (higher numbers-more complexity). Complexity factor considers number of parameters to be adjusted and the amount of accessory equipment needed.

which a polarizing voltage is applied; and a supply of alkali salt, CsBr or Rb_2SO_4 being the most commonly used. It is in the salt placement that the variety ensues. The salt has been coated on wires and screens or pressed into pellets or cups which have been placed in the flame, above the flame, or in the collector electrode. Alternatively, it may be placed in platinum diffusion capillaries near the electrodes. Apparently the alkali salt need only be positioned in the general vicinity of the detector chamber to obtain some sensitivity to organophosphates or other particular species.

Physical Basis of Operation. Although the AFID's theoretical basis of operation is uncertain at this time, there are a few statements that can be made about its physical basis of operation. A controlled vaporization of an alkali salt is accomplished in a hydrogen-rich flame. Certain particular molecules or their fragments formed in the burning process are exposed to this vapor. Unique ion species are formed which are then collected at the electrode. Scolnick (7) has presented evidence which suggests that a major contribution to ion formation is chemiionization. Alkanes do not readily produce collectable ion species, so they remain undetected by the AFID.

Operating Characteristics.

• The stability of the hydrogen flow rate is critical to baseline stability. Flow perturbations of 0.01 ml/min are detectable.
• The AFID is a specific detector primarily for phosphorus although halogens, sulfur, and nitrogen organics have also been detected by some designs.
• The stability of air flow is about 1/10 as critical as that of hydrogen flow.
• The AFID is a mass flow rate-dependent detector.
• Subpicogram quantities of phosphorus compounds are detectable.

Limitations.

• The biggest disadvantage in using an AFID is that rate of vaporization of the alkali salt cannot be controlled easily over long periods of time. This results in changing sensitivities, drifting baseline, and the need to calibrate frequently. Early designs were usable for only a few hours, whereas recent designs are usable for months (8).
• Large amounts of solvent (over 1 μl) usually cause serious disturbances to the detector which require minutes to hours for recovery, depending on the design.
• Although the AFID has a good specificity factor for phosphorus of about 10,000 relative to alkanes, other types of compounds can also be detected, such as halogens and nitrogen-containing organics. The sensitivity difference between halogens, nitrogen, and sulfur-containing organics compared to organophosphates is about 1000-fold.

Flame Photometric Detector (FPD)

Description. The FPD has received considerable interest in the few years since its introduction in 1966 (9). The FPD is the first inexpensive spectroscopic detector for sulfur- and phosphorus-containing compounds to be coupled with a gc, which is the primary reason for its popularity. Despite its potential as a specific detector, there are special considerations which require attention by the discerning analyst. Figure 6 shows a schematic representation of the essential components: burner jet; reflective mirror; narrow band pass filter (394 nm for S, 526 nm for P); and photomultiplier tube.

Physical Basis of Operation. When substances containing phosphorus or sulfur are burned in the hydrogen-rich flame, they emit characteristic light at 526 nm and 394 nm, respectively. The advantageous place to detect this light is directly above the flame. The light from this area is collected with a mirror, passed through the appropriate filter, and monitored by a photomultiplier tube. Typically —750 volts direct current is supplied to the photomultiplier tube, the output of which is fed to an electrometer.

Operating Characteristics.

- The FPD is a mass flow rate-dependent detector.
- It is a specific detector with pertinent limitations in specificity.
- It has low ppm detectability for pesticide residue analysis; but ppb detectability when specially tuned, such as for sulfur gas in air pollution analysis.

Limitations.

- The FPD exhibits a peculiar calibration curve. Every compound yields a unique calibration—*i.e.*, it has no linear relationship at all (only a straight line on a log-log plot with a slope of 1.5 to 2.0 which is constant only over a range of about 500-fold).
- Specificity to hydrocarbons is good; but with the sulfur filter, phosphorus is also detected—*i.e.*, with about a 100-fold lower sensitivity than for sulfur. With the phosphorus filter, sulfur is detected with a sensitivity only 4-fold lower than for phosphorus. A special laborious technique of response ratioing can minimize this ambiguity (*10*).
- The hydrogen-rich flame is not very stable and the solvent usually extinguishes the flame, requiring the operator to proceed through a simple but bothersome reignition procedure.

Other Detectors

Table I lists 17 additional detectors which have been used successfully with a gas chromatograph. The most significant performance characteristics of each have been summarized therein. Generally these detectors are not being placed into service at the rate of the previous six for one or more of the following reasons: high complexity; high cost; poor detectability; non-unique performance; or poor reliability.

For those interested in additional information, at least one key reference is given for each of the detectors.

Future Developments

In conclusion, this article has presented, in current terminology, the state of gc detectors today. Indications are that future development will lie in the area of spectroscopic detectors for higher specificity and qualitative analysis. However, better interfacing, better performance specifications, and simpler operation will be required before these detectors become an accepted reality. Also expected in the future will be gradual improvements in the more popular detectors such as the TCD, FID, and ECD. For optimum performance with any detector, attention must be paid to the entire system, including flow and temperature control and removal or minimization of contaminant sources.

References

(1) I. Halasz, ANAL. CHEM. **36**, 1429 (1964).
(2) E. Lawson and J. Miller, *J. Gas Chromatogr.* **4**, 273 (1966).
(3) J. C. Sternberg, W. S. Gallaway, and D. T. L. Jones, "Gas Chromatography, 3rd International Symposium," 1961. N. Brenner, Ed., Academic Press, New York, 1962, p 231.
(4) D. C. Fenimore and C. M. Davis, *J. Chromatogr. Sci.* **8**, 519 (1970).
(5) C. H. Hartmann and K. P. Dimick, *J. Gas Chromatogr.* **4** (5), 163 (1966).
(6) B. O. Prescott and H. L. Wise, *J. Gas Chromatogr.* **4** (2), 80 (1966).
(7) M. Scolnick, 6th International Symposium—Advances in Gas Chromatography, Miami, Florida, June, 1970.
(8) C. H. Hartmann, Aerograph Tech. Bull. **136-69**, Varian Aerograph, Walnut Creek, Calif., 1969.
(9) S. S. Brody and S. E. Chaney, *J. Gas Chromatogr.* **4** (2), 42 (1966).
10. H. W. Grice, M. L. Yates, and D. J.

David, *J. Chromatogr. Sci.* **8**, 90 (1970).

(11) D. Buhl, ANAL. CHEM. **40**, 715 (1968).

(12) J. T. Walsh, *J. Gas Chromatogr.* **5** (5), 233 (1967).

(13) J. E. Lovelock, G. R. Shoemake, and A. Zlatkis, ANAL. CHEM. **36**, 1410 (1964).

(14) J. E. Lovelock, "Gas Chromatography," R. P. W. Scott, Ed., Butterworths, London, 1960, p 16.

(15) H. A. Moye, ANAL. CHEM. **38**, 1441 (1967).

(16) J. G. W. Price, D. C. Fenimore, P. G. Simmonds, and A. Zlatkis, ANAL. CHEM. **3**, 541 (1968).

(17) R. Villalobos, *ISA Trans.* **7**, 38 (1968).

(18) H. P. Williams and J. D. Winefordner, *J. Gas Chromatogr.* **6**, 11 (1968).

(19) J. R. Penton and C. H. Hartmann, submitted for publication in *J. Chromatogr.*

(20) H. W. Grice and D. J. David, *J. Gas Chromatogr.* **7**, 239 (1969).

(21) D. M. Coulson, *Amer. Lab.* 1969, (May), p 22.

(22) D. M. Coulson, *J. Gas Chromatogr.* **3** (4), 134 (1965).

(23) A. B. Littlewood and W. A. Wiseman, *J. Gas Chromatogr.* **5** (7), 334 (1967).

(24) R. P. W. Scott, "Gas Chromatography 1966," A. B. Littlewood, Ed., Institute of Petroleum, London, 1967, p 318.

(25) J. T. Watson, "Ancillary Techniques of Gas Chromatography," L. S. Ettre and W. H. McFadden, Eds., Wiley, New York, 1969, p 745.

(26) C. Merritt, *Appl. Spectrosc. Rev.* **3** (2), 263 (1970).

(27) W. Kaye, ANAL. CHEM. **34**, 287 (1962).

(28) M. C. Bowan and M. Beroza, ANAL. CHEM. **40**, 535 (1968).

Countercurrent Chromatography

Development of support-free liquid-liquid partition techniques holds promise for preparative and analytical chromatographic separation of compounds without complications arising from solid supports

YOICHIRO ITO and ROBERT L. BOWMAN
Laboratory of Technical Development
National Heart and Lung Institute
Bethesda, Md. 20014

W HEN two immiscible solvents containing solutes are shaken in a separatory funnel and then separated, the solutes are partitioned between the two phases. The ratio of solute concentration in the upper phase to the lower is called the partition coefficient. If the partition coefficients of two substances differ greatly, the only process needed for separation is a one-step operation called extraction. As the nature of the substances becomes more similar, the difference of their partition coefficient decreases, hence requiring multistep extraction for separation. This technique, called the countercurrent distribution method (1), can be performed with a Craig apparatus. Since the method is time-consuming and tends to dilute the sample, it is best suited for large preparative separation and is usually performed with no more than a few hundred glass partition tubes or "plates."

On the other hand, high-efficiency microscale partition techniques, termed partition chromatography, involve a continuous partition process between the moving and the stationary phases. The variety of methods developed elsewhere so far have employed solid supports (cellulose, silica, alumina, glass) to hold one phase stationary. The granular and porous nature of the solid support provides an enormous surface area in relation to the liquid volume and divides the free space into thousands of plates. In each plate, the partition process is theoretically completed, thus yielding an efficiency as high as thousands of theoretical plates. [Although plate concept in chromatography is not equivalent to that in the countercurrent distribution method, analogy may help to compare the efficiency between these methods (2)]. However, the affinity of the solid supports for the solute can add an undesirable adsorption effect evidenced by tailing of the elution curves of the solutes. When one deals with a minute amount of biological components, ad-

213

sorption can result in a significant loss or denaturation of the sample in addition to contamination by foreign materials eluted from the support.

General Features of Countercurrent Chromatography

We have called our method countercurrent chromatography to distinguish it from the countercurrent distribution method or conventional liquid partition chromatography. Our system is similar to the countercurrent distribution method in that the two immiscible phases pass through each other in a tubular space. However, it involves a continuous nonequilibrium partition process comparable to chromatography. It was developed to achieve a high-efficiency chromatographic separation on both a preparative and analytical scale in the absence of solid supports. However, elimination of the solid support creates a number of problems as listed below:

(1) How to keep the stationary phase in the column as the moving phase is steadily eluted

(2) How to divide the column space into numerous partition units and reduce laminar flow spreading of sample bands

(3) How to increase interfacial area

(4) How to mix each phase to reduce mass transfer resistance.

Several arrangements have been developed in our laboratory, each with a specific or potential advantage. In each system, a tubular column is made to form multiple traps to hold the stationary phase in a segmented pattern while a gravitational or centrifugal force maintains the two phase states. The relative interface area is increased by decreasing the tubular diameter and/or increasing the number of the phase segments per unit length of the column. In some cases, effective mixing is accomplished by rotational or gyrational motion of the column, while the interface is held stable by gravity or centrifugal force.

We will describe these arrangements and report their performance to separate a test set of nine DNP (dinitrophenyl) amino acids having partition coefficients ranging from >100 to 0.18, using a two-phase system composed of $CHCl_3$, CH_3COOH, and $0.1N$ HCl (2:2:1).

Helix Countercurrent Chromatography (Helix CCC) *(3, 4)*

The method is so called because the two phases are formed in a helical tube, each phase occupying half the volume of each turn (Figure 1). The continued injection of one phase causes the moving phase to percolate through the stationary phase which is trapped by the effect of gravity. Solute locally introduced into the moving phase is exposed to each stationary segment, at-

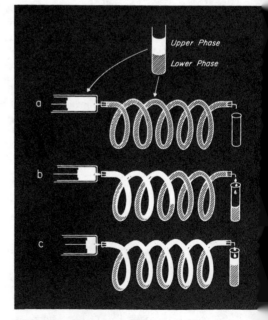

Figure 1. Model of helix countercurrent chromatography

taining a degree of equilibration dependent on the degree of mixing that results from the percolation, filming, and surface tension changes as the solute is partitioned between the phases.

To be effective in the separation of materials with similar partition coefficients, a large number of partitioning exposures is required. This system has been tested with up to 17,000 turns of Teflon tubing of 0.2-mm i.d. When small tubes are used, the cohesive forces in the liquid exceed the gravitational forces tending to separate the phases. This necessitates the use of a centrifugal force field to maintain the phase separation and permit the percolation process to occur in fine tubes in a manner similar to that demonstrated in the large helix.

In our prototype, the separation tube is supported at the periphery of a cylindrical centrifuge head and is fed from a coaxially rotating syringe. The plunger of the special high-pressure syringe is pushed through a thrust bearing by a multispeed syringe pusher. Pressures of up to 20 atm were used in these experiments, and we found it easier to push a rotating syringe than to make a high-pressure rotating seal. Effluent from the helix at approximately atmospheric pressure is conducted from the rotating system through a rotating seal situated at the center of the thrust bearing. We have used two methods of coil preparation. In the first, the Teflon tubing is wound tightly on a flexible rod support which is subsequently coiled in a number of turns around the inside of the centrifuge head. Alternatively, the tubing is folded in two and twisted along its length to provide a rope-like structure in which the individual strands have the appearance of a stretched helix of small diameter.

Our experiments have shown that the efficiency increases with tube length, smaller bore, and slower flow. Also, with a given length of tubing, we can increase efficiency by increasing the number of coil units by decreasing the helix diameter. This may be reasonably explained by the fact that the number of the interfaces situated at the top and the bottom of each coil unit (Figure 1c) increases with the number of turns. Such interfaces not only contribute to the partition process but also prevent laminar flow spreading of sample bands. Hydrostatic pressure accumulated in the coil is found to be the limiting factor in this method and calculated from the equation:

$$P = R\omega^2(\rho_1 - \rho_u)nd \qquad (1)$$

where R denotes the distance between the center of the rotation and the axis of the helix; ω, the angular velocity of rotation; n, the number of coil units; d, the helical diameter; and ρ_1 and ρ_u, the density of lower and upper phases, respectively. Since the twisted configuration of the column described earlier gives a smaller helix diameter, d, than that of the ordinary coil, the twisted column has the potential for the highest efficiency by reducing the column pressure. Efficiency of over 8000 TP (theoretical plates) has been obtained on a twisted column of 17,000 turns prepared from 80 meters of 0.2-mm i.d. tubing. Seven DNP amino acids (each component 20 to 50 μg) were eluted out within 30 hr at a flow rate of 125 μl/hr under a feed pressure of 16 atm.

Recently, this technique has been substantially improved by eliminating the use of both the rotating syringe and rotating seal, thus enabling gradient or stepwise elution by adopting the conventional elution system. The new system, called "the elution centrifuge" (5), is capable of passing the fluids in and out through fine lead tubes from a high-speed centrifuge head without rotating seals. In this system, a cylindrical column holder containing a separation column is supported radially with a pair of bearings on one side of the centrifuge head, a counterbalance being applied on the other side. When the holder revolves around the central axis of the apparatus, it simultaneously counterrotates about its own axis at the same angular velocity so as to avoid twisting the lead tubes. This synchronous rotation is achieved by coupling a pulley fixed to the holder by means of a toothed belt to a stationary pulley of equal diameter on the axis of the centrifuge drive.

The effect of the synchronous rotation on the centrifugal acceleration field has been analyzed to determine configuration, bore size, and location of the column suitable to this system. With a column consisting of 20,000 twisted turns of a 60-meter long, 0.2-mm i.d. Teflon tubing, nine DNP amino acids elute within 54 hr at efficiencies ranging between 10,000 and 6400 TP.

Droplet Countercurrent Chromatography (4, 6)

For preparative purposes, direct application of the model system (Figure 1) was found to be time-consuming, for the flow rate has to be low to prevent plug flow. When the helical configuration was flattened by winding a tube onto a plate, introduction of a light phase with low-wall surface affinity formed discrete droplets of the light phase at the bottom of each coil unit. These droplets rose rapidly through the heavy phase at a regular interval with visible evidence of very active interfacial motion. Under ideal conditions, each droplet could become a plate if kept more or less discrete throughout the system.

A system to exploit these ideas was developed by making long vertical columns of silanized glass tubing with fine capillary Teflon tubes to interconnect the wider bore glass tubes. Discrete droplets, formed at the tip of the finer tube inserted at the bottom of the long glass tube, were made to follow one another with minimal space between and a diameter near that of the internal bore of the column (Figure 2). These droplets divide the column into discrete segments that prevent laminar flow along the length of the column as they mix locally near equilibrium. The fine Teflon tubing, interconnecting the individual columns, preserves the integrity of the partitioning with minimal longitudinal diffusion and forms new droplets at the bottom of the next column. The regularity of droplet size and close spacing, a droplet size nearly filling the bore, and thin return tubing, are important for best results.

Our present column assembly consists of 300 glass tubes, 60-cm long and 1.8-mm i.d., and has a total capacity of 540 ml. At a flow rate of 16 ml/hr, several hundred milligrams of seven DNP amino acids can be eluted out within 80 hr at an efficiency of 900 TP.

Locular Countercurrent Chromatography (LCCC) (4)

This method includes two varieties, rotation LCCC and gyration LCCC. Both systems employ a column pre-

Figure 2.   Droplets in the column of droplet countercurrent chromatography

Droplets formed at the column junction can travel through the column length of 60 cm without accident at a rate of about 2 cm/sec

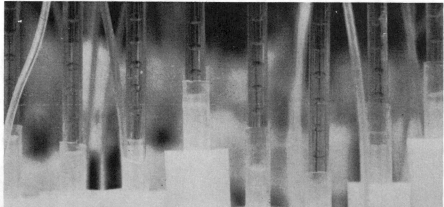

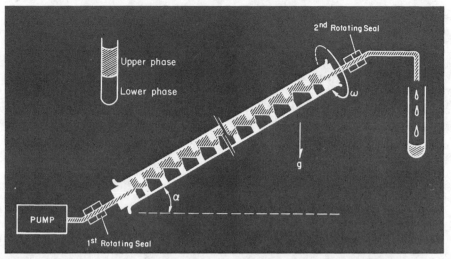

Figure 3. Mechanism of rotation locular countercurrent chromatography

pared by placing multiple centrally perforated partitions across the diameter of a tubular column which divide the space into multiple cells called locules. In each locule, the liquids form an interface while the solute partitioning is promoted by stirring of each phase induced by either rotation or gyration of the column.

Rotation LCCC: Figure 3 illustrates the mechanism used for rotation LCCC. The column inclined at angle α from the horizontal position is filled with the lower phase, and the upper phase is fed through the first rotating-seal connection as the column is rotated around its own axis. The upper phase then displaces the lower phase in each locule down to the level of the hole which leads to the next locule, the process being continued throughout the column. If the relationship between the upper and the lower phases is reversed at the beginning, the countercurrent process similarly proceeds, leaving the stationary upper phase in the column. Consequently, solute introduced into the column is subjected to a multistep partition process, promoted by the rotation of the column, and finally eluted out through the second

rotating-seal connection. Here, the gravity contributes to the phase separation, the column angle determines the phase volume ratio in the locule, and the rotation of the column produces circular stirring of the liquids to promote partition. In practice, we use a modified system in which multiple column units interconnected by fine Teflon tubings are mounted lengthwise on a rotating shaft to produce a similar effect.

Preliminary studies indicate that the efficiency increases with column length, and decreases with locule length and fast flow. The efficiency also increases with rotation speed up to 180 rpm or until the centrifugal force induced by rotation approaches the gravitational force and the liquids are about to rotate with the column. The column angle determines both volume ratio and interface area between two phases in the locule. The best result appears at 30° where the volume ratio of the two phases is nearly one, providing the maximal interface area. With a column consisting of 5000 locules, each 2.6 mm in diam and 3-mm long, and with a total capacity of about 100 ml, 1 ml of DNP amino acid sample is eluted out within 70 hr at an efficiency of 3000 TP or slightly over 50% partition efficiency/locule.

Gyration LCCC: The second variety employs a vertical locular column which gyrates on a horizontal plane without rotational motion (nonrotational gyration). The circular gyrational motion causes the interface to rotate inside the locule to produce necessary mixing. As the gyrating vertical column does not actually rotate, no rotating seals are necessary to permit flow in and out of the column during separation.

Figure 4A shows a diagram of the apparatus which provides a nonrotational gyration effect. When a pair of large wheels rotates synchronously at angular velocity ω, a plate eccentrically bridging these wheels conveys the motion to the column secured on the plate. Radius r of the column gyration is equal to the distance on the wheel between the center and the eccentric connection. In our recent prototype, r has a fixed value of 2.5 cm and ω is continuously adjustable up to 1500 rpm. The apparatus is installed in such a way that the axis of the gyration is held vertical.

Figures 4B and C show the effect of

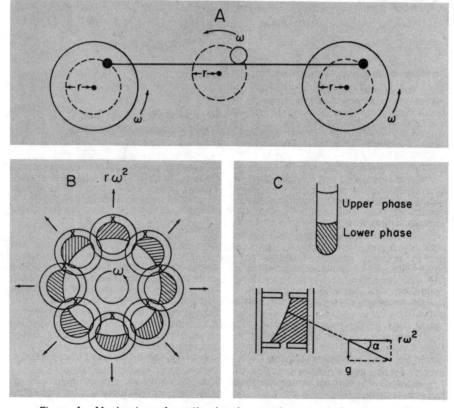

Figure 4. Mechanism of gyration locular countercurrent chromatography

A. Diagram of the apparatus
B. Gyration effect on liquid interface in locule on cross section of successive positions of the same locule
C. Gyration effect on liquid interface in locule on vertical section

218

the nonrotational gyration on the liquids and the interface present in an individual locule. The cross section (Figure 4B) through a middle portion of the locule shows successive positions of one locule as it is gyrated about a central point. The upper and lower phases are seen separated by centrifugal force forming a circular interface perpendicular to the action of the force. With the steadily changing direction of the centrifugal force, both liquids and interface rotate with respect to the "x" fixed to the column wall. Thus, friction between the liquids and the internal surface of the locule induce stirring in each phase to accelerate the partition process. On the vertical section (Figure 4C) through the center of the locule, the liquids form a parabolic interface perpendicular to the vector given by summation of the centrifugal and gravitational acceleration. The angle α here again determines both volume ratio and interfacial area between the two phases in the locule. Since the moving phase usually covers the entire area of the exit hole in the locule, the maximum interface area is given by moving either the upper phase upward or the lower phase downward through the column.

Preliminary experiments have shown that, in general, the efficiency rises with column length, number of locules, slower flow, and faster gyration. Carry-over of the stationary phase, which tends to increase with vibration of the apparatus at a high-gyration speed, is found to be the limiting factor

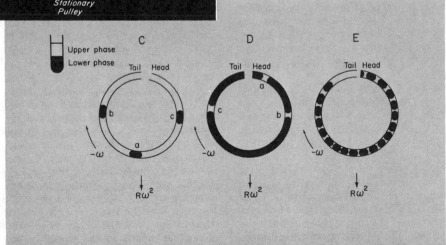

Figure 5 (A thru E). **Principle of countercurrent chromatography with flow-through coil planet centrifuge**

The figures schematically illustrate the general principle and the motion of the two phases in a coiled tube undergoing planetary motion

in this method. Two devices have been successful in eliminating this complication. The use of a column with threaded partition holes prevents the carry-over, if the direction of gyration is chosen to screw the carried-over droplets back to the original locule. Also, we have found that the surface nature of the partition hole greatly relates to this phenomenon. When the exit hole of the locule is made of a material wettable by the moving phase, the bore becomes smaller, there is less tendency of carry-over, and higher efficiency is expected. To make the locular column more convenient for the choice of the moving phase (either aqueous or nonaqueous phase may be chosen), we embedded small glass beads (1.2-mm o.d., 0.5-mm i.d., 0.7 mm long) partially into each Teflon partition hole to make one side of the hole wettable to the aqueous phase and the other wettable to the organic phase. Use of such partitions completely eliminates carry-over of the stationary phase at 800 rpm and yields a high partition efficiency from 100% to 40% in each locule.

Countercurrent Chromatography with Flow-Through Coil Planet Centrifuge (7)

This method stems from a partition technique utilizing a coil planet centrifuge (8, 9) which induces a planetary motion to a helical tube in a manner such that the rotation of the helical tube is extremely slow in comparison with the revolution. In such a system, however, flow becomes difficult because of the need for the rotating seals; therefore, the separation is usually performed within a closed helical tube. Thus, problems of sample introduction and fractionation limit the practical use of the method. On the other hand, the present method introduces a continuous flow-through system to the coil planet centrifuge without rotating seals.

Figure 5A, B illustrates the general principle of the method. The separation column is a helix, as indicated in Figure 5A. Flexible feed and return tubes are supported by the moving

disc at the top of the helix and the stationary disc fixed to the center of the upper frame of the centrifuge. The entire helix and the moving disc revolve around the axis of the centrifuge, but they are not permitted to rotate with respect to the stationary disc. The fixed orientation of the helix is maintained by coupling a pulley on the helix holder through a toothed belt to a stationary pulley of equal diameter on the axis of the centrifuge drive. This coupling causes a counter rotation ($-\omega$) of the helix to cancel out the rotation (ω) of the helix induced by its revolution. The feed and return tubes do not twist because the moving disc does not rotate with respect to the stationary disc as indicated in Figure 5B by the position of the "x" marks shown in successive positions of the relation between the stationary and moving discs. Because the helical column maintains a fixed orientation while it revolves, the radially directed centrifugal force rotates with respect to the column as in the gyrating system described earlier.

Let us consider the motion of the two immiscible phases, the upper and lower phases, confined in such a tube without the external introduction of flow (Figure 5C, D, and E). For the convenience of illustration, the direction of the centrifugal force which actually rotates is fixed at the bottom and, instead, the rotation of the coil units is substituted as indicated by the curved arrows. Figure 5C shows the motion of a small amount of the lower (heavier) phase in the upper phase. If the relative centrifugal force, $R\omega^2$, is strong enough to keep the lower phase near the bottom of the coil unit (a) at all times, the latter will move steadily toward one end of the coil, the head (the other end is called the tail), at angular velocity ω. In the present system, the centrifugal force usually fails to fix the lower phase which subsequently appears at any portion of the coil unit. When the lower phase appears at the left half of the coil unit (b), it tends to move toward the head, while, at the right half (c), it moves toward the tail. Because the lower phase tends to spend more time in the left half, these two motions do not cancel out and the lower phase moves toward the head

with an oscillatory motion at a mean angular velocity smaller than ω. Figure 5D illustrates the motion of a small amount of the upper phase in the lower phase. With a strong centrifugal force, the upper phase could be fixed near the top of the coil unit (a), constantly moving toward the head of the coil. When the centrifugal force fails to fix the upper phase, the latter moves toward the head at the right half (b) and toward the tail at the left half (c) of the coil unit. However, these motions again do not cancel out, and the upper phase also moves toward the head. When a similar volume of the two phases is introduced into the tube, the motion of the phases becomes quite complex but finally reaches an equilibrium state, illustrated in Figure 5E. At the equilibrium, the multiple segments of the two phases are alternately arranged from the head to the tail side, and any excess of either phase remains at the tail. (With a relatively large-bore tubing, a strong centrifugal force field may produce multiple droplets of either phase in the other. The picture

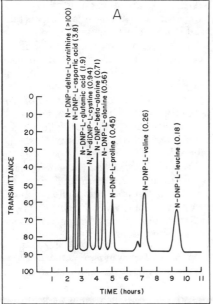

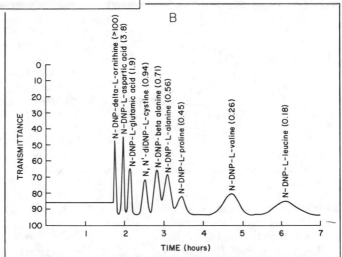

Figure 6.   Results of DNP amino acid separation.
Recording was originally made by LKB uv monitor (Uvicord II) at 280 nm. Identification of the peaks and their partition coefficients are given on the chart
A. Analytical chromatogram   B. Preparative chromatogram

of this hydrodynamic phase equilibrium may be modified by various factors such as viscosity, interfacial tension, wall surface affinity, density difference of the two phases, etc., and for some solvent systems, the ideal condition may be obtained by using an even greater radius of revolution than that used in the present test system.) Consequently, this phenomenon determines the proper direction of elution. When the tube is filled with either phase and the other is introduced from the head, the equilibrium process is quickly established from the head through the tail as elution proceeds and a given volume of the stationary phase is held within the equilibrated coil (Figure 5A). On the other hand, introduction of the flow through the tail results in a steady carry-over of the stationary phase until the moving phase fills the entire column. Thus, the moving phase should be fed from the head end, determined by both handedness of the coil and direction of the planetary motion. Consequently, a sample solution introduced into the tube is subjected to a partition process between the oscillating alternate segments of the two phases and finally eluted out through the tail end of the coil.

To form the two phases into multiple segments in a small-bore tubing without plug flow, an adequate centrifugal force is provided by the relatively long radius of revolution. We have constructed a test system with radii adjustable to 30.7, 20.2, 13.5, and 8.6 cm by modifying a conventional centrifuge. The motor shaft is extended through the top-end bearing of a stationary tube (60 cm long) mounted to the motor housing and then connected to a rotating tube that fits freely over the stationary tube with another bearing at the bottom. A pair of arms fixed to the top and bottom of the rotating tube holds the rotating rods or coil holders with bearings. When a pair of toothed pulleys of the same size, one fixed to the bottom of the stationary tube and the other at the bottom of the coil holder, are connected by a toothed belt, revolution introduces the desired planetary motion to the coil holder—i.e., one rotation per revolution in the opposite direction. The column is made of Teflon tubing either by winding the tube onto the coil holder or by arranging multiple column units interconnected in a series (tail-head connection) or several separate parallel columns around the holder. Both feed and return tubes are passed through the center hole at the top of the holder and then supported at the level of 25 cm above the center of the apparatus. These tubes, protected with a piece of silicone rubber tube at the hole, have not failed or stretched appreciably for many runs.

Figure 6A shows separation for nine DNP amino acids on an analytical column of 0.30-mm bore Teflon tubing, 100 meters long with a helix diameter of 5 mm and a total capacity of about 8 ml. Sample size is 10 μl containing each component at about 1% where solubility permits. The upper phase is fed with a metering pump at a rate of 2.4 ml/hr while the apparatus is spun at 550 rpm with a radius of 30.7 cm, the equilibrium feed pressure being approximately 200 psi. The efficiency ranges between 10,000 and 3000 TP, showing a tendency to decrease with increased retention time. Figure 6B shows a chromatogram obtained on a preparative column prepared from 1.4-mm bore tubing, 100 meters long with a helix diameter of 1 cm and with a total capacity of about 140 ml. At 520 rpm with an 8.6-cm radius of revolution and at a flow rate of 60 ml/hr, a sample size of 2 ml can be eluted out within 7 hr at an efficiency ranging between 2000 and 500 TP.

Conclusions

The various systems developed for countercurrent chromatography show high reproducibility and meet preparative and analytical needs. The advantages of countercurrent chromatog-

222

raphy over conventional liquid partition techniques are briefly discussed below.

Compared with the countercurrent distribution method (CCD), countercurrent chromatography practically yields more theoretical plates and reduces the degree of sample dilution occurring during separation. The time required for separation is also lessened. The values of countercurrent chromatography analogous to one transfer time in CCD, obtained by dividing the retention time of the solvent front by the yielded theoretical plate number, range from 40 sec (droplet CCC) to 1 sec (flow-through coil planet centrifuge technique), whereas in CCD, one transfer requires several minutes. Thus, the present method is suitable for separation of samples on the order of 10 mg, which is just enough to complete most biochemical investigations.

On the other hand, countercurrent chromatography eliminates complications arising from the use of solid supports in conventional partition chromatography, hence tailing of the solute peaks, sample loss, denaturation, and contamination are minimized. Also, this method of countercurrent chromatography enables us to predict locations of the eluted solute peaks once their partition coefficients are known.

Thus, it may be extremely useful for purification of a minute amount of biological material from a crude mixture.

Countercurrent chromatography, as introduced here, is a new development. When the instruments and techniques are further refined, the method will be even more useful. Even now, the technique offers many unique advantages over other separation methods.

References

(1) L. C. Craig, "Comprehensive Biochemistry," Vol 4, Elsevier, Amsterdam, London, and New York, N.Y., 1962.

(2) A. I. M. Keulemans, "Gas Chromatography," Chapt. 4, Reinhold, New York, N.Y., 1957, pp 96–129.

(3) Y. Ito and R. L. Bowman, *Science*, 167, 281 (1970).

(4) Y. Ito and R. L. Bowman, *J. Chromatogr. Sci.*, 8, 315 (1970).

(5) Y. Ito, R. L. Bowman, and F. W. Noble, *Anal. Biochem.*, in press (1972).

(6) T. Tanimura, J. J. Pisano, Y. Ito, and R. L. Bowman, *Science*, 169, 54 (1970).

(7) Y. Ito and R. L. Bowman, *Science*, 173, 420 (1971).

(8) Y. Ito, M. A. Weinstein, I. Aoki, R. Harada, E. Kimura, and K. Nunogaki, *Nature*, 212, 985 (1966).

(9) Y. Ito, I. Aoki, E. Kimura, K. Nunogaki, and Y. Nunogaki, ANAL. CHEM., 41, 1579 (1969).

New Directions for Ion-Selective Electrodes

Garry A. Rechnitz

Department of Chemistry, State University of New York, Buffalo, N. Y. 14214

Electrode development and applications mutually stimulate one another. Outlook for the future indicates that ion-selective electrodes will play an important part in measurement/science

THE IMPACT of ion-selective electrodes on solution chemistry is comparable to that of the laser on optical physics. Since solutions are part of the make-up of all living organisms and cover a majority of the earth's surface, it is not surprising that a new experimental tool capable of measuring the ionic composition of solutions should be of major importance. Moreover, ion-selective electrodes are most effective for the measurement of exactly those ions— e.g., Na^+, K^+, Ca^{2+}, F^-, SO_4^{2-}, S^{2-}, NO_3^-, ClO_4^-—which are most difficult to measure by other techniques.

Ion-selective electrodes measure the activities of ions in solution with considerable sensitivity (often to below one part per billion) and selectivity (selectivity ratios in excess of 10,000 are not uncommon). Such measurements are rapid, nondestructive, and can be carried out on a continuous, automated basis. Because of these desirable characteristics ion-selective electrodes are being used widely for chemical studies, biomedical measurements, pollution and oceanographic monitoring, and industrial control. About two dozen different ion-selective electrodes are already available, but more are being developed as the fundamentals of electrode operation and selectivity are elucidated (1).

Despite this notable progress, the development and exploitation of ion-selective electrodes are still in a state of infancy—perhaps comparable to the state of polarography in the early 1930's. Entirely aside from the possibility of a major fundamental breakthrough, which cannot be predicted, several new directions for the future development of new types of electrodes have become apparent this year. These are:

Immobilized liquid membrane electrodes
Mixed-crystal membrane electrodes
Enzyme electrodes
Antibiotic electrodes

The first two types are useful and important extensions of contemporary electrodes while the latter two are radical departures arising from the impact of the biological sciences upon analytical chemistry and chemical instrumentation.

Historically, the development of ion-selective electrodes began with glass electrodes and a major forward step was taken just 10 years ago when researchers (2) succeeded in correlating glass compositions with electrode selectivity. Up to that time, selection of compositions had been made largely on an empirical basis. These correlations and accompanying theoretical advances immediately resulted in the development of practical cation-sensitive glass electrodes. It was soon realized, however, that ion-selective electrodes consisting of glass membranes would be primarily limited to electrodes responsive to univalent cations and that electrodes for other ions would require new types of materials.

During the last three or four years two classes of new membrane materials have been found to be highly useful—ionically conducting inorganic solids and organic liquid ion exchange resins (3). The first category includes salt crystals, such as LaF_3, Ag_2S, and the silver halides, which can be used directly as membranes or in combination with some inert matrix material; the second consists of viscous solutions of exchangers such as organophosphorus compounds which require the use of support membranes to construct stable electrodes. In practice, the range of pure crystalline and other inorganic materials suitable as electrode membranes is limited but highly effective in specific, favorable cases. The liquid ion exchangers, on the other hand, offer a wide range of possibilities for almost any ion but usually yield electrodes which are only moderately selective for the ion of interest.

At the present time, the available range and degree of effectiveness of ion-selective electrodes still offer plenty of scope for improvement and innovation. The field is a highly competitive one, of course, with substantial commercial potential; thus, much valuable research may be going on in secrecy. Nevertheless, several possibilities of promise are discernible:

Immobilized (Matrix) Liquid Membranes

One of the disadvantages of the current types of liquid membrane electrodes is that the sample (itself a liquid) comes into contact with the liquid exchanger resin, which is restrained by some inert support material at a rather poorly defined interface. This liquid-liquid interface is subject to stirring and pressure effects, has poor mechanical stability, and can readily lead to mutual contamination of the two liquid phases. It would be advantageous, indeed, if the fundamental advantages of liquid membrane electrodes could be retained with some improvement in the mechanical and physical characteristics of these electrodes.

Now, the main property of liquid membrane electrodes which makes them so useful as ion-selective electrodes is that their exchange sites are "mobile" (4, 5). Because of this property, liquid membrane electrodes can be made to respond to cations and anions of varying charge. However, the mobile charge carriers (sites, ions or ion-site pairs) are of molecular size and there is no reason why the bulk membrane phase must be mobile—i.e., a flowing, if viscous, liquid.

This realization suggests that the exchanger liquid could be immobilized in a bulk matrix permeable to the microscopic charge carriers. One such matrix material is collodion and an early prototype electrode (6), selective for calcium, was indeed prepared by immobilizing a solution of the calcium salt of a dialkylphosphoric acid in collodion. In this manner, a solid membrane electrode is achieved which retains the desirable characteristics of the liquid membrane system. Recently, an improved version has been described (7) which has important handling and selectivity advantages over the liquid membrane calcium electrode, especially for biochemical application.

The principle involved here is perfectly general and it should be possible to prepare a wide variety of such matrix electrodes. At the present time, several new anion- and cation-selective electrodes of this type are being evaluated in the author's laboratory.

Mixed Crystal Membrane Electrodes

A useful extension of crystal membrane electrodes to additional ions can be achieved through the employment of suitable salt mixtures as the membrane phase (8). The well-established (9) sulfide ion-selective electrode, responsive to sulfide and silver ions, has as its

225

active element a polycrystalline Ag$_2$S membrane and functions by the transport of silver ions in the membrane. If this membrane is altered from pure Ag$_2$S to a CuS–Ag$_2$S mixture (see Figure 1), a cupric ion-selective electrode results. Similarly, if the membrane is made of a CdS–Ag$_2$S mixture or a PbS–Ag$_2$S mixture, one obtains cadmium-selective or lead-selective electrodes, respectively. These electrodes still transport charge by the movement of silver ions, but their potential is determined indirectly by the availability of S^{2-} which, in turn, is fixed by the activity of the divalent metal in contact with the membrane. Thus, an important requirement is that the solubility of the divalent metal sulfide be greater than that of Ag$_2$S. Practical electrodes need not have the configuration shown in Figure 1; a silver wire could be fastened directly to the inside surface of the membrane to yield a completely solid-state electrode.

Because they function well in aqueous and nonaqueous media, these mixed crystal electrodes are of considerable importance to analytical chemists. The CuS–Ag$_2$S electrode has been successfully employed for potentiometric titrations (10) in nonaqueous media (see Figure 2) while the PbS–Ag$_2$S electrode has already been found useful for the determination of sulfate by potentiometric titration with Pb^{2+} (11).

It should be kept in mind that the mixed crystal concept is not limited to sulfides; indeed, there should be a variety of two- and three-component crystal mixtures which meet the chemical and electrical requirements of electrode membrane systems. A recent review (12) of ionic conduction and diffusion in solids points the way toward new ion-selective electrodes of this type.

Enzyme Electrodes

Ion-selective electrodes need not be limited to inorganic substances or, even, to ionic species, provided a means can be found to provide charge transport and selectivity. Enzymes appear to be promising in this connection. The action of enzymes is very highly selective

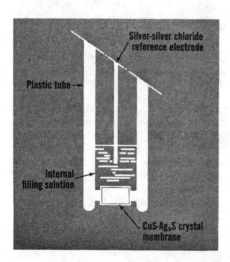

Figure 1. Cupric ion-selective electrode

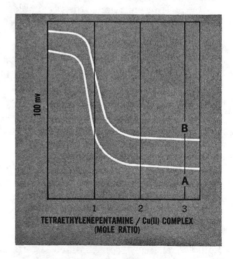

Figure 2. Potentiometric titration using CuS-Ag$_2$S electrode

A. Acetone
B. Methanol

Table I.[a] K_{MK}^{Pot} for Nonactin (25 °C)

(Nujol/2-octanol)

	NH$_4^+$	K$^+$	Rb$^+$	Cs$^+$	H$^+$	Na$^+$	Li$^+$
				M			
K_{MK}^{Pot}	0.4	1	2.4	32	55	150	1780

[a] Courtesy Prof. W. Simon

Table II.[a] K_{MK}^{Pot} for Valinomycin (25 °C)

(Diphenyl ether)

	Rb$^+$	K$^+$	Cs$^+$	NH$_4^+$	Na$^+$	Li$^+$	H$^+$
				M			
K_{MK}^{Pot}	0.52	1	2.6	84	3800	4700	18,000

[a] Courtesy Prof. W. Simon

and, furthermore, the product of enzymatic reactions is often a simple ion to which conventional ion-selective electrodes respond.

Thus, if the enzyme system can be interposed between the sample solution and the final sensing electrode, a highly selective measurement system should result. Guilbault and Montalvo (13, 14) have accomplished just this by immobilizing an enzyme in a matrix which coats a conventional cation-sensitive glass electrode. The enzyme urease is fixed in a layer of acrylamide gel held in place around the glass electrode bulb by porous nylon netting or thin cellophane film. The urease acts specifically upon urea in the sample solution to yield ammonium ions which diffuse through the gel to give rise to a potential at the glass electrode proportional to the original urea concentration in the sample. The electrode could be used continuously for three weeks without loss of activity and, since the catalytic efficiency of enzymes is very high, the overall system is essentially a nondestructive, urea-specific sensor.

The exciting possibilities of this approach must be obvious. There are literally thousands of enzymes with high activity and selectivity. Many of these yield products measurable with existing ion-selective electrodes; thus, it should be feasible to prepare electrodes of the immobilized enzyme type with selectivity for a variety of organic and biological substances such as glutamine, asparagine, and amino acids (15).

Antibiotic Electrodes

Some recent studies on biological membranes are directly applicable to the development of ion-selective electrodes as analytical sensors. Indeed, there are formal similarities between ion electrodes and biological membrane systems which virtually ensure that progress in one of these areas will also advance the other.

Recently, it has been shown (16) that certain antibiotics display notable selectivity in their interaction with alkali metal cations. Nonactin and valinomycin, for example, preferentially associate with K$^+$ rather than Na$^+$. Since the construction of a successful potassium-selective electrode by conventional means has so far eluded workers in the field, this observation has important practical implications and led Simon (17) to the construction of workable antibiotic-based membrane electrodes.

The antibiotics nonactin and valinomycin were suspended in solvents consisting of Nujol/2-octanol and diphenylether, respectively, and incorporated into liquid membrane electrodes similar to the conventional versions described above. The potentiometric selectivity constants for the resulting electrodes are given in Tables I and II and show most remarkable electrode characteristics. The valinomycin electrode, for example, displays a selectivity for K^+ over Na^+ of 3800:1. This may be compared with a maximum K^+ to Na^+ selectivity of about 30:1 for the best available glass electrode. Furthermore, the valinomycin electrode has an 18000:1 selectivity for K^+ with respect to H^+; this means that the electrode should be usable in strongly acidic media, where cation-sensitive glass electrodes lose their effectiveness. The nonactin electrode, on the other hand, shows an interesting selectivity for NH_4^+ over hydrogen ion and the alkali metal ions and may be of considerable practical value in this connection.

It is too early to say whether antibiotic electrodes will be of general utility and give rise to a broad new class of ion electrodes; however, there is no question but that electrodes based upon the synthetic organic analogs of such compounds are worthy of serious investigation. Pedersen (18) recently synthesized a whole series of cyclic polyethers, so-called "Crown" compounds, which bind (19) alkali metal ions selectively. These compounds can be tailor-made to display desired ion binding and transport properties; thus, they should play a major role in the development of new ion-selective electrodes. Two U. S. manufacturers have recently announced potassium ion–selective, liquid-membrane electrodes. A K^+ to Na^+ selectivity of about 5000:1 is claimed for one of these.

New directions for ion-selective electrodes are not limited, of course, to the development of electrodes. Novel and imaginative applications are of equal importance. Electrode development and application mutually stimulate one another, however, so that the present vigorous pace of research in this area assures ion electrodes a major place in modern measurement science.

Literature Cited

(1) G. A. Rechnitz, *Chem. Eng. News,* 43 (25), 146 (1967).
(2) G. Eisenman, (Editor) "Glass Electrodes for Hydrogen and other Cations," Marcel Dekker, New York, N. Y., 1966.
(3) R. A. Durst, (Editor) U. S. Bureau of Standards Monograph on Ion-Selective Electrodes, Government Printing Office, Washington, D. C., 1969.
(4) G. Eisenman, ANAL. CHEM., 40, 310 (1968).
(5) M. J. Brand and G. A. Rechnitz, *ibid.,* 41, 1185 (1969).
(6) F. A. Schultz, A. J. Petersen, C. A. Mask, and R. P. Buck, *Science,* 162, 267 (1968).
(7) G. A. Rechnitz and T. M. Hseu, ANAL. CHEM., 41, 111 (1969).
(8) J. W. Ross, paper presented at meeting of the Electrochemical Society, New York, May 1969.
(9) T. M. Hseu and G. A. Rechnitz, ANAL. CHEM., 40, 1054 and 1661 (1968).
(10) G. A. Rechnitz and N. C. Kenny, *Anal. Letters,* 2, 395 (1969).
(11) J. W. Ross and M. S. Frant, ANAL. CHEM., 41, 967 (1969).
(12) J. Kummer and M. E. Milberg, *Chem. Eng. News,* 47 (20), 90 (1969).
(13) G. G. Guilbault and J. G. Montalvo, *J. Am. Chem. Soc.,* 91, 2164 (1969).
(14) G. G. Guilbault and J. G. Montalvo, *Anal. Letters,* 2, 283 (1969).
(15) G. G. Guilbault, R. K. Smith, and J. G. Montalvo, ANAL. CHEM., 41, 600 (1969).
(16) L. A. R. Pioda and W. Simon, *Chimia,* 23, 72 (1969).
(17) W. Simon, paper presented at meeting of the Electrochemical Society, New York, N. Y., May 1969.
(18) C. J. Pedersen, *J. Am. Chem. Soc.,* 89, 7017 (1967).
(19) R. M. Izatt, J. H. Rytting, D. P. Nelson, B. L. Haymore, and J. J. Christensen, *Science,* 164, 443 (1969).

BASED on a lecture presented at the Analytical Summer Symposium, Athens, Ga., June 1969.

Addendum

The promise shown by ion-selective membrane electrodes in 1969, when the preceding article was prepared, has not only been met but exceeded by 1972. Indeed, skeptics have been confounded by the rapid acceptance of ion electrodes as measurement tools and by the continuing vigorous progress in fundamental understanding and practical application. Developments in this area have excited the interest of not just analytical chemists but of experimental and theoretical scientists in the biological, physical, and engineering fields.

There is, moreover, no evidence that the rate of progress is slowing; certainly, the number of publications in this field is increasing every year. Most of the pertinent literature for the 1969–71 period has been well covered in the excellent new review by Buck (1). A small book emphasizing the practical side of ion electrode measurements (2), a self-teaching audio course (3), and an analytical methods guide (4) have also appeared recently.

Numerous new and improved electrodes have been described; a trend toward the use of biological materials in membranes or coupled to membranes is especially noteworthy (5). The reader is referred to Buck's review (1) for critical listings of laboratory and commercial electrodes and their application, but mention must be made of subsequent developments dealing with nitrate (6), sulfate (7), semiconducting salt-based (8), coated wire type (9), and ammonia (10) electrodes.

From the standpoint of practical application, ion electrodes are now ubiquitous. Main application areas include process control, clinical measurement, pollution monitoring, industrial analysis, fundamental equilibrium and kinetic studies, and even medical diagnosis. Additional applications, perhaps involving miniaturization of electrodes, elimination of liquid internals, the design of new configurations (e.g., flow-through), and high-precision or differential measurements, are likely to gain wider acceptance as more potential users overcome their distrust of electrochemical measurements and as academic institutions introduce new generations of scientists to ion electrodes and their potentialities as part of the technical curriculum.

References

(1) R. P. Buck, ANAL. CHEM., 44, 270R (1972).
(2) G. J. Moody and J. D. R. Thomas, "Selective Ion-Sensitive Electrodes," Merrow, Waterford, England, 1971.
(3) G. A. Rechnitz, "Ion Selective Membrane Electrodes—Audio Course," American Chemical Society, Washington, D.C., 1971.
(4) "Analytical Methods Guides," 2nd ed., Orion Research, Inc., Cambridge, Mass., 1972.
(5) W. Simon, Pure Appl. Chem., 25, 811 (1971).
(6) J. E. W. Davies, G. J. Moody, and J. D. R. Thomas, Analyst, 97, 87 (1972).
(7) G. A. Rechnitz, G. H. Fricke, and M. S. Mohan, ANAL. CHEM., 44, 1098 (1972).
(8) M. Sharp, Anal. Chim. Acta, 59, 137 (1972).
(9) H. James, G. Carmack, and H. Freiser, ANAL. CHEM., 44, 856 (1972).
(10) Product Bulletin, Model 95-10 electrode, Orion Research, Inc., Cambridge, Mass., 1971.

COMMENTARY

by Ralph H. Müller

D<small>R. RECHNITZ'S ESSAY</small> on this subject is most stimulating and should convince analysts that continued research and development of ion selective electrodes is a fertile field. What has been accomplished so far is impressive and very useful, but, as he has pointed out, the possibilities are almost unlimited. Dr. Rechnitz and his associates continue to contribute heavily to the subject and this discussion combines enthusiasm with extensive experience.

It seems quite certain that a host of useful systems can be developed for inorganic, organic, or biological systems. We are not too happy about the present state of knowledge of the electrical behavior of selective ion electrodes. For example, what is the equivalent circuit of such systems? How are potential, current, capacitance, and resistance related and how do they combine to account for the observed behavior? If this query reeks too much of the electrical engineer's "black box," it still seeks to get a practical answer.

Knowledge about the attainment of equilibrium at the electrode is unsatisfactory. In practically all cases, no data are given and it is difficult to analyze the emf *vs.* time curves in any precise fashion. Dr. Rechnitz has made repeated pleas for more information (data), to which we would add, "always give the final value of the equilibrium potential, even if you have to wait hours or days for the answer." There is the temptation to quote the half-time ($t_{1/2}$)—*i.e.*, the time required for the emf to attain half the final equilibrium value. In the usual analytical techniques, this is an illusory constant, because the measurement, in the early stages, is a fruity mélange of mixing time and electrode equilibration. It is not unusual to encounter in the literature a statement that "the electrode had fast response with a half time of 1 second" and then followed by "the po-

tentials were recorded after an interval of 2.5 minutes." This still gives no indication of the true potential equilibrium or what fraction thereof is attained in 2.5 minutes. There is an empirical approach which affords an accurate prediction of the equilibrium value, obtainable from a few values taken at moderately short times. We speak of this briefly because of its utility, particularly in kinetic studies. Basically, the object is not to delineate the entire emf *vs.* time curve, but to get an accurate prediction of the equilibrium value. But in simpler terms, "one is interested in where he is going rather than where he has been."

For some time, here at Baton Rouge, we have looked into this matter in collaboration with Dr. Doris Müller and Dr. Philip W. West. Measurements of emf to ± 0.1 mV as a function of elapsed time were made with an Orion Ag_2S membrane electrode for "jump" increments of Ag^+ ion. The data were accurately represented by the equation: $E = t/(a + bt)$ where E is the increase in potential caused by the sudden increment of Ag^+ concentration. The curve is a hyperbola and a plot of t/E *vs.* t yields an excellent straight line, the slope of which is b and the intercept is a. The maximum value of E (for $t = \infty$) is equal to $1/b$, the reciprocal slope. As expected, there is a small but progressive deviation from linearity for the lower values of t in which range the experimental conditions are ill-defined. The prediction of the equilibrium potential E_e does not involve any dubious extrapolation procedure because it is given by the reciprocal slope of a straight line and the degree of linearity of this line is a reliable criterion of the confidence which can be placed in the calculation of E_e.

For a given value of constant b, which defines E_e, an infinite number of hyperbolas can be drawn, all converging to E_e at $t = \infty$, depending upon the choice of constant a. These would range from almost instantaneous response to almost infinitely slow or sluggish response. The ratio of these constants a (intercept) and b (slope) is a useful quantity; for example, the half-time, $t_{1/2} = a/b$. The time required to attain any desired fraction f of the equilibrium value E_e is simply: $t_f = [f/(1 - f)] \times a/b$. Beyond this, constant a is a reliable response speed measure of the electrode, but more useful when used in the ratio a/b.

Some doubt may be expressed that E_e is attained only at infinite time. The apparent stabilization at finite and relatively short times is a function of the sensitivity of the detecting system, and simple "eyeballing" of a recorder trace is not reliable. At very long time intervals after apparent equilibrium had been attained, we could detect increments in potential of the order of 10–50 μV. We have examined several curves reported in the literature, as far as our patience in disinterring the data permitted, and they were found to yield well to this treatment.

The empirical approach to this problem is useful because the calculated values agree with the observations with the same precision that the data are known. For want of better information, such results are preferable to plausible theoretical derivations yielding results in which the agreement between calculated and observed is called "encouraging."

The hyperbolic relationship is not without significance for an ultimate understanding of the time-response phenomenon. It bears a formal, if not accidental, similarity to the Langmuir adsorption isotherm and is, at least circuitwise, akin to the equivalent resistance of two resistors in parallel, one of which is changing uniformly with time. In the latter case, it is conceivable that such an arrangement could be set up as a compensator to turn out an emf value corresponding to the true E_0 value over a reasonable period of elapsed time. One might even drive it with an alarm clock.

Data supporting these conclusions are being submitted to this Journal. We obtrude these few generalizations upon this commentary merely in the sense that there are other questions relating to selective ion electrodes worthy of detailed inquiry. They should not detract from the larger design and vista which Dr. Rechnitz has outlined for us.

Ion Specific Liquid Ion Exchanger Microelectrodes

Miniature ion specific electrodes with liquid ion exchanger membranes provide a convenient means of measuring intracellular ionic activities in living cells

JOHN L. WALKER, JR., Department of Physiology,
University of Utah College of Medicine,
Salt Lake City, Utah 84112

NERVE CELLS convey information to other nerve cells and to effector organs by means of electrical impulses propagated along the cell membrane. In many types of muscle cells, an electrical impulse propagated over the cell membrane is a necessary prerequisite to mechanical contraction. These electrical events are dependent upon an asymmetrical distribution of inorganic ions between the extracellular and intracellular spaces. In addition, the electrical-mechanical coupling in muscle involves changes in the intracellular activity of calcium. To understand the electrical and mechanical activities of nerves and muscles, an accurate knowledge of the intracellular activities of the ions involved is necessary.

To measure the intracellular activities of ions is a difficult technical problem for two reasons. First, the cells and, therefore, the intracellular volumes are very small. Table I contains the dimensions and volumes of some cells of interest. The first two entries in the table represent the upper size limit, with few exceptions, while the third is approaching the lower size limit. Second, the measurements must be made without appreciable damage to the cell membrane. If the membrane is damaged to the extent that the membrane potential is not normal, the intracellular ionic activities are certain to be different from their normal values.

It is possible to make estimates of the intracellular concentrations by measuring the amount of an ion in whole tissue, the whole tissue volume, the extracellular space concentration, and the extracellular volume. However, in addi-

---

Table I. Dimensions and Volumes of Some Electrically Excitable Cells

Kind of cell	Cell shape	Diameter	Length	Volume
Molluscan neuron	Spherical	1 mm		0.52 μl
Frog skeletal muscle	Cylindrical	0.1 mm	30 mm	0.25 μl
Frog ventricle	Cylindrical	0.01 mm	0.13 mm	1×10^{-5} μl

tion to the uncertainties of the method, it gives only concentrations, which are not very meaningful (1–3). One needs to know the intracellular activities of the ions.

Since the development of ion specific glasses for hydrogen and the alkali cations, microelectrodes have been fabricated from these glasses (4, 5). Chloride microelectrodes have been made by coating a platinum wire protruding from the end of a glass micropipet with Ag–AgCl (4) or by depositing silver chloride inside the tip of a micropipet (6). These electrodes all have the drawback of having a large ion sensitive surface which must be entirely within the cell and their use is, therefore, limited to cells the size of skeletal muscle or larger. While these electrodes may have a tip diameter of one micron or less, the sensitive area is on the order of ten microns in length with a diameter of five microns or more at the top of this area. In addition to the problem of size, they are difficult to fabricate.

Liquid Ion Exchangers

Liquid ion exchangers have been used in liquid-liquid ion extraction processes in industry for some time and, from time to time, as models for biological membranes (7, 8). In the past decade, there was renewed interest in liquid ion exchangers on the part of biologists, and the theory of their behavior as ion selective membranes was developed (9, 10). At the same time, a practical calcium electrode was developed which utilizes a liquid ion exchanger membrane as the sensitive element of the electrode (11). Since then, several other ion specific liquid ion exchanger electrodes have become commercially available. This paper discusses the miniaturization of these electrodes for the express purpose of measuring intracellular ionic activities in living cells.

A liquid ion exchanger is composed of an organic electrolyte dissolved in a water-immiscible solvent, usually an organic solvent with a low dielectric constant. The organic ion should have a low water solubility and is often highly branched to prevent micelle formation (12). Owing to the low dielectric constant of the exchanger, inorganic ions have a very low solubility in the exchanger and, consequently, a membrane made of a liquid ion exchanger is much more permeable to ions whose valence sign is opposite to that of the organic ion than to ions of the same valence sign because of ion pair formation with the organic ion. For example, negatively charged phosphate esters are cation exchangers while positively charged amines are anion exchangers. Furthermore, some exchangers exhibit marked selectivity within a group of ions of the same valence sign, the selectivity being a function of the strength of interaction between the organic and inorganic ions.

An ion specific electrode is made by forming a liquid ion exchanger membrane in a suitable holder so that one side of the membrane is in contact with an aqueous reference solution of constant composition while the other side can be brought into contact with the solution to be analyzed. Electrical contact is made with the internal reference solution by means of an electrode which is reversible to one of the ions in the reference solution. The electrical potential difference across the membrane changes in proportion to the activity in the test solution of the ion(s) for which the liquid ion exchanger is selective.

A quantitative theory has been developed (9, 10) which describes the electrode potential, but it is cumbersome and contains parameters which are difficult to measure. It is, therefore, more convenient to use the following empirical equation:

$$E = E_o + \frac{nRT}{F} \log_e (a_i + K_{ij} a_j^{z_i/z_j}) \quad (1)$$

E is the electric potential (volts); E_o is a constant (volts); R is the gas constant

233

(8.3 joules deg^{-1}mole^{-1}); T is the temperature (°K); F is Faraday's number (96,500 coulomb equivalent^{-1}); n is an empirical constant (dimensionless) chosen so that nRT/F is the slope of the line when E is plotted as a function of $\log_e a_i$ when $K_{ij}a_j = 0$; z_i and z_j are the valences of the ith and jth ions, respectively; a_i is the activity (arbitrary concentration units) of the ion the electrode is expected to measure; the a_j's are the activities (same units as a_i) of interfering ions whose valence sign is the same as that of the principal ion and the K_{ij}'s are the selectivity constants for the jth ions with respect to the ith ion. When $K_{ij} < 1$ the electrode has a higher selectivity for the ith ion than for the jth ion. K_{ij} is an empirical number and should not be assigned a strict physical interpretation.

The problem, then, is to form a liquid ion exchanger membrane in a holder small enough to insert into a cell so that one surface of the membrane is in contact with the intracellular fluid.

Electrode Fabrication

A standard technique for making intracellular electric potential measurements has been to pull a glass micropipet with a tip diameter of about 0.5 micron and fill it with 3M KCl. Extensive research with this technique has shown that when the tip of such a pipet is inserted into a cell, it does not disrupt the normal functioning of the cell (13) and, therefore, if a liquid ion exchanger membrane could be formed in the tip of such a pipet, the problem would be solved.

The primary difficulty is that a clean glass surface is hydrophilic, so an organic liquid membrane will not remain in place in the tip of the pipet. Instead, it will be displaced by one of the aqueous phases with which it is in contact. To prevent this, it is necessary to render hydrophobic the portion of the pipet which is to contain the organic liquid. The method used to do this is to apply an organic silicone compound to the terminal 200 microns of the pipet tip.

A successful method for making potassium and chloride microelectrodes involves the following steps. Borosilicate glass capillary tubing (1.2 mm od, 0.3-mm thick wall) is cleaned with hot ethanol vapor and dried. Pipets are pulled from the clean tubing with a commercially available pipet puller and have a tip diameter of 0.5 to 1.0 micron. Immediately after pulling, the pipet tips are dipped in a fresh solution of 1% Siliclad (Clay-Adams) in 1-

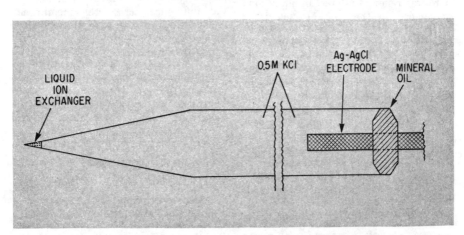

Figure 1. Schematic diagram of an ion specific liquid ion exchanger microelectrode

chloronaphthalene until there is a column of the solution about 200 microns long inside the tip. This takes approximately 15 sec. The pipets are then placed tip up in a metal block and when the desired number (one to two dozen) have been prepared, they are placed in a 250°C oven for one hour. After being removed from the oven and allowed to cool, the pipets are ready for filling but can be held in this condition for at least one week before being filled.

To fill a pipet, the tip is dipped into the ion exchanger (Corning code 477317 potassium exchanger or Corning code 477315 chloride exchanger) until the terminal 200 microns (approximately) of the tip is filled with the exchanger (one to two minutes). One-half molar KCl is then injected as far down into the top of the pipet as possible using a three-in., 30-gauge needle attached to a syringe. The pipet is placed horizontally on the movable stage of a compound microscope and a hand-drawn, solid glass needle mounted on a micromanipulator is advanced down the inside of the pipet while viewing under 100X magnification. The tip of the glass needle is brought to the center of the field and the microscope stage is carefully moved until the tip of the needle touches the meniscus of the ion exchanger. The KCl solution flows down to the liquid ion exchanger, displacing the air toward the top of the pipet where it can be flushed out using the 30-gauge needle. Air bubbles of 100 microns or less in length may be disregarded since they will be quickly absorbed. With practice, electrodes can be dipped and filled in an average of five min. A small amount of mineral oil is then injected into the top of the pipet to prevent water evaporation, and the pipet is stored with the tip in a solution of 0.5M KCl. Best results are obtained when the filled electrodes are allowed to stand for at least two hr before being used. Figure 1 shows a schematic diagram of an electrode ready for use, and Figure 2 is a photograph of

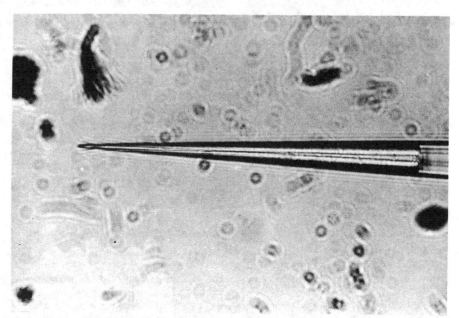

Figure 2.   Photograph of a potassium liquid ion exchanger microelectrode with 125 microns of ion exchanger inside the tip.   Magnification, 400X

a potassium microelectrode with 125 microns of liquid ion exchanger in the tip.

Two difficulties have been encountered with these electrodes. Sometimes the tips of the electrodes will not fill with the ion exchanger, or take it up very slowly. This is owing, presumably, to the tips being plugged by the Siliclad. If a pipet does not take up the desired amount of exchanger within two min, it is discarded. An alternative to this is to break the tip slightly, but this is only useful if a tip diameter of two microns or larger can be tolerated. The other difficulty is that the electrodes will sometimes lose the exchanger within two to three hr after being filled. The cause for this difficulty has not been determined. To alleviate these problems, other siliconizing procedures using a variety of organic chlorosilanes in different solvents are being investigated.

Although both kinds of electrodes can be made in the same way, there is another method of siliconizing which works better for the potassium electrodes but is not satisfactory for the chloride electrodes. After being pulled, the pipet tip is dipped in a 5% solution of tri-n-butylchlorosilane in 1-chloronaphthalene until there is a column about 200 microns long inside the tip (about 15 sec). The pipet is then allowed to air dry for at least 24 hr before being filled as described above. The tips of pipets siliconized in this way do not plug up as they sometimes do with the Siliclad solution.

Electrode Resistance and Selectivity

Because of the small size of the tip of the electrodes and the low conductivity of the liquid ion exchangers, the resistance of the electrodes is high and a voltage measuring device with an input impedance of at least 10^{13} ohms is required. A satisfactory solution has been to use a varactor bridge operational amplifier (Analog Devices, model 311K) which has an input impedance of 10^{14} ohms with an input capacitance of two picofarads. The microelectrode is connected to the input of the amplifier via a chloridized silver wire inserted into its top, and the amplifier output is connected to a digital voltmeter. The resistance of the microelectrodes, as measured from the charging time of the input of the operational amplifier, is in the range of 10^9 to 10^{10} ohms.

The response time of the electrodes when the concentration of potassium or chloride is changed has not yet been measured carefully because of the technical problems involved. However, when one of the electrodes is dipped into a solution and the amplifier input is opened as rapidly as possible, the new steady-state potential is reached within five sec, which indicates that the time constant of the response is not more than one sec. Once the steady-state potential is attained, it remains constant to within $\pm$ 1 mV for several hours. In Table II, some data from experiments with $Aplysia$ neurons (14) are presented showing the long-term stability of both the potassium and chloride electrodes.

The selectivities of the electrodes for interfering ions with respect to the principal ion has been determined for some interfering ions commonly found in biological preparations. These have been determined by measuring the electrode potentials in mixtures of constant ionic strength and finding the K_{ij} which makes the data fit Equation 1. The results of these measurements are presented in Table III for the chloride electrode and for potassium electrodes made by both of the methods described above.

Making Intracellular Measurements

Immediately before making an intracellular measurement with one of the electrodes, the electrode must be calibrated. This is done by measuring its potential with respect to a $3M$ KCl-filled micropipet in a series of KCl solutions of known concentrations.

Table II. Data from Experiments on *Aplysia* Neurons with Chloride and Potassium Microelectrodes[1] *(14)*

Slope, mV		Potential in artificial seawater, mV		Time in
Before[2]	After[2]	Before[2]	After[2]	cell(s), min
58	58	34	34	335
58	58	31	30	486
58	58	48	48	137
58	58	44	44	118
58	58	35	36	347
54	54	46	44	495
56	56	50	51	140
54	54	49	49	380
52	52	42	40	144
52	52	40	40	384

[1]First five measurements are with potassium electrodes and last five are with chloride electrodes. In most experiments, more than one cell penetration was made with the electrode during the indicated time. [2]Before and after intracellular measurement(s).

Table III. Selectivities for Interfering Ions with Respect to Principal Ion for Chloride and Potassium Microelectrodes

Interfering ion	K_{ij}		
	0.1M[1]		1.0M[1]
Bicarbonate[2]	0.05		0.05
Isethionate[2]	0.2		0.2
Propionate[2]	0.5		0.7
Calcium[3]	0.002	a[4]	0.03
	0.002	b	0.03
Hydrogen[3]	0.02	a	0.03
	0.025	b	0.016
Sodium[3]	0.02	a	0.02
	0.02	b	0.014

[1] Ionic strength. [2] Ions interfering with chloride. [3] Ions interfering with potassium. [4] *a* and *b* refer to Siliclad and tri-*n*-butylchlorosilane potassium microelectrodes, respectively.

The KCl concentrations used should bracket the expected concentration of the unknown solution. The electrode is then moved to the solution bathing the cell in which the measurement is to be made, which normally contains some of the ion to be measured, potassium and/or chloride. The potential of the electrode in that solution should agree with the potential predicted for the electrode from the calibration curve, taking into account any interfering ions which may be present in the solution. The tip of the electrode is then inserted into the cell. Since it is not possible to see the tip of the electrode enter the cell, the criterion used to determine cell entry is an abrupt shift in the potential of the electrode as it is slowly advanced toward the cell. This shift in potential is due to two factors: (1) the difference in activity of the ion being measured between the extracellular and intracellular fluids; and (2) the cell membrane potential. The shift of the electric potential as the electrode tip enters the cell can be written as:

$$\Delta E = E_m + \frac{nRT}{z_i F} \log_e \left[\frac{a_i^o + K_{ij} a_j^o}{a_i^i + K_{ij} a_j^i} \right] \quad (2)$$

where ΔE (volts) is the difference in electric potential between the inside and the outside of the cell; E_m (volts) is the cell membrane potential; the superscripts on the activities, o and i, refer to outside and inside of the cell; and n, determined from the calibration curve for each electrode, is 1.0 for the potassium electrodes and in the range of 0.90 to 0.97 for the chloride electrodes. While n varies from one chloride electrode to another, it is constant for any one electrode and constant over the range of at least $1.0 \times 10^{-3}M$ KCl to $1.0M$ KCl for both potassium and chloride electrodes. E_m is measured by impaling the cell with a $3M$ KCl-filled micropipet.

When Equation 2 is solved for the denominator of the logarithmic term, it takes the form of Equation 3.

$$a_i{}^i + K_{ij}a_j{}^i =$$

$$(a_i{}^o + K_{ij}a_j{}^o) \exp\left[\frac{(\Delta E - E_m)z_i F}{nRT}\right] \quad (3)$$

It is now a simple matter, knowing all the factors on the right side of Equation 3, to calculate a numerical value for the left side. The only remaining problem is to determine the value of the term, $K_{ij}a_j{}^i$. If microelectrodes specific for the interfering ions are available, the $a_j{}^i$'s can be measured directly. Unfortunately, this is not usually possible and, therefore, it is necessary to use estimates arrived at by other means. When whole tissue analyses have been done, the values of the concentrations can be used by assuming values for the intracellular activity coefficients (2). Intracellular sodium activity can be estimated from the overshoot of the action potential (15), and in the case of ions which appear to be passively distributed across the cell membrane, the intracellular activity can be calculated by using the extracellular activity of that ion and assuming a Donnan distribution (2).

When the cells are large, as in the case of *Aplysia* neurons or frog skeletal muscle, the KCl-filled micropipet can be seen in the same cell as the ion specific electrode, but with smaller cells this is not possible. When the cells are too small to see if both electrodes are

in the same cell, the membrane potential and ion specific electrode measurements can be made separately in several cells and the average values of the measurements used to calculate intracellular activity. A better solution to the problem is to make double-barrel electrodes where one barrel is the ion specific electrode and the other barrel is the reference electrode. Attempts to fabricate such electrodes are currently being made.

As shown in Table II, both the potassium and chloride microelectrodes are stable over a period of several hours even when several cell penetrations are made with the same electrode. When an electrode has shown a DC drift, it has usually been traced to a change in the potential of the Ag–AgCl electrode making contact with the internal reference solution of the electrode. Most important is the fact that the slopes of the electrodes do not change during the course of the experiments. Since the intracellular measurements are made with respect to the extracellular solution, which is of known composition, a slow DC drift is of little consequence as long as the slope of the response curve remains constant.

As is the case with any new technique, it is necessary to establish that the data obtained with the technique are valid. In measuring intracellular activities with ion specific electrodes, this is not easy to do because of the lack of accurate information concerning intracellular activities of ions, both organic and inorganic, which may interfere with the measurement. This is especially true of the chloride electrode because of the presence of organic anions in the intracellular fluid. Some experimental results bearing on this point have been presented and discussed by Cornwall *et al.* (16).

At the present time, only the potassium and chloride ion exchanger microelectrodes are of practical use. Attempts to make a calcium microelectrode have not been completely successful, although some progress has been made. They can be made to exhibit a Nernst slope in solutions of $CaCl_2$ from $1.0 \times 10^{-4}M$ to $1.0M$ with a time constant and stability comparable to the potassium and chloride microelectrodes. However, when potassium, in concentrations approximating those found inside cells, is added to the $CaCl_2$ solutions, it reduces the calcium response drastically. Work is continuing on the calcium microelectrode and as liquid ion exchangers for other ions of interest to biologists become available, they will be used to make ion specific microelectrodes of the type described above.

References

(1) J. A. Johnson, *Amer. J. Physiol.* **181**, 263 (1955).

(2) E. J. Conway, *Physiol. Rev.* **37**, 84 (1957).

(3) E. Page, *J. Gen. Physiol.* **46**, 201 (1962).

(4) J. A. M. Hinke in "Glass Microelectrodes," Lavalee *et al.*, Eds., Wiley, New York, N. Y. 1969.

(5) N. W. Carter, F. C. Rector, Jr., D. E. Campion, and D. W. Seldin, *J. Clin. Invest.* **46**, 920 (1967).

(6) G. A. Kerkut and R. W. Meech, *Life Sci.* **5**, 453 (1966).

(7) R. Beutner, *Amer. J. Physiol.* **31**, 343 (1913).

(8) G. Eisenman, J. P. Sandblom, and J. L. Walker, Jr., *Science* **155**, 965 (1967).

(9) J. P. Sandblom, G. Eisenman, and J. L. Walker, Jr., *J. Phys. Chem.* **71**, 3862 (1967).

(10) J. P. Sandblom, G. Eisenman, and J. L. Walker, Jr., *J. Phys. Chem.* **71**, 3871 (1967).

(11) J. Ross, *Science* **156**, 1378 (1967).

(12) R. Kunin and A. G. Winger, *Angew. Chem., Int. Ed. Engl.* **1**, 149 (1962).

(13) W. L. Nastuk and A. L. Hodgkin, *J. Cell. Comp. Physiol.* **35**, 39 (1950).

(14) D. L. Kunze and A. M. Brown, unpublished results, 1970.

(15) A. L. Hodgkin and B. Katz, *J. Physiol.* **108**, 37 (1949).

(16) M. C. Cornwall, D. F. Peterson, D. L. Kunze, J. L. Walker, Jr., and A. M. Brown, *Brain Res.* **23**, 433 (1970).

This work was supported by American Heart Association Grant #70 887, Utah Heart Association Grant 1T, and the United States Public Health Service Grant GM 14328.

Electrofocusing in Gels

DANIEL WELLNER

Department of Biochemistry, Cornell University Medical College
New York, N. Y. 10021

Gel electrofocusing provides the enzymologist, the protein chemist, the geneticist, the immunologist, and the clinician with a versatile supplementary research tool for the analysis and characterization of proteins and peptides. Its simplicity, its high resolving power, and its ease of combination with other techniques indicate that the number of applications to which it will be put will greatly increase in the near future

THE TECHNIQUE of gel electrofocusing is one of the most promising new methods for the analysis and characterization of enzymes, hormones, and other ampholytes of biological interest. It is equal or superior in sensitivity, resolution, and simplicity to other analytical methods commonly employed for the study of proteins. Furthermore, because it separates compounds on the basis of differences in isoelectric point, it provides additional information not readily obtainable by other methods. Therefore, while electrofocusing is not likely to displace such methods as electrophoresis, chromatography, or ultracentrifugation, it represents a valuable supplementary research tool. In this article, the principle of the method will be outlined briefly, the experimental technique will be described, and some examples of its use will be given.

Principle of Method

In an electric field, proteins and other ampholytes move toward the cathode in solutions more acidic than their isoelectric point and toward the anode in solutions more basic than their isoelectric point. At their isoelectric point (when pH = pI) they carry no net charge and therefore do not migrate. When a protein is added to a solution containing a pH gradient, and when an electric potential is applied to such a solution, with the anode on the acidic side and the cathode on the basic side, the protein molecules migrate toward that region of the gradient where the pH equals their isoelectric point. If the pH gradient were stable and if convection and diffusion were absent, the protein would become concentrated in a sharp stationary zone, its position depending on the isoelectric point of the protein.

240

An early attempt to exploit this principle for the separation and concentration of proteins was made by Kolin (*1*) in 1954. He used a gradient formed between two buffers of different pH and, at the same time, a sucrose density gradient to prevent convective remixing of the separated proteins. Although good separations were achieved in that system, the sharpness of the focusing was limited by the migration of the buffer ions in the electric field and the consequent instability of the pH gradient.

This problem was overcome by Svensson (*2*) and Vesterberg and Svensson (*3*) by the use of a "natural" pH gradient—i.e., a gradient formed under the influence of an electric current and stable in the presence of the current. This was accomplished by using "carrier ampholytes" consisting of a mixture of many compounds of low molecular weight (<1000) containing both basic and acidic groups. Because of differences in the number and dissociation constants of these groups, the mixture included compounds covering a range of isoelectric points. When a potential difference is applied to a solution of such a mixture of ampholytes, initially uniformly distributed, the negatively charged (acidic) ampholytes migrate electrophoretically toward the anode and the positively charged (basic) ampholytes migrate toward the cathode.

To prevent the ampholytes from coming into contact with the electrodes, the anode is immersed in a strong acid and the cathode in a strong base. Thus, whenever an ampholyte molecule enters the electrode solution, its charge is reversed and it is forced in the opposite direction. In addition, a sucrose density gradient is used to prevent convection.

After the current has been applied for a sufficient time, any ions such as Na^+, Cl^-, or SO_4^{2-}, which carry only positive or negative charges, will have migrated into the electrode solutions. In the central region, the only ions present in addition to the ampholytes will then be hydrogen ions and hydroxide ions. Since the acidic ampholytes have migrated to the anodic end and the basic ampholytes to the cathodic end, a pH gradient will now exist in the solution.

An equilibrium or steady state is established, as elegantly demonstrated by Svensson (*4*), in which all the ampholytes are ordered in a series of overlapping zones, each having its maximum concentration in the region of its isoelectric point. The pH gradient is stable as long as the voltage is maintained, diffusion of the ampholytes away from their isoelectric position being exactly counteracted by electrophoretic migration in the opposite direction.

When proteins are introduced in such a system, either before or after the gradient is established, they are concentrated or focused in sharp zones. Their isoelectric points may be determined with accuracy simply by measuring the pH of these zones. In this way, proteins differing in isoelectric point by as little as 0.02 pH unit can be resolved from one another (*3*).

In 1968, not long after suitable carrier ampholytes became commercially available (several ranges of pI are available from LKB Produkter AB, Bromma, Sweden), procedures for performing electrofocusing in polyacrylamide gels were developed independently in many laboratories (*5–14*). Polyacrylamide gel has several important advantages as an anticonvection medium over sucrose density gradients. One of these is that it allows separations to be carried out on a much smaller scale, thereby making it possible to analyze smaller samples and to work with smaller amounts of the relatively expensive ampholytes. Another advantage is that it makes unnecessary the use of elaborate jacketed glass columns. Instead, experiments can be carried out in a very simple

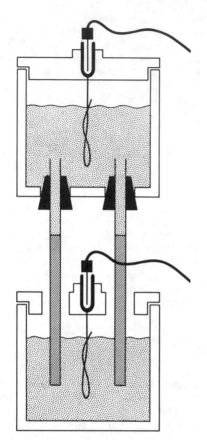

Figure 1. Apparatus for gel electro-focusing

taneously and under identical conditions for comparative purposes.

Experimental Procedure

A simple apparatus which has been used successfully for gel electrofocusing in this laboratory is shown diagrammatically in Figure 1. It is similar to that described by Davis (*15*) for disc electrophoresis. The apparatus consists of two cylindrical plexiglas containers whose covers are fitted with platinum wire electrodes. The bottom of the upper container has a number of circular openings symmetrically arranged along its circumference. These are fitted with rubber stoppers through which a 6-mm i.d. glass tube has been inserted. When an experiment involves only one or a few samples, the unused openings may be closed with solid stoppers. The bottom container has a corresponding number of openings in its cover. Glass tubes of various lengths may be used, the height of the upper container being varied accordingly by raising or lowering its support on a ring stand.

Since electrofocusing concentrates proteins into sharp zones, the sample may be uniformly distributed throughout the gel at the start of the experiment. Thus, it is possible to polymerize the acrylamide in a solution containing both the carrier ampholytes and the sample to be analyzed. When this procedure is adopted, photopolymerization with riboflavin as catalyst is the preferred method of forming the gel. This avoids the possibility of alteration of protein samples which may occur in the presence of persulfate. Photopolymerization is a mild procedure, 30 min at room temperature under ordinary laboratory overhead lights usually being sufficient for the gel to form. Gel formation is sometimes facilitated by removing dissolved oxygen from the solution under reduced pressure.

In some cases, however, it is not

apparatus of the type used for gel electrophoresis, as described below.

Furthermore, less time is required for focusing to be complete on the small scale used in gels than in the column procedure. Greater resolution may also be achieved in gels because it is possible to avoid the mixing of closely spaced bands which takes place when gradient solutions are transferred to a fraction collector. In addition, the convective remixing which occurs in sucrose gradients when focusing produces excessive local concentrations of proteins does not take place in gels. Finally, gel electrofocusing makes it simple to analyze many samples simul-

possible to prepare satisfactory gels by photopolymerization. For example, some samples of carrier ampholytes inhibit the riboflavin-catalyzed reaction. Some protein samples are also inhibitory. It is then preferable to use an alternative procedure in which persulfate-catalyzed polymerization is allowed to proceed with the carrier ampholytes incorporated into the solution, and to add the sample subsequently on top of the gel, as is done for disc electrophoresis (14).

Contact between persulfate and the sample is avoided by a short prelimi-

nary focusing. If this is done with the cathode at the top and the anode at the bottom, it is not necessary for all of the persulfate to be out of the gel before the sample is applied, because any remaining persulfate will move down ahead of the protein when the current is turned on. After the preliminary focusing, the sample, in a solution of ampholytes and 10% sucrose, is carefully introduced above the gel under a layer of ampholytes in 5% sucrose which separates it from the electrode solution.

The resolution obtainable with gel electrofocusing may be illustrated by describing an experiment carried out with crystalline L-amino acid oxidase, an enzyme purified from the venom of the Eastern Diamondback rattlesnake (*Crotalus adamanteus*) (10). The result of this experiment is shown in Figure 2. For comparison, a disc electrophoresis experiment performed with the same sample of enzyme is also shown. The gel used in the electrofocusing experiment was made by photopolymerization of a solution containing the following components: acrylamide, 7.5%; N,N'-methylene bisacrylamide, 0.2%; N,N,N',N'-tetramethylethylenediamine, 0.058%; carrier ampholytes ("Ampholine," pH 3–10), 2%; enzyme, about 50 μg/ml; and riboflavin, 0.0015%. After deaerating about 3 ml of this solution by placing it under reduced pressure for 1 to 2 min, it was pipetted into a glass tube, 6 × 120 mm, closed off at the bottom with Parafilm. The solution was covered with a 1-cm layer of water and allowed to gel at room temperature for 30 min.

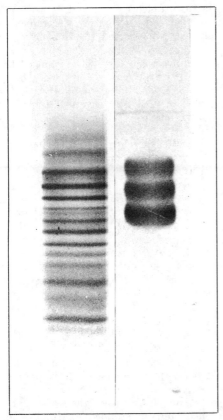

Figure 2. Isozymes of L-amino acid oxidase separated by gel electrofocusing (left) and by gel electrophoresis (right) (from ref 10). For experimental details, see text

Electrofocusing was carried out at 4° with a final voltage of 120 V for 9 hr in the apparatus shown in Figure 1. The anode (top) was immersed in 1% H_2SO_4 and the cathode in 2.9% ethanolamine. To prevent overheating, the voltage was increased gradually to its final value, the power output always being kept below 0.2 W per gel. The optimum focusing time and final voltage used depend on a number of factors, such as the length of the gel, the pH range of the carrier ampholytes, and the concentration of salts or buffer initially present in the sample.

When electrofocusing is complete, the gel is removed from the tube. It may then be fixed and stained for protein. Alternatively it may be stained for a specific enzymatic activity or analyzed in a variety of other ways. Because amido black and some other protein stains form complexes with the carrier ampholytes, they do not give satisfactory results unless the fixed gel is first extensively washed. Such washing is not necessary, however, when the protein is stained with Coomassie blue. A method by which the gel may be simultaneously fixed and stained is to immerse it overnight in a solution containing 5% trichloroacetic acid, 5% sulfosalicylic acid, 18% methanol, and 0.02% Coomassie blue. Excess stain is then removed by washing the gel in water.

A convenient method of verifying that a pH gradient has been established and of measuring the pH in various regions of the gel is to touch the surface of the gel at various points with a microelectrode (combination glass and reference electrode, Cat. No. 14153, available from Instrumentation Laboratory, Inc., Watertown, Mass.). This should be done immediately after removing the gel from the tube. The gel is laid alongside a ruler and touched at 1-cm intervals with the electrode in such a way that both the glass membrane and the reference junction are in contact with the surface. These measurements take less than 5 min per gel.

Comparison of Gel Electrofocusing and Gel Electrophoresis

As shown in Figure 2, gel electrofocusing revealed the presence of at least 18 protein components in crystalline L-amino acid oxidase, whereas only three zones could be separated by gel electrophoresis. It could be shown, by comparing the gels stained for protein with Coomassie blue with similar gels stained specifically for L-amino acid oxidase activity using a staining solution containing L-leucine, phenazine methosulfate, and nitroblue tetrazolium, that all of the protein bands possessed enzymatic activity. The possibility was considered that these results could represent an artifact arising from an interaction between the enzyme and the gel or from an alteration of the enzyme in the course of purification. However, such artifacts were ruled out by showing that a similar pattern of zones was obtained when electrofocusing was conducted in a sucrose density gradient or when untreated snake venom was electrofocused in gels and stained for enzymatic activity (10).

It may therefore be concluded that there are at least 18 isozymes of L-amino acid oxidase, perhaps a larger number than has been found for any other enzyme. This raises many interesting questions concerning the nature of the chemical differences between the isozymes, their origin, and their physiological significance. It seems probable that when other enzymes are examined by this technique, there will be other cases where a greater multiplicity of forms will be found than were previously known to exist. Several examples of such heterogeneity have already been discovered by using the sucrose gradient technique (16).

These results indicate that the resolution obtainable from gel electrofocusing is considerably greater than that of gel electrophoresis. There are several reasons for this. One is that, whereas in the course of electrophoresis protein bands become increasingly diffuse, in electrofusing they are sharpened and maintained sharp by the electric field. Another is that differences of a fractional charge, which may not contribute an appreciable difference in electrophoretic mobility, may result in a sufficient difference in isoelectric point to yield a separation by electrofocusing. Such a difference may arise from a different environment, and consequently an altered dissociation constant, of an ionizable group such as a carboxyl group or an amino group. The changed environment may be due to a difference in conformation between the two proteins or perhaps to the substitution of a neighboring amino acid residue—e.g., leucine for alanine—which by itself would not be expected to alter the charge on the protein.

Another reason may be illustrated by the following example. If protein A differs from protein B by having a group of pK 4 instead of one of pK 6, no separation of A from B would be expected by electrophoresis at pH 8.5, although at pH 5 such a separation might be feasible. On the other hand, at pH 5, proteins C and D might not be separated if they differ by groups having pK's of 8 and 9, respectively. Electrophoresis at pH 7 might not separate either of the protein pairs. Electrofocusing, on the other hand, would be expected to separate all four in a single experiment.

In view of these considerations, and because electrophoretic mobility depends on additional factors such as binding of buffer ions by the protein, frictional forces, and interactions between the protein and the gel matrix, it is evident that in general there is no direct correlation between electrophoretic mobility and isoelectric point. Thus, a combination of the two techniques provides more information than either one alone.

Two-Dimensional Separations

A mixture of proteins may be separated in two dimensions by a combination of electrofocusing and electrophoresis to produce a "fingerprint," in analogy to the well-known two-dimensional separation of peptides from a protein digest. This was done with the L-amino acid oxidase isozymes in the following way. After the gel was removed from the tube, immediately following electrofocusing, it was embedded in a rectangular gel slab containing buffer. A current was then applied at right angles to the direction of the original separation (10). In this way, it could be shown that some components not separated by electrofocusing were separable by electrophoresis.

The two-dimensional technique thus allows greater resolution than either electrofocusing or electrophoresis alone. This technique was also used by Dale and Latner (17) with serum proteins. The results they obtained suggest the possible value of this method in the diagnosis of disease. Such a two-dimensional mapping of the soluble proteins of potato tubers has been shown to be capable of differentiating between several plant varieties, each having its own characteristic pattern (18). This suggests that the method may be of considerable value in genetic studies. In a variation of the technique, in which starch gel was used for electrophoresis in the second dimension (19), it was possible to separate about 40 different proteins from wheat grains, while under similar conditions electrophoresis or electrofocusing alone could separate only about 20.

It is also possible to combine gel electrofocusing with gradient gel electrophoresis (20). In the latter technique, proteins migrate into a gel with decreasing pore size until the size of the molecules in relation to the pore size of the gel makes further migration impossible. By use of such a technique, it is possible in a single experiment to obtain an estimate both of a protein's isoelectric point and molecular weight. Electrofocusing combined with immunodiffusion has also been carried out and named immunoelectrofocusing (21).

Possible Future Developments

Since electrofocusing has achieved its popularity only recently, it seems inevitable that significant improvements and many new applications will soon appear in the literature. Among the developments to be expected are improvements in resolution, decrease in the time required to achieve equilibrium, and more sensitive and rapid methods for detecting proteins, enzymes, or hormones in the gel. A spectrophotometric scanning procedure for detecting proteins after gel electrofocusing was recently described (*22*). Although not as sensitive as some of the staining procedures, it has the advantage of allowing the separation and focusing to be observed as they progress. Good resolution is obtained, since it is possible to scan the gel before the current has been turned off and hence before the occurrence of band spreading through diffusion.

Better resolution than has been achieved thus far may possibly be attained by the use of longer gels, by the use of ampholytes with narrower ranges of pI, or perhaps by other methods of creating stable pH gradients. One novel and promising method has recently been suggested by Luner and Kolin (*23*). Although it was performed in free solution, there appears to be no reason why it could not be carried out equally well in gels. In this method, advantage is taken of the variation in pH of tris buffers as a function of temperature. A pH gradient may thus be formed simply by creating a temperature gradient. Luner and Kolin (*23*) report the possibility of fractionating proteins in about 15 min by this procedure and of determining their isoelectric points simply by measuring the temperature of the protein zone (the dependence of the buffer pH on temperature having been determined previously).

In summary, gel electrofocusing appears to be a versatile new technique for the analysis of proteins and peptides. Because of its simplicity, its high resolving power, and the ease with which it can be combined with other techniques, it may be confidently predicted that the extent of its use and the number of applications to which it will be put will greatly increase in the near future. This method provides a valuable new tool for the enzymologist, the protein chemist, the geneticist, the immunologist, and the clinician.

Acknowledgment

Gel electrofocusing techniques were developed in this laboratory in collaboration with Melvin B. Hayes. Many experiments were also performed by Mark Sivakoff and Linda Winsor. This work was supported by Grant #AM-12068 from the National Institutes of Health, U. S. Public Health Service.

References

(1) A. Kolin, *J. Chem. Phys.*, **22**, 1628 (1954).

(2) H. Svensson, *Arch. Biochem. Biophys., Suppl. 1,* 132 (1962).

(3) O. Vesterberg, and H. Svensson, *Acta Chem. Scand.*, **20**, 829 (1966).

(4) H. Svensson, *ibid.*, **15**, 325 1961).

(5) Z. L. Awdeh, A. R. Williamson, and B. A. Askonas, *Nature*, **219**, 66 (1968).

(6) N. Catsimpoolas, *Anal. Biochem.*, **26**, 480 (1968).

(7) G. Dale, and A. L. Latner, *Lancet*, **1**, 847 (1968).

(8) J. S. Fawcett, *FEBS Letters*, **1**, 81 (1968).

(9) M. B. Hayes, and D. Wellner, Abstract #151, 156th American Chemical Society Meeting, Division of Biological Chemistry, September 1968.

(10) M. B. Hayes, and D. Wellner, *J. Biol. Chem.*, **244**, 6636 (1969).

(11) D. H. Leaback, and A. C. Rutter, *Biochem. Biophys. Res. Commun.*, **32**, 447 (1968).

(12) R. F. Riley, and M. K. Coleman, *J. Lab. Clin. Med.* **72**, 714 (1968).

(13) C. W. Wrigley, *J. Chromatog.*, **36**, 362 (1968).

(14) C. W. Wrigley, *Sci. Tools*, **15**, 17 (1968).

(15) B. J. Davis, *Ann. N. Y. Acad. Sci.*, **121**, 404 (1964).

(16) W. A. Susor, M. Kochman, and

W. J. Rutter, *Science,* **165** 1260 (1969).

(17) G. Dale, and A. L. Latner, *Clin. Chim. Acta,* **24,** 61 (1969).

(18) V. Macko, and H. Stegemann, *Hoppe-Seyler's Z. Physiol. Chem.,* **350,** 917 (1969).

(19) C. W. Wrigley, *Biochem. Genetics,* **4,** 509 (1970).

(20) K. G. Henrick, and J. Margolis, *Anal. Biochem.,* **33,** 204 (1970).

(21) N. Catsimpoolas, *Immunochemistry,* **6,** 501 (1969).

(22) N. Catsimpoolas, and J. Wang, *Anal. Biochem.,* **39,** 141 (1971).

(23) S. J. Luner, and A. Kolin, *Proc. Nat. Acad. Sci.,* **66,** 898 (1970).

High-Speed Current Measurements

PIETER G. CATH
ALAN M. PEABODY
Keithley Instruments, Inc., 28775 Aurora Rd., Cleveland, Ohio 44139

Electrometers provide the means of detection and measurement of small electrical currents in many important instrumental methods utilized in chromatography, electrochemistry, and spectroscopy. A consideration of several approaches to high-speed current measurements can prove helpful in designing or selecting an optimum detector for a particular instrumental application.

THE MEASUREMENT of small electrical currents has been the basis for a number of instrumental methods used by the analyst. Ion chambers, high-impedance electrodes, many forms of chromatographic detectors, phototubes, and multipliers are commonly used transducers which require the measurement of small currents. Devices used for this measurement are often called electrometers. It is the purpose of this article to point out the trade offs that one makes to obtain desired characteristics and to present in some detail the design techniques for a new type of electrometer which is optimized for measurements in the 1 Hz to 5 kHz region (10^{-14} to 10^{-11} A resolution).

Current-Detection Limitations. In any measurement, if source noise greatly exceeds that added by the instrumentation, optimization of instrumentation is unimportant. When source noise approaches the theoretical minimum, optimization of instrumentation characteristics becomes imperative. To determine the category into which his measurement falls, the researcher needs to be familiar with the characteristics which impose theoretical and practical limitations on his measurement.

Most researchers are familiar with the theoretical limitations present in voltage measurements. The noise increases with source resistance, and the familiar equation for the mean-square noise voltage is

$$\overline{e_n{}^2} = 4\,kTR\Delta f \qquad (1)$$

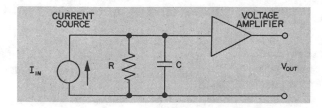

Figure 1. In the shunt method, current is measured by the voltage drop across a resistor

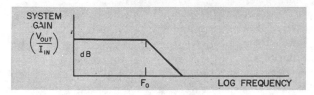

Figure 2. The frequency response of the shunt method is limited by omnipresent shunt capacitance

where k is the Boltzmann constant, T is the absolute temperature of the source resistance, R, and Δf is the noise bandwidth ($\pi/2$ times the 3 dB bandwidth for a single RC roll-off).

In the case of current measurements, it is more appropriate to consider the noise current generated by the source and load resistances. The mean-square noise current generated by a resistor is given by Equation 2.

$$\overline{i_n{}^2} = \frac{4\,kT\Delta f}{R} \qquad (2)$$

From this equation, it is immediately apparent that the measurement of a small current requires large values of R—i.e., high impedance levels.

However, this presents difficulties for measurements requiring wide bandwidths because of the RC time constant associated with a high-megohm resistor and even a few picofarads of circuit capacitance. Figure 1 shows a current source generating a voltage across a parallel RC. The frequency response of this current measurement is limited by the RC time constant. Figure 2

shows this response and the —3 dB point occurs at a frequency

$$F_0 = \frac{1}{2\,\pi RC} \qquad (3)$$

Low noise and high speed, therefore, are contradictory requirements. To optimize a current-measuring system, techniques must be used which obtain high speed using high-impedance devices.

High-Speed Methods

High speed can, of course, be obtained in a shunt-type measurement by using a low value for the shunt resistor. As pointed out above, such a small resistor value introduces excessive noise into the measurement.

A second method to achieve bandwidth is to keep R large, to accept the frequency roll-off starting at F_0, and to change the frequency response of the voltage amplifier as shown in Figure 3a. The combined effects of the RC time constant followed by this amplifier are shown in Figure 3b, and it is seen that the frequency response of the current measurement has been extended to

249

F_1. The frequency at which the amplifier gain starts to increase must be exactly equal to the frequency F_0 determined by the RC time constant in order for this approach to result in a flat frequency response. Therefore, this method is useful only for applications where the shunt capacitance, C, is constant. Aside from this drawback, this is a legitimate approach being used in low-noise, high-speed current-measuring applications.

In addition to current noise in the shunt and in the amplifier input stage, a major source of noise in this system arises from the voltage-noise generator associated with the input stage (reflected as current noise in the shunt resistor) caused by the high-frequency peaking in the following stages of amplification. More will be said about this in the discussion on noise behavior.

A third method used for speeding up a current measurement employs guarding techniques to eliminate the effects of capacitances. Unfortunately, only certain types of capacitances, such as cable capacitance, can be conveniently eliminated in this manner. Eliminating the effect of parasitic capacitances associated with the source itself becomes very cumbersome and may not be feasible in many instances. The major sources of noise in this system are identical to those mentioned in the second system.

A fourth circuit configuration combines the capability of low-noise and high-speed performance with tolerance for varying input C and eliminates the need for a separate guard by making the ground plane an effective guard. This is the current-feedback technique. This technique gives a typical improvement of a factor of three over shunt techniques. Again, the major sources of noise are identical to those mentioned in the second system.

Current-Feedback Technique

The basic circuit configuration used in the current-feedback technique is shown in Figure 4. In this configuration, the current-measuring resistor, R, is placed in the feedback loop of an inverting amplifier

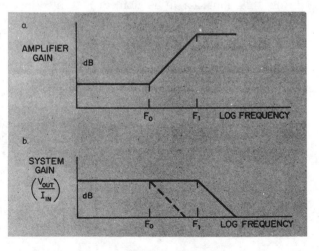

Figure 3. By tailoring the frequency response of the amplifier (3a), the frequency response of the shunt method can be extended (3b)

250

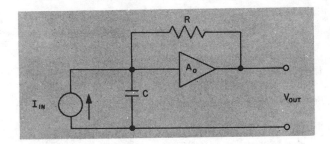

Figure 4. Basic circuit configuration for the feedback method

with a gain of A_0. The frequency response obtained with this circuit is identical to that shown in Figure 3b. F_0 again is the frequency associated with the RC time constant

$$F_0 = \frac{1}{2\,\pi RC} \qquad (4)$$

The frequency response of the system is extended to a frequency F_1 where

$$F_1 = A_0 F_0 \qquad (5)$$

Note that the frequency response is automatically flat without having to match break points. However, the total bandwidth of the system, F_1, is still limited by the value of the shunt capacitance, C, across the input.

This improved frequency response of the feedback technique avoids the use of low values for R which could generate excessive current noise.

Refinements of the Feedback System. A major difficulty of the feedback system arises from shunt capacitance associated with the high-megohm resistor, R, in the feedback path. If the shunt capacitance across this resistor is C_F, then the bandwith, F_F, of the system is determined by the time constant, RC_F

$$F_F = \frac{1}{2\,\pi RC_F} \qquad (6)$$

A slight modification of the feedback loop can correct this problem, as shown in Figure 5. If the time constant, R_1C_1, is made equal to the time constant, RC_F, it can be shown that the circuit within the dotted line behaves exactly as a resistance R. The matching of time constants in this case does not become a drawback because the capacitances involved are all constant and not affected by input capacitance.

Noise in Current Measurements

Noise forms a basic limitation in any high-speed current-measuring system. The shunt system gives the simplest current measurement but does not give low-noise performance. A properly designed feedback system gives superior noise-bandwidth performance.

Noise Behavior of the Shunt System. High speed and low noise are contradictory requirements in any current measurement because some capacitance is always present. The theoretical performance limitation of the shunt system can be calculated as follows:

The rms thermal noise current, i_n, generated by a resistance R is given by

$$i_n = \sqrt{\frac{4\,kT}{R}\Delta f} \qquad (7)$$

The equivalent noise bandwidth, Δf,

251

of a parallel RC combination is $\Delta f = 1/(4\ RC)$ and the signal bandwidth (3 dB bandwidth) is $F_0 = 1/(2\ \pi RC)$. For practical purposes, peak-to-peak noise is taken as five times the rms value.

The peak-to-peak noise current can now be written as

$$i_{npp} = 2 \times 10^{-9}\ F_0\ \sqrt{C} \qquad (8)$$

In practice, a typical value for shunt capacitance is 100 pF. With this value, the following rule of thumb is obtained:

The lowest ratio of detectable current divided by signal bandwidth using shunt techniques is 2×10^{-14} A/Hz for a peak-to-peak signal-to-noise ratio equal to 1.

A corollary for this rule of thumb expresses the noise current in terms of obtainable rise time (10–90% rise time $t_r = 2.2\ RC$).

The lowest product of detectable current and rise time using shunt techniques is 7×10^{-15} Asec.

In this derivation it has been assumed that the voltage amplifier does not contribute noise to the measurement.

Noise Behavior of the Feedback System. There are three sources of noise in the feedback system that have to be looked at closely. The first two, input-stage shot noise and current noise from the measuring resistor, are rather straightforward. The third, voltage noise from the input device of the amplifier, causes some peculiar difficulties in the measurement.

Any resistor connected to the input injects white current noise (Equation 7). In the circuit of Figure 4, the only resistor that is connected to the input is the feedback

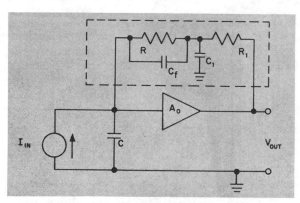

Figure 5. Frequency compensation in the feedback method removes the effect of shunt capacitance across the high-impedance measuring resistor

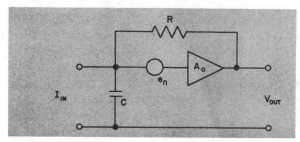

Figure 6. The voltage noise associated with the amplifier input device is an important source of noise in the high-speed feedback system

252

resistor, R. As in the shunt system, R must be made large for lowest noise. Because this noise is white, the total contribution can be calculated by equating Δf to the equivalent noise bandwidth of the system.

The second source of noise is the current noise from the amplifier input. This component is essentially the shot noise associated with the gate leakage current, i_0, of the input device. Its rms value equals

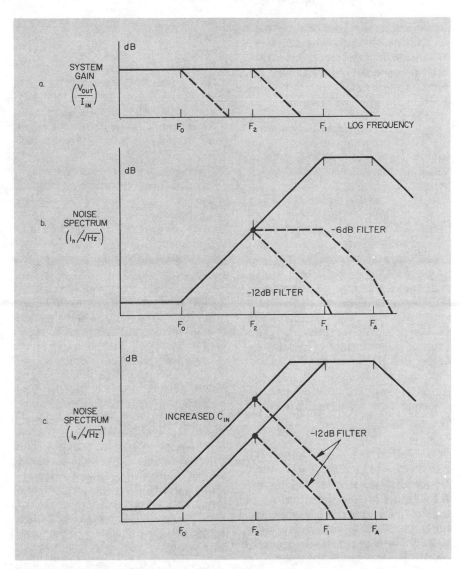

Figure 7. The bandwidth of the high-speed feedback system (a) can be limited by using a filter with either a —6 dB/octave or a —12 dB/octave roll-off. The effect of the filter on the noise spectrum is shown in b. Effect of input capacitance on noise is shown in c

$$\overline{i_n} = \sqrt{2\ ei_o\Delta f} \qquad (9)$$

where e is the electronic charge. The contribution of this noise generator is also white. Not only do these two noise sources generate white current noise, the noise in a given bandwidth is also independent of the input capacitance, C.

The major source of noise in a feedback current measurement is the noise contribution associated with the voltage noise of the input amplifier. The voltage noise can be represented by a voltage noise generator, e_n, at the amplifier input as shown in Figure 6. This noise generator itself is assumed to be white. However, its total noise contribution to the current-measuring system is not white.

Inspection of Figure 6 will reveal that at low frequencies a large amount of feedback is applied around the voltage noise source, e_n. However, the RC combination attenuates the high-frequency components of V_{out} so that no feedback is present at high frequencies. Thus, the noise contribution to the output voltage, V_{out}, from the voltage noise source, e_n, is no longer independent of frequency. The noise is "colored" and increases in intensity for all frequencies higher than F_0. The resulting noise spectrum is shown in Figure 7b. The total system noise is related to the area under this curve. Because the logarithm of frequency is plotted on the horizontal axis, the area under the curve at higher frequencies represents a significantly larger amount of noise than a similar area at low frequencies.

For comparison, Figure 7a shows the frequency response of the current-measuring system. Figure 7a is identical to Figure 3b.

It is interesting at this point to compare this noise spectrum with the frequency response of the voltage amplifier in Figure 1 as shown in Figure 3a. A voltage noise source at the input of the amplifier would generate a noise spectrum according to the amplifier frequency response as shown in Figure 3a. The noise spectrum of such a system, then, is identical to the noise spectrum of the feedback system as given in Figure 7b. This illustrates the well-known fact that signal-to-noise performance of a measurement cannot be improved by feedback techniques.

At the high-frequency end, the voltage noise is limited by the frequency, F_A, which is the high-frequency roll-off point of the operational amplifier. It should be noted that even though the useful bandwidth of the system extends only to F_1, there are noise components of higher frequency present. To obtain best wideband noise performance, these high-frequency noise components have to be removed. This can be achieved by adding a low-pass filter section following the feedback input stage. If the bandpass of this low-pass filter is made adjustable, this filter can serve the dual purpose of removing high-frequency noise and of limiting the signal bandwidth of the system.

To obtain optimum wideband noise performance under these conditions, a filter with a single high-frequency roll-off—i.e., -6 dB/octave—is not sufficient and -12 dB/octave is required. The effect of a -6 dB-filter is shown in Figure 7a, b. The filter is used to limit the system bandwidth to a frequency F_2, smaller than F_1. The effect of this filter on the noise spectrum is shown in Figure 7b. It can be seen that there are again high-frequency noise components above F_2, the useable bandwidth of the system. These can be eliminated by using a filter with a -12 dB/octave roll-off. The result of such a filter on noise performance is also shown in Figure 7b.

The reason that a -12 dB-filter gives superior performance over a -6 dB-filter in this application arises from the unusual manner in which the noise is colored. In most other cases, the noise is reasonably white across the bandwidth of interest and a -12 dB-filter then gives only marginal improvement.

One parameter affecting the noise behavior remains to be discussed. This is the effect of the input capacitance, C, on the voltage noise. An increase in the input capacitance will lower the frequency, F_0, and also F_1 since $F_1 = A_0 F_0$. Figure 7c shows how an increase in input capacitance changes the noise spectrum. Because total-system wideband noise is related to the area under the noise-spectrum curve, this increase in input capacitance results in more wideband noise.

For the same reason, increased noise also results from adding capacitance across the feedback resistor, R, which is often done to limit the signal bandwidth and is referred to as damping. For low-noise wideband performance, signal bandwidth should be limited by use of a filter amplifier.

Narrowband signal-noise performance, as when using the amplifier in a lock-in system, is dependent only on the net system noise at the chosen operating frequency and is independent of any bandwidth limiting in the input amplifier. As can be seen in Figure 7c, increased input capacitance increases narrowband noise if F_0 falls below the operating frequency.

Event-Counting Techniques

An entirely different approach to the measurement of small currents is possible with event-counting techniques. In these techniques, individual electrons are counted in a fashion similar to the counting of radiation quanta in photon counting. These techniques are limited to measuring small currents in a vacuum.

To be detectable, electrons must be detected with an electron multiplier or photo-multiplier. Very small currents can be measured, but circuit speed limits the maximum detectable current to the range of 10^{-12} to 10^{-13} A. Resolution in this technique is determined by statistical considerations, and the product of resolution and rise time for this method is 4×10^{-18} Asec.

High Performance Current Amplifier

As an example of what can be achieved with the feedback technique, the performance of a commercially available current amplifier will be described. This amplifier (Keithley Model 427) is designed to incorporate the principles described above. Figure 8 shows a block diagram of this high-speed current-measuring system. The input amplifier is a wideband, high-gain feedback system using a field-effect transistor input.

The choice of input device is determined by the trade-offs involving the desired sensitivity, stability, and frequency response. Figure 9 shows the frequency and sensitivity areas for which different input devices have proved optimum, taking into account that most of the devices are limited by the practical rule of thumb of 10^{-14} A/Hz. Electron counting is the exception with a practical limit of 10^{-17} A/Hz.

The input amplifier is followed by an adjustable low-pass filter having a -12 dB/octave roll-off and a voltage gain of $10\times$. The voltage gain in the low-pass filter avoids premature overloading in the input amplifier which can be seen as follows. The maximum output voltage V_{out} is ± 10 V. The maximum signal level at the input of the low-pass filter is, therefore, ± 1 V. At this point in the circuit, wideband noise could still be present and exceed the 1-V signal level. The voltage gain of 10 in the filter allows the total pre-filter wideband noise to exceed the full-scale signal by a fac-

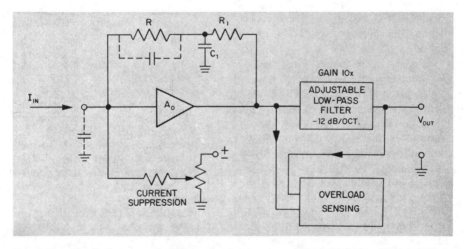

Figure 8.   Block diagram of a high-speed current amplifier (Keithley Model 427)

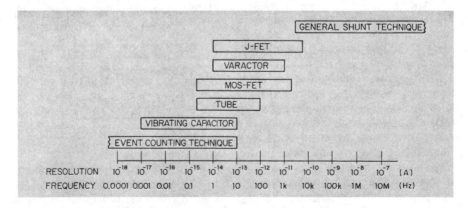

Figure 9.   Optimum performance area for different input devices depends on desired frequency response and sensitivity

tor of 10 (20 dB). The frequency response of this filter is adjustable for variable damping control.

To complete the 427 Current Amplifier a current-suppression circuit is added and overload-sensing circuits monitor pre- and post-filter overloads.

Noise performance must be examined for two methods of measurement. First, dc techniques for real-time measurements with bandwidth equal to the current-amplifier frequency response must be considered. Second, we must examine ac techniques (lock-in, Boxcar) where the current-amplifier frequency response must extend to the operating or center frequency while bandwidth is limited to a much lower value by the demodulator time constant.

It should be noted that use of ac techniques is to be avoided if the prime noise contributors are white, that is, if they have a constant noise per unit bandwidth. First, by definition, white noise increases with bandwidth regardless of center frequency, even if that frequency is dc.

Second, a current-amplifier—shunt or feedback—requires lower shunt resistors as operating frequency is raised and noise per unit bandwidth increases with operating frequency. Thus, translating the operating frequency from dc provides no reduction in noise and may actually increase it.

The sensitivity and speed of the 427 Current Amplifier for either dc or ac measurements can be compared to the best performance obtainable with the shunt method. The best noise–rise time product that can be achieved for dc measurements with a 100-pF shunt capacitance in a shunt system is equal to 7×10^{-15} Asec. The feedback amplifier achieves 2×10^{-15} Asec with a 100-pF input shunt capacitance, which is approximately a 10-dB improvement. This 10-dB improvement over a theoretically perfect shunt system typically covers the span of bandwidths from 1 Hz to 3 kHz, with input shunt capacitance between 10 and 1000 pF.

When used in ac narrowband systems, the degree of improvement de-

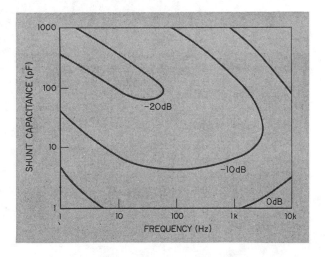

Figure 10. Plot of noise-improvement contours illustrating the improvement that can be obtained with the feedback method over the best that can be achieved with the shunt method at different operating frequencies

pends on the amount of input capacitance and the operating frequency. The achievable improvement over the shunt method can be plotted in a graph similar to a set of noise contours. Figure 10 shows the measured improvement (negative dB) that can be obtained with the 427 amplifier at any given frequency and input capacitance compared to a shunt method using an ideal (noiseless) voltage amplifier.

Acknowledgment

The contributions of G. Angeline, J. Maisel, and R. Babinec are gratefully acknowledged.

Novel Superconducting Laboratory Devices

Part I. Introduction to Superconductivity, the Josephson Effect, and Tunneling

By Juri Matisoo

SUPERCONDUCTIVITY has been a source of many interesting and useful devices. However, until the advent of superconducting magnets and the Stanford superconducting linear accelerator, the technological impact of superconducting devices has been small, primarily because of the high cost of adequate refrigeration. However, the low temperature physicist has often taken advantage of the properties of superconductors to design devices which he can use in his experiments.

Typical of such devices, which have been in use for twenty years or more, are current switches, thermal switches, and an application which is familiar to many chemists, the superconductive bolometer.

All of these are based on the rather dramatic changes in the properties of a metal as it becomes superconducting. Thus, the current switch depends for its operation on the resistance being zero in the superconducting state as long as the current density, the magnetic field, temperature, and the photon energy of any incident radiation all are sufficiently small. These are the usual quantities which destroy superconductivity. If any of these become large enough, the metal again looks like an ordinary conductor; that is, it has a fairly small but nonzero resistance, R_N. The ratio of resistances in the normal and superconducting states is infinite, however, and since the normal state can be restored by applying a magnetic field or by raising the temperature, current flow can be interrupted in one path and steered into another.

The thermal switch operates similarly with the heat flow being turned on or off. This possibility arises from a large difference in thermal conductivities of the superconducting and normal states. This time, however, things are turned about with the superconducting state being the poorer conductor. Switching action can be obtained by turning on a magnetic field to quench the superconducting state.

In the superconductive bolometer, the incident radiation (whose intensity is to be measured) creates a temperature rise in a superconducting film. This temperature rise can be measured very accurately by monitoring the resistance of the film whose temperature is just below the critical temperature, so that a small temperature change produces a large change in resistance.

These devices take advantage of one of the two major manifestations of the superconducting state, zero resistance and complete magnetic field exclusion. This exclusion of field exists regardless of how the

259

superconducting state is brought about.

Until rather recently all superconducting devices were based on these two properties and on the change in properties which accompanies the transition from the superconducting state to the normal state. To build these devices one really needed to know very little about the nature of the superconducting state. However, in 1962 Josephson made some startling predictions about what would happen if two pieces of superconductor were separated by a very thin (15Å) nonsuperconducting region. His predictions, which were based on understanding the details of what happens when a metal becomes superconducting, have led to a new class of superconducting devices, many of them with remarkable performance capabilities.

In what follows, Josephson's predictions will be outlined, then three laboratory instruments will be de-

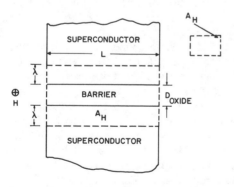

Figure 1. This shows (schematically) a tunnel junction. Two superconductors are separated by a barrier of oxide of thickness, d_{oxide}. The barrier is of area A and the area of field penetration, A_H, is indicated by the dotted lines ($A_H = L \times (2\lambda + d_{oxide})$)

scribed. These should be of interest to anyone wanting to measure very small magnetic fields ($\sim 10^{-10}$ Gauss), very small voltages ($\sim 10^{-15}$ V) or to anyone interested in a faster response, sensitive infrared detector ($\tau < 10^{-8}$ S; 5×10^{-13} W).

Superconductivity, Tunneling, and Josephson Effect

As in other co-operative phenomena, when a metal becomes superconducting, an ordering takes place. An easily visualized case of ordering occurs in ferromagnetism, in which electron spins align below the Curie temperature. The ordering in the superconducting state, however, is in electron momentum, not in their position. The process was finally understood by Bardeen, Cooper, and Schrieffer in 1957 (superconductivity was discovered in 1911). When a metal becomes superconducting the conduction electrons condense into a state of energy which is lower than the normal state. This comes about as follows: Under appropriate circumstances the force between two electrons can become attractive so that the electrons form bound pairs. Each electron is paired with another electron of equal but opposite momentum and spin. The force or energy which ties them together is weak (T_c, after all, is $< 20°$K), which the average rather far (~ 500Å) or so apart, so millions of other electrons are in between. If such a weak attractive interaction can exist with all those other electrons around, these other electrons must move in a very special way, in unison with the given pair; in fact, the center of mass momentum of *all* the pairs must be the same. This, then, is the order which exists in the superconducting state.

The number of paired electrons is a very sensitive function of temperature; it is largest at $T = 0$ and vanishes at $T = T_c$, the transition temperature. As the temperature is raised from $T = 0$, more and more of the pairs are broken by thermal energy until at T_c all the electrons are unpaired, the metal is again in its normal state.

From the behavior of the electron wavefunctions in elementary quantum mechanics, we can get an idea of how superconducting order is modified by electromagnetic potentials; that is, by electric and magnetic fields.

We know that, since the pairs have equal momenta, the waves associated with these pairs have the same wavelength (since $p\lambda = h$). But there is another quantity associated with waves, the phase of the waves, which determines how the waves overlap in space and time. It turns out that for maximum pair binding energy (which corresponds to the minimum energy state of the superconductor) the phases, as well as the wavelengths, must be the same. If by some means we attempt to impose phase differences, the superconductor will resist because a phase difference represents a condition of higher energy. Indeed, if one were able to momentarily impose a phase difference between two parts of the superconductor, a current will flow in response to this phase difference, ϕ, wiping it out.

Since phases are defined only modulo 2π, the current will be a periodic function of ϕ. Indeed, it turns out to be

$$j = j_1 \sin \phi \qquad (1)$$

where j is the current density and j_1 is a constant which is the maximum current density possible; it depends on the properties of the superconductor and its environment (such as T, etc.)

We can learn what the phase depends on from the properties of the elementary wavefunction. We know that an electron in a stationary state has the wavefunction

$$\psi(r,t) = \psi(r)e^{-i(E/\hbar)t}$$

so the phase is $-\dfrac{Et}{\hbar}$ where E is the energy of the state and $\hbar$ is Planck's constant.

The phase of the pair wavefunction turns out to be of the same form with $E = 2\mu$, where μ is the chemical potential (2 because there are 2 electrons forming the pair). Thus, even in the absence of any applied potentials the phase is a linear function of time. If the pairs see a potential (voltage), E will be $2\mu - 2eV$ and an additional time dependence is introduced.

We also recall that momentum in quantum mechanics depends on magnetic fields, or specifically on vector potential. If the fields are not too strong, the predominant effect of a magnetic field will be to give a spatial variation to the phase.

From this point of view, the zero resistance and perfect diamagnetism of a superconductor results from the successful attempt of the superconductor to exclude all phase differences.

Thus, for example, suppose that a superconductor carrying transport current develops a voltage drop along its length. This means that the chemical potential at the two ends (and along the entire length) is different, and thus the time variations of phases are different. Thus a phase difference between the ends will develop. Rather than go to this higher energy state, the superconductor carries the current without voltage (or exhibits zero resistance). The center of mass momentum for the pairs is still the same for all pairs; however, now it is not zero, but has a value corresponding to the impressed current. Eventually, however, as the current is increased the kinetic energy in this motion becomes comparable with the pair binding energy, pairs break up, and the superconducting state is destroyed. This maximum current density is j_c, the critical current.

The situation with respect to magnetic fields is similar. An impressed magnetic field leads to spatial phase differences which lead to supercurrents on the surface via Equation 1. These set up a magnetic field which cancels the applied field in the interior, having it free of fields and currents. Again, if the field is large enough, the superconductor can't resist and the superconducting state is destroyed.

On the basis of these arguments one would say that there is little hope of being able to introduce phase differences in a superconductor, because by the time you do, you have destroyed the superconductivity.

Josephson, however, pointed out that there exists a situation in which one could modify the phase externally, without destroying superconductivity altogether. He suggested that a barrier be interspersed between two pieces of superconductor. If the barrier is sufficiently thin, the superconducting order extends through the barrier, but is very weak in the barrier region. The barrier (empty space, a dielectric, or a nonsuperconducting metal) acts then as a region which is superconducting, but very, very weakly. But potential differences can be established across the barrier and magnetic fields can penetrate the barrier.

The basis for the difference between the response of the "weakly superconducting" region and the bulk superconductor lies precisely in the strength of the superconductivity. The bulk superconductor (as we have seen) resists all attempts to introduce phase differences until it catastrophically gives in. By that time superconducting order is gone too. The "weakly superconducting" region doesn't have the energy resources to fight back; it meekly accepts the introduced phase differences.

Obviously the barrier must be properly chosen; if it is too thick, the superconductivity of the barrier becomes so weak it disappears altogether; if it is very thin (zero in the limit), we are back to the bulk case. An intermediate situation can be devised, however, in which the necessary correlations are maintained, and yet these are not so strong that they become limiting. There are several ways in which this can be done in practice. One is to make a sandwich of superconductor, insulator, superconductor in which the insulator is $\sim 15\text{Å}$ thickness of oxide of one of the superconductors (Figure 1). Such a barrier thickness permits electrons to tunnel from one superconductor to the other without loss of their superconducting properties. Tunneling is a phenomenon which is common to all wave phenomena, and results from the fact that when a propagating wave strikes a barrier its amplitude does not drop to zero discontinuously at the barrier edge, but rather dies to zero in a characteristic way over a distance which, although small, is not zero (what this distance is depends, of course, on the barrier height, etc.). If the barrier separates two propagation regions and is thin enough that there is nonzero amplitude on the other side of the barrier, the wave can pass through and propagate freely on the other side. In the case of the electron, it is said to have *tunneled*.

Returning to the sandwich, called a tunnel junction, what is experimentally observed? The current through the junction depends on the phase difference across the barrier, as in Equation 1. In practice the current is controlled and the phase difference adjusts itself accordingly, so what is seen is zero voltage current through the barrier up to a critical value, I_c, when ϕ has become $\pi/2$. $I_c = j_1 A$, where A is the area of junction. If a magnetic field is applied, one observes that this critical current, I_c, depends on the value of the field. This follows from the spatial dependence of ϕ on H. Because of the field, different parts of the junction have different values of ϕ; hence the currents in different parts of the junction can interfere, reducing the *total* supercurrent the junction can carry. These changes in I_c are brought about by very small fields, and thus lead to the possibility of using a junction to *detect* very small fields. This is the operating principle of the magnetometer and the voltmeter.

A steady potential V_{dc} across the barrier leads to a time variation of ϕ

$$\phi = \frac{2e}{\hbar} V_{dc}\, t + \phi_0 \qquad (2)$$

and thus the current density is of the form

$$j = j_1 \sin (\omega_0 t + \phi_0) \qquad (3)$$

where the frequency ω_0 is $\frac{2e}{\hbar} V_{dc}$.

Since $\frac{2e}{\hbar} = 483$ MHz/applied microvolt, these currents are of very high frequency for very modest voltages. As is obvious from (3) the time average value of the oscillating supercurrents is zero in this case. If, however, an ac voltage is also applied, *i.e.*

$$V = V_{dc} + V_1 \cos \omega t$$

then j takes the form

$$j = j_1 \times$$
$$\sin \left(\omega_0 t + \phi_0 + \frac{2eV_1}{\hbar \omega} \sin \omega t \right) \qquad (4)$$

This is just standard frequency modulation and whenever $\omega_0 = n\omega$ $\times$ $(n = 1, 2 \ldots)$ a d.c. or zero frequency sideband is obtained. This means that if one traces out the d.c. current *vs.* voltage characteristic of the junction, increases in current are

observed whenever V is $\frac{n\hbar}{2e}$ ω. The presence of this step structure in the I-V characteristic constitutes detection of the alternating voltage (which can be radiation in practice). This is the principle of the radiation detector.

NEXT MONTH: The operating principles and characteristics of a magnetometer, a voltmeter, and a radiation detector which are based on the Josephson Effect will be discussed.

Dr. Ralph H. Müller, Contributing Editor of ANALYTICAL CHEMISTRY, will present timely comments on both Parts I and II of this feature in the February issue.

Selected Reading for Part I

A. *A recent book reviews superconductivity and applications.* M. H. Cohen (Ed.) "Superconductivity in Science and Technology," University of Chicago Press, Chicago, 1968.

B. *Superconductivity General:*
  (a) E. A. Lynton, "Superconductivity," Methuen & Co., Ltd., London, 1962.
  (b) P. G. De Gennes, "Superconductivity of Metals and Alloys," W. A. Benjamin, Inc., New York, N.Y. 1966.

C. *Josephson Effect:*
  (a) D. N. Langenberg, D. J. Scalapino, and B. N. Taylor, "The Josephson Effects," *Scientific American,* **214,** 30 (1966).
  (b) B. D. Josephson "Supercurrents Through Barriers," *Advances in Physics,* **14,** 419 (1965).
  (c) P. W. Anderson, "The Josephson Effect and Quantum Coherence Measurements in Superconductors and Superfluids" in: "Progress in Low Temperature Physics," C. J. Gorter, Ed., Vol. V, North Holland Publishing Co., Amsterdam, 1967.

D. *Superconducting Devices:*
  (a) V. L. Newhouse, "Applied Superconductivity," John Wiley & Sons, Inc., New York, 1964.
  (b) "Proceedings of the Symposium on the Physics of Superconducting Devices," University of Virginia, Charlottesville, Va. [copies available through Office of Naval Research, Physics Branch, Code 421, Main Navy Building, Washington, D. C. *Attention:* E. A. Edelsack].

Dr. Juri Matisoo is associated with the Thomas J. Watson Research Center, International Business Machines Corporation, Yorktown Heights, New York.

Novel Superconducting Laboratory Devices

Part II. Operating characteristics of a magnetometer, voltmeter, and radiation detector based on the Josephson Effect

By Juri Matisoo

LAST MONTH the general effects of superconductivity, the Josephson Effect, and tunneling were discussed as a prelude to the understanding of the operation of certain laboratory devices. The operating principles and characteristics of a magnetometer, a voltmeter, and a radiation detector will now be discussed.

Devices

A: MAGNETOMETER. The magnetometer and voltmeter, operate on the same principle; that is, the voltmeter is also basically a magnetic field detector. The magnetometer is largely the development of Zimmerman and Silver and their colleagues at Ford (*1*) while the voltmeter was developed by Clarke at Cambridge (*2*).

Let us review briefly the properties of a Josephson tunnel junction. First, there exists a critical current value, I_c, below which current can flow through the junction with no voltage drop and above which a voltage appears. The value of I_c depends strongly on applied magnetic fields, but in a rather unexpected way; rather than decreasing monotonically with increasing field, it is periodic in field and (ideally) is given by

$$I_J(H) = I_J(0) \left| \frac{\sin \dfrac{\pi A_H H}{\Phi_0}}{\dfrac{\pi A_H H}{\Phi_0}} \right| \qquad (5)$$

where H is the applied field; Φ_0 is a constant (2×10^{-7} G-cm^2 or 2×10^{-15} Wb) and A_H is the effective area of the field penetration, $L(2\lambda + d_{ox})$ (Figure 1).

This dependence is the result of summing the current density over the area of the junction. Because of the magnetic field (as discussed)

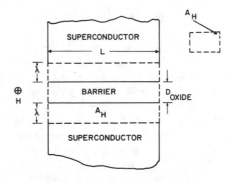

Figure 1. This shows (schematically) a tunnel junction. Two superconductors are separated by a barrier of oxide of thickness, d_{oxide}. The barrier is of area A and the area of field penetration, A_H, is indicated by the dotted lines ($A_H = L \times (2\lambda + d_{oxide})$)

there is a spatial dependence to Φ. If the field is just equal to $n\Phi_0/A_H$ ($n = 1, 2, \ldots$), one half of the current in the junction flows in one direction and half in the other, and the externally measured I_c is zero. If the field has some other value, a nonzero I_c is measured.

If one could make a junction of sufficient length so that $A_H = 1$ cm^2, then the period of I_c in field is 2×10^{-7}G. Since one can certainly detect a change of a fraction of a period, magnetic fields on the order of 10^{-9}G could be detected.

Unfortunately, it is difficult in practice to make such a long junction ($L \simeq 10^5$ cm), but this problem can be overcome by using the geometry shown in Figure 2. This shows two junctions as a part of a superconducting ring.

The magnetic field (to be measured) is oriented as shown in the plane of the paper. The idea in this geometry is the following. Recall again that in the superconducting state the phase of the pair wavefunctions is the same for all pairs and that when the superconductor sees a magnetic field, a spatial variation of the phase results. In particular the flux through the ring leads to a phase difference between the ends of the semicircles. If the flux is such that this difference, $\Delta_{AB} = n2\pi$ ($n = 0, 1, 2 \ldots$), the ends A and B have the same phase, and the critical current for the pair of junctions is twice that for one alone

$$I_{cr}(\Phi_r) = 2 \times I_c(\Phi) \left| \frac{\sin \frac{\pi n\Phi}{\Phi_0}}{\frac{n\pi\Phi}{\Phi_0}} \right|$$

However, if $\Delta_{AB} = m \left(\dfrac{\pi}{2}\right)$ ($m =$

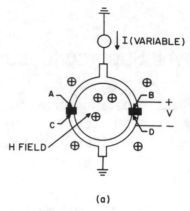

(a)

Figure 2. Schematic of the Zimmerman-Silver magnetometer consisting of a superconducting ring. The junctions lie between A-C and B-D, respectively. The orientation of the magnetic field is as shown. The leads for monitoring the value of I_{cr} are shown as well

$1, 3, 5, \ldots$) then the current at B is flowing opposite to that at A and the externally measured $I_{cr}(\Phi_r)$ is zero. Thus,

$$I_{cr}(\Phi_r) = 2I_c(\Phi) \left| \frac{\sin \frac{n\pi\Phi}{\Phi_0}}{\frac{n\pi\Phi}{\Phi}} \right| \left| \cos \frac{m\pi\Phi_r}{\Phi_0} \right|$$

(This is completely analogous to double slit interference in optics.)

If we for the moment ignore the interference effects in the junctions themselves (as can be approximately done in practice), then

$$I_{cr} = 2I_c \left| \cos \frac{n\pi\Phi_r}{\Phi_0} \right|$$

Since it is easy to make a ring with the area of the hole ~ 1 cm^2 we now have the situation in which the measured critical current for the ring, I_{cr} undergoes a complete oscillation whenever the field is changed by 2×10^{-7}G (2×10^{-15} Wb/m^2).

266

In order to measure magnetic fields, all we have to do is monitor the value of I_{cr} (which can be done as shown in Figure 2; one simply runs up the current I from zero and measures the value of I at which V goes from zero to some nonzero value. The value is I_{cr}).

To obtain an absolute value of H one must also know the number of complete oscillations I_{cr} has undergone.

Zimmerman and Silver have realized the structure shown in Figure 2 where the ring is made of niobium ($A \sim 1$ cm²) and the weakly superconducting regions have been made by splitting the ring, interspersing a layer of Mylar, and using Nb screws with very sharp points to form the tunneling junctions. In their magnetometer they indeed achieved a field period of 2×10^{-7}G and have detected field changes of 10^{-9}G.

There is a complication which limits the ultimate sensitivity of their device. The area of the ring cannot be made indefinitely large, as it turns out, because the amplitude of the oscillations ΔI_{cr} turns out to be inversely proportional to the inductance of the ring. In their structure, $\Delta I_{cr} \sim 0.2$ μA for an $L \sim 5 \times 10^{-9} H$.

This magnetometer is suited, obviously, to very low field measurements and particularly incremental ones; in practice one must shield very carefully against stray fields and noise. Also, the mechanical stability has been a problem in the past; recently, however, Zimmerman has succeeded in devising a structure which is mechanically very stable.

Compared with conventional (nonsuperconducting) magnetom-

eters, this has unprecedented sensitivity; however, other magnetometer schemes involving superconductors, but not the quantum interference phenomena, exist. These have comparable sensitivity.

B. VOLTMETER. Clarke's voltmeter is based on the same interference phenomenon as the ring. The basic idea is very simple (although his practical realization is unique). If one couples a superconducting loop into the magnetometer (as shown in Figure 3) and passes current, I, through the loop, this will create a field through the double junction ring (which is measured as before). The field is directly a measure of current in the loop. For practical configurations, this does not make a particularly sensitive ammeter but because of the zero resistance of the loop (it is superconducting) one can make a very sensitive voltmeter.

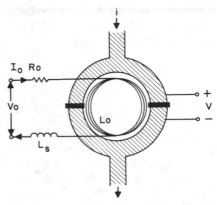

Figure 3. Schematic of the Clarke voltmeter. A superconducting coil of inductance L_o is tightly coupled to the magnetometer. The voltage to be measured is connected at the terminals of the loop. A fixed resistor R_0 converts the ammeter to a voltmeter. The other terminals monitor the critical current I_{cr}

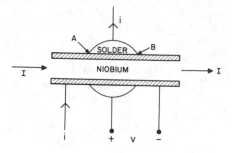

Figure 4. Clarke's realized structure. The center portion (where *I* flows) is a solid circular Nb wire. *A* and *B* indicate the junctions

One can easily find the sensitivity from Figure 3. The critical current period of the double junction ring is Φ_0, or in terms of current through the loop

$$I_0 = \frac{\Phi_0}{L_0}$$

Thus the input voltage required to change I_{cr} by one period is $I_0 R_0$ which is $\Phi_0 R_0 / L_0$. If one is willing to accept a one-sec time constant (L_0/R_0), then

$$V_0 = 2 \times 10^{-15} \text{ Volt}$$

I_{cr} will undergo one complete oscillation when the voltage is changed by $V_0 = 2 \times 10^{-15}$ volts. Clarke's realization of the configuration is shown in Figure 4. He took a piece of Nb wire and dipped it into a tin-lead solder, forming a blob around the wire as shown. This forms a double junction. Apparently the solder only makes contact to the Nb at the edges A and B in Figure 4. (The barrier presumably is the natural oxide of niobium); the superconducting path is completed by the Pb-Sn solder; the loop which creates the

magnetic field is the Nb wire itself. The current to be measured is represented by I, and the terminals for monitoring I_{cr} are attached as shown.

The characteristics of his achieved structure are: I_0, the current required for a change in I_{cr} by one period is $\sim 1\mu A$ ($L_0 = 10^{-8} H$, the inductance of Nb wire). He added a series resistance $R_0 = 10^{-8}$ Ω with a resulting $V_0 \cong 10^{-14}$ V and a time constant of 1 second.

In practice, this galvanometer is used as a null detector; *i.e.*, one attaches the galvanometer to the voltage to be measured and then by passing current through R_0 balances the total voltage seen by the loop to zero; by monitoring the change in I_{cr}, one can determine the value of the voltage which was to be measured.

The arrangement can be made as fancy as one likes, with various electronic schemes possible for balancing and monitoring I_{cr}.

As used by Clarke, the device has a sensitivity of $\sim 10^{-15}$V (with the Johnson noise $\sim 5 \times 10^{-16}$V). It is a simple rugged device of very low impedance. The impedance levels, however, are changeable by using transformer schemes.

Compared with other instruments, the usable voltage sensitivity of this device is orders of magnitude greater.

C. INFRARED DETECTOR. The infrared detector, which is useful even in the far infrared, is the development of Grimes, Richards, and Shapiro (*3*). The operating principle is based on the response of a weak superconductor to applied electric fields (potentials), rather than on the magnetic field response, which was the basis of the magnetometers. We have seen that a voltage applied across a barrier leads to oscillating supercurrents and that if an ac voltage is present, this leads to structure in the dc i-v characteristic of the junction. The amplitude of these current increases $\left(\text{at voltages } \frac{nh\nu}{2e}, \text{ where} \right.$ ν is the frequency of the incident radiation$\left. \right)$, varies with the power level of this radiation, as $J_n \left(\frac{2ev}{nh\nu}, \right.$ where v is the equivalent voltage amplitude of the incident radiation and J_n is the nth Bessel function$\left. \right)$

Thus, in principle a junction can be used to detect not only the power level but also the frequency (unambiguously). In the presence of monochromatic radiation well defined structure exists. If, however, the incident radiation is broadband, no steps appear; however, a change does occur, the critical current I_c decreases, which thus reflects the presence of radiation.

The structures devised by Shapiro et al. are point contacts and are fabricated by pressing together the ends of two pieces of wire, one of which is flattened and the other sharpened (by filing, for example). The wires are ~ 30 mils in diameter and a variety of superconductors have been used, Nb, Ta, In, and Pb.

Such a point contact with leads for measuring the i-v curve is immersed in helium and contact pressure adjusted until the desired i-v characteristic is obtained. The radiation to be detected is incident on the junction (via a waveguide or light pipe, depending upon frequency), and the i-v characteristics of the point-contact junction is monitored (for presence of step structure if the radiation is monochromatic or for a decrease in I_c, if broadened).

Shapiro et al. showed that: (1) They could detect a noise equivalent power of 5×10^{-13} W (1-sec time constant) at 2.5 cm^{-1} with an Nb-Nb junction, (2) The response of the Nb-Nb junction extended to $\nu > 40$ cm^{-1}, (3) The junction could follow a pulsed signal which had a 10-nsec rise time, (4) Current steps appeared in the dc i-v curve when the junction was irradiated with a monochromatic laser at 32.2 cm^{-1}.

These properties are quite remarkable when compared with other helium temperature far infrared detectors.

Conclusion

The devices which have been described by no means exhaust the possible applications of the Josephson Effect. At our laboratory, for example, we have concentrated on the possibility of utilizing the effect in a computer memory or logic element which appears to have high performance capabilities.

Langenberg and others at the University of Pennsylvania, as well as Dayem and Grimes at Bell Labs, have used junctions to generate microwave radiation (simply by applying a dc voltage across the junction) of high spectral purity. Unfortunately the power levels are low ($\sim 10^{-10}$ W) because of impedance matching problems. Zimmerman and Silver have constructed a spectrometer which they used to detect NMR in ^{59}Co, and so on.

It should be emphasized that the contributions of Josephson have led not only to the devices described but to a greater appreciation of the nature of the superconducting state. Furthermore, in the study of tunneling which was spurred by the work of Josephson, other devices of interest to chemists have been developed. One, in particular, is the molecular spectrometer of Thompson in which tunneling electrons interact with molecules in the barrier region and give information about molecular levels (4). It is probable that other ingenious and useful laboratory devices will result from a study of superconductivity, phase coherence, and tunneling.

Literature Cited

(1) J. E. Zimmerman and A. H. Silver, *Phys. Rev.*, **141**, 367 (1966).
(2) J. Clarke, *Phil. Mag.*, **13**, 115 (1966).
(3) C. C. Grimes, P. L. Richards, and S. Shapiro, *J. Appl. Phys.*, **39**, 3905 (1968).
(4) W. A. Thompson, *Phys. Rev. Letters*, **20**, 1085 (1968).

COMMENTARY

by Ralph H. Müller

IT IS QUITE apparent from the lucid and interesting discussion by Dr. Matisoo on "Some Novel Superconducting Laboratory Devices" that chemists have a lot of homework cut out for them to acquaint themselves with the attendant aspects of the Josephson Effect and its manifold applications. Dr. Matisoo is indeed correct in stating that these developments "should be of interest to anyone wanting to measure very small magnetic fields ($\sim 10^{-10}$ gauss), very small voltages ($\sim 10^{-15}$ V) or to anyone interested in a faster response, sensitive infrared detector ($\tau < 10^{-8}$ sec.) at a power level of 5×10^{-13} watt)." These fantastically small values open many vistas for measurements heretofore impossible. We possess the necessary tribal prejudices to believe that the analysts will be among the first to seek useful applications. This, coupled with their native reverence for precision and reproducibility and, above all, to know when they are right and when they are wrong, should add zest and purpose to an otherwise permissive age.

It may be some time before these devices can become generally available because few investigators are familiar with cryogenic techniques, particularly in the superconductivity range. Then, too, numerous problems will have to be solved in coupling systems to the cryogenic detector. Typical of the latter would seem to be the case of using the superconducting bolometer in the 1–40 μ range in the infrared where wave guides or light pipes are hardly feasible. These temporary difficulties will no doubt be solved when instrument companies will design equipment to utilize these effects in an efficient and practical manner. In the meantime the physicists concerned with the Josephson Effect and its manifold aspects will continue to explore new possibilities and some experimental simplifications. There is ample precedent for this ordered development. A case in point is the use of lithium-drifted silicon and germanium detectors for gamma and X-rays in which very high resolution is obtained by operating them at liquid nitrogen temperatures (77.3 °K.). Although this is well above the superconducting range, it turned out that, in a practical instrument, a liquid nitrogen reservoir required recharging only once in a week or ten days.

In any case, if we remember that superconductivity was discovered in 1911 and we are now blessed with the Josephson prediction and discovery a half century later, the residual difficulties should be solved very quickly because our technology is vastly improved. It is related that, when Professor H. Kamerlingh-Onnes was summoned from his laboratory at Leyden to address the Royal Society of London on his cryogenic studies, he decided to speak about superconductivity. His approach was decidedly old fashioned. He did not arrive with an array of higher order differential equations or computerized values, but simply immersed a lead ring in liquid helium and induced a current in the ring. This he carried across the English Channel and up to London. Before his audience, he waved a search coil, coupled to a ballistic galvanometer and clearly demonstrated the continuous and unabated flow of current in the lead ring. It required no great argument to convince his listeners that the resistance of the lead was, indeed, zero, inasmuch as the current was induced in the ring more than 24 hours before in Holland.

Dozens of questions arise in one's mind when these extraordinary sensitivities are considered. Shall we be able to look into the "structure" of Johnson noise, the magnitude of which is such that the root mean square voltage across a 1000-ohm resistor at room temperature is of the order of a few microvolts? It is given by:

$$\overline{V^2} = 4\pi kT\Delta f$$ where Δf is the band width in cycles per second. Noise thermometers employing these principles have been used for more than 20 years, but the technique is very difficult at these low voltages.

Can a magnetic switch arrangement be devised to improve the measurement of magnetic susceptibility of substances? This is a subject of great theoretical interest but experimental methods have been notoriously unimpressive and primitive. The measurement of magnetostrictive effects (shrinking or expansion of a conductor in a magnetic field) is beset with great difficulties because thermal expansion coefficients are orders of magnitude greater than the magnetic effect. Despite this, early work led to the development of alloys (permalloy, perminvar, etc.) of extremely high permeability. These are of great commercial importance.

A physicist once remarked that our knowledge of magnetic phenomena was disgraceful. In the hands of Francis Bitter and others, this is no longer true, but much remains to be done in extending the techniques to the solution of the chemists' problems. Theory has far outrun practice and we make relatively little use of the Faraday effect, the Hall effect and its many cousins, the Nernst, von Ettingshausen, Leduc, and Corbino effects. When inquiring minds get back to phenomenological studies, something new and useful almost always pops up.

The historical record is clear and we still have a vast store of minute second-order or feeble effects awaiting application. In the past, such strange effects were observed and interpreted, but prevailing technology was not adequate for their exploitation. We no longer have this excuse.

Tunable Lasers

by R. G. Smith

Bell Telephone Laboratories, Incorporated, Murray Hill, N. J.

Several techniques achieve tunable laser action, some of which yield large tuning ranges. Possible analytical uses will result when instrumentation of dye laser and parametric oscillator systems become readily available

IN THE FIELDS of physics and chemistry the interaction of radiation with matter plays a key role in the research of most workers. Fundamental to these studies is the availability of an appropriate source of electromagnetic radiation. In many cases the development of sources has opened up fruitful new areas of research or rejuvenated old ones. For example, following World War II developments in microwave sources made possible great advances in microwave spectroscopy, and recently the laser has revived and extended the field of Raman spectroscopy. Sources in the microwave range and below are characterized by high spectral purity and high power levels, and have the desirable property of being tunable. It is the tunability, for example, which makes possible microwave absorption spectroscopy. In the optical and infrared regions of the spectrum a combination of monochromator and broadband light source is used to obtain light of various wavelengths. The primary drawback of the monochromator-light source combination is that the power available in a given frequency range is low, except for a small number of resonance lines. The power per unit wavelength range becomes smaller as one moves to the infrared. Consequently, as one looks for higher resolution, for example in absorption spectroscopy, signals become weaker, ultimately causing detection problems. The desirability of high power tunable optical and infrared sources is clear. The methods for achieving such tunable sources are at last in hand, and recent developments in tunable lasers suggest that their evolution into available laboratory apparatus may not be too far off. In this article several techniques for achieving tunable laser action will be described, with emphasis on those techniques which yield large tuning ranges. They include the stimulated Raman oscillator, the optical parametric oscillator, and the dye laser. Other techniques will be briefly mentioned.

Stimulated Raman Oscillator (1)

The Raman Effect, familiar to most chemists, involves the inelastic scattering of radiation by matter and is used as an analytic tool to permit the study of vibrational modes of molecules. In Raman scattering the difference in frequency between the incident and scattered radiation is characteristic of the scattering medium. Under normal illumination conditions the amount of scattered radiation is a small fraction, perhaps one part in 10^6, of the incident beam and is emitted in all directions. However, for sufficiently intense illumination, such as is afforded by high power pulsed lasers, the scattered radiation becomes stimulated rather than spontaneous; it reaches power levels comparable to the incident pumping beam and it is emitted as a collimated beam. As such, it has the basic characteristics of laser radiation, with the important fact that it is shifted in frequency from the incident laser. Since large numbers of materials—including gases, liquids, and solids—have been made to emit stimulated Raman radiation, the number of discrete shifted frequencies is large. Further, since radiation both up-shifted and down-shifted in frequency by integral multiples of the Raman frequency is usually observed, a given Raman oscillator will produce a picket fence of frequencies given by the relation

$$v = v_l \pm n\, v_r \tag{1}$$

where v_l is the frequency of the laser, v_r is the Raman frequency of the molecule, and n is an integer. Useful powers are obtained for values of n up to 3 or so. By using the output of one Raman oscillator to pump another oscillator using a different material, large numbers of frequencies can be obtained. By employing various combinations of materials, as well as by using the pulsed ruby or neodymium lasers and their

second harmonics, an extremely large number of discrete frequencies are possible. The primary drawback of this technique for achieving tunable laser action is that continuous tunability is not possible (one must in general change materials to obtain a new wavelength). Further, some regions of the spectrum are difficult to cover with other than a large number of iterations, and operation has been achieved to date only on a pulsed basis. Its advantages are that a Raman oscillator is easy to build, requiring only a pulsed laser, and in the case of Raman active liquids, a simple absorption cell and a few cubic centimeters of chemicals, and it provides high powers reproducibly at a large number of discrete wavelengths.

A recent advance reported only this year has produced a continuously tunable Raman oscillator with a tuning bandwidth of 200 cm^{-1}. Continuous tuning was made possible by using materials which had transitions which were both Raman and infrared active. Because the transitions are infrared active, they radiate the infrared, providing a powerful source in this frequency range. Observation of stimulated radiation extending from 40 to 200 microns has recently been reported.

Optical Parametric Oscillator (2)

Another device for the generation of tunable coherent light is the optical parametric oscillator which also makes use of optical nonlinearities. The basic configuration of the optical parametric oscillator consists of a suitable nonlinear crystal surrounded by a pair of mirrors to provide feedback for the oscillation. The source of energy for this oscillator is usually derived from a single frequency laser, called the pump, and the oscillation consists of two electromagnetic waves called the signal and idler. The frequencies of the pump, signal, and idler are related by

$$v_p = v_s + v_i \tag{2}$$

The nonlinear crystal in which the interaction between the three waves takes

Figure 1. An example of one of the continuous optical parametric oscillators

place must have several properties: it must have a second-order optical nonlinear coefficient relating the polarization to the electric field by a term $P = dE^2$; it must be transmitting at the three frequencies involved, and it must possess sufficient birefringence to allow synchronous propagation of the three waves. The latter condition is expressed mathematically as

$$n_p v_p = n_s v_s + n_i v_i \qquad (3)$$

where the n's are the indexes of refraction at the three frequencies. Birefringence is required of the crystal to overcome the effects of dispersion, limiting the number of suitable nonlinear materials. Tunability of the optical parametric oscillator is accomplished by varying the indexes of refraction of the crystal in a controlled manner, altering the propagation constants and ultimately the frequencies v_s and v_i which satisfy Equations 2 and 3. Tuning has been achieved by varying the angle of propagation through the crystal, its temperature, or by applying an external electric field. Materials in which optical parametric oscillation has been achieved are ADP (ammonium dihy-

drogen phosphate), KDP (potassium dihydrogen phosphate), $LiNbO_3$, (lithium niobate) and $Ba_2NaNb_5O_{15}$ (barium sodium niobate) nicknamed "bananas."

A number of optical parametric oscillators have by now been constructed. Some of the salient features of oscillators reported to date are tunability from 0.7 μ to 2 μ, and efficiencies of up to 45%. Most of the oscillator work has been pulsed and peak powers of several hundred kilowatts have been reported. One of the features that makes the optical parametric oscillator particularly interesting is the fact that it is capable of continuous operation. Two continuously pumped tunable oscillators, one using $LiNbO_3$ and the other $Ba_2NaNb_5O_{15}$ were constructed last year, one operating in the visible and the other in the infrared. An example of one of the continuous optical parametric oscillators is shown in Figure 1. In very recent work, an efficiency of 30% with an output power of 45 MW has been achieved using a "banana" crystal only 5 mm long. These results clearly point to the practicality of tunable oscillators. Another

274

feature of the optical parametric oscillator is the fact that it is capable of operation in the infrared out as far as the nonlinear material used is transmitting.

Although great strides have been made in the development of optical parametric oscillators during the past year all the problems are not solved. Particularly troublesome with both pulsed and continuous oscillators is the lack of wavelength reproducibility, in some cases excessively broad emission spectra, and amplitude instabilities. Considerable improvements have recently been made both in amplitude stability and spectral width and further improvements appear feasible with improved designs. It is not unreasonable to expect that progress in optical parametric oscillators will parallel that of the laser in the past several years where reasonably-priced reliable devices have resulted.

Dye Lasers

In contrast to the Raman oscillator and the optical parametric oscillator which use nonlinear optical effects, the dye laser makes use of the basic laser technique of inverted population to achieve gain. The active species of the dye laser is an organic dye in solution. It is pumped optically, either with the output of another laser or by a flashlamp. The laser configuration consists of the optically-pumped dye cell and a pair of reflectors to provide feedback for the oscillation. Its simplicity is one of the main features of the dye laser.

The wavelength range covered by dye lasers is large, extending from 4000 Å to 1 micron. This broad range is covered because a large number of dyes can be made to exhibit laser action, with different dyes being used to cover different regions of the spectrum. Familiar compounds such as Rhodamine 6G and Fluorescein are examples of dyes which have been made to lase. The output spectrum of a dye laser is usually broad, varying from the order of 50 to several hundred angstroms. The central wavelength can be shifted by varying such parameters as the solvent, the concentration, the cavity length, and cavity losses. The most successful method of tuning that has been found to date employs a diffraction grating as one of the reflectors. When used as a Littrow reflector (reflected wave directed back toward incident wave) the output wavelength can be adjusted by simply rotating the grating. With different dyes and a grating, most of the visible can be covered. Further narrowing of the spectrum to the order of 10^{-2} Å bandwidth can be done through the use of other frequency selective elements within the cavity. The features of narrow linewidth, ease of tunability, simplicity, and high efficiency (50% has been reported for laser pumping) make the dye laser an attractive choice as a tunable source. Although to date dye lasers have been made to lase only on a pulsed basis, it is the opinion of many workers in the field that continuous operation can be achieved by using high-intensity arc lamps as the pumping source. At present the major limitation of the dye laser is that it does not appear suitable for operation in the infrared due to the lack of dyes which have efficient fluorescence bands in that region.

Three principal types of tunable laser have been described. The simplest, because they make use of readily available organic liquids, are the stimulated Raman oscillator and the dye laser. Provided one has a high power pulsed laser (and perhaps the ability to generate the second harmonic of this laser) either of these two types of laser is simple to construct—the Raman oscillator giving a large number of fixed frequencies and the dye laser a more continuous coverage. Continuous operation of neither has been demonstrated. The optical parametric oscillator is the most difficult to construct, primarily because it requires crystals of high optical quality which are not yet readily available. The characteristics of the optical parametric oscillator which make it interesting are continuous tunability, operation in the infrared, and operation on a cw basis. We have mentioned only three methods of achieving tunable laser action; a number of others exist (4). They include the direct tuning of lasers by the application of electric or magnetic fields, by variation of temperature, or by the application of pressure. An interesting example of the last technique was the demonstration of tunable laser action from 7.5 to 25 microns by the application of hydrostatic pressure up to 14 kbar to a lead selenide diode laser. By using any of the techniques described to generate new wavelengths it is possible to further extend the spectrum covered by mixing various combinations of frequencies in suitable non-linear media to obtain sum and difference frequencies. As we have seen, a number of techniques exist to achieve tunable laser action. At present, product oriented companies are working on dye lasers and optical parametric oscillators and, in fact, one company is advertising a pulsed dye laser for sale. In the next year or two many competitive systems should be available, and it is fair to assume that a large number of interesting experiments will result.

References

(1) A good summary of the Stimulated Raman Effect is given by R. W. Terhune and P. D. Maker in "Lasers," **Vol. II,** ed Albert K. Levine, Marcel Dekker, Inc., New York, 1968.

(2) A review of optical parametric oscillator theory is given by S. A. Akhmanov and R. V. Khokhlov in *Soviet Physics Uspekhi,* 9 (2), 210–222, September–October 1966.

(3) For details of dye lasers see "Organic Dye Lasers" by M. R. Kagan, G. I. Farmer, and B. G. Huth, Laser Focus, pp 26–33, Sept., 1968. and "Organic Lasers" by P. P. Sorokin, *Scientific American* 220, pp 30–40, February, 1969.

(4) A more complete paper by the author including a detailed bibliography will appear in the *Annals of the New York Academy of Sciences.*

Addendum

Since this article was first written, many advances have been made in the field of tunable lasers. In the field of stimulated Raman scattering, the greatest excitement has centered around the tunable "spin flip" Raman oscillator. These oscillators, with indium antimonide as the nonlinear material, have been pumped in the infrared by the CO_2 laser in the 10-micrometer region and by the CO laser in the 5-micrometer region. Operating at cryogenic temperatures, these devices can be tuned by varying an applied magnetic field, thereby giving rise to continuous tuning. The tuning range is further extended by using different pump frequencies. Line widths less than 1 kHz have been reported for the output of such devices. Continuous operation has been obtained when pumping with the CO laser, and with further material development, continuous operation in the 10-micrometer region appears feasible.

The field of optical parametric oscillators has also advanced. An oscillator has been built which tunes over the full visible spectrum, and, recently, operation in the interesting 10-micrometer region of the infrared has been achieved. One company is currently selling an optical parametric oscillator which covers the range from 0.55 to 3.7 micrometers. Further wavelength coverage has been obtained by harmonic generation and sum frequency mixing in nonlinear crystals.

Progress in dye lasers has been extremely rapid in the past three years. Most important is the achievement of continuous operation of dye lasers by use of the argon laser as the pump source. Continuous dye lasers presently emit in wavelength range from about 0.55 to 0.65 micrometers, but extension of this tuning range cannot be ruled out. Linewidths as narrow as 0.001 cm^{-1} have been reported. Nonlinear mixing has also been employed to extend the range of tunability of dye lasers. At the present time there are a number of companies offering tunable dye lasers, both continuous and pulsed.

Advances in semiconductor lasers make them an important contender in the tunable laser field. At the present time coarse wavelength tuning is achieved by alloying the various semiconductor materials. Fine tuning is achieved by varying the temperature of the device. Semiconductor lasers can cover the wavelength range from 1 to 30 micrometers. By use of semiconductor lasers chosen to emit in particular wavelength bands, extremely high-resolution spectroscopy has been demonstrated.

At the present time work is continuing on all phases of tunable laser development. Efforts are focused on achieving higher power levels, narrower linewidths, and attaining broader tuning ranges. With the present state of the art in devices, many workers are turning to new and exciting applications of these sources. Among the areas being investigated are very high-resolution spectroscopy, particularly in the area of air pollution detection, and the study of chemical reactions by the selective excitation of atoms and molecules. The prospect of new and innovative work with these tunable sources seems promising indeed, and in the next few years many new and fruitful areas of research will undoubtedly evolve.

COMMENTARY

by Ralph H. Müller

WE FIND Dr. Smith's discussion of Tunable Lasers most provocative, not solely as a promise for new, versatile and useful light sources but for the multitude of electro-optical phenomena which will undoubtedly be investigated in the course of these studies.

In the comparatively short time that the laser has been used, it has found important uses in communications radar, holography, surgery, time standards, metal cutting and welding, alignment instrumentation, high temperature studies, and in ignition systems. In the forty-one years since its discovery, the Raman effect has been a source of information on molecular structure supplementary and complementary to infrared spectroscopy, but until the advent of laser excitation of Raman spectra, it was rarely used by the analytical chemist, largely because it was too insensitive. Now that excitation by a high power pulsed laser can cause the scattered radiation to become stimulated rather than spontaneous, we have the high power Raman oscillator. This is an accomplishment that transcends mere improvement in detection sensitivity—it presents innumerable possibilities. The number of systems which can be used and the multiplicity of frequencies which can be generated would seem to be limited only by the number of substances which the chemist can find on his reagent shelf. It is of major importance that the radiation so produced is coherent and collimated.

Dye lasers are likely to be of greatest interest to chemists even though, as the author points out, they are not particularly useful in the infrared. Fluorescence spectrophotometry is a highly developed field and elaborate instruments are available for the accurate delineation of both the absorption (excitation) and fluorescence spectrum. Sensitivities are of the order of parts per million or parts per billion and hundreds of substances have been precisely characterized, many of them of biochemical and medical importance. Is it not reasonable to assume that many of these substances can be excited to the point where laser action ensues? If high levels of output can be attained, it would seem that detection and identification could be achieved by relatively simple optical methods of abridged spectrophotometry.

The physicist and engineer will continue to study and improve tunable lasers and, at each stage of development, the chemist will benefit from the results, but he may do well by considering the implications of Raman and dye laser action. He should not merely regard them solely as high intensity, tunable light sources to be employed in conventional spectrophotometric techniques.

In these thoughts, we may do well to keep the meaning of the acronym LASER, constantly in mind—"light amplifications by stimulated emission of radiation." To the extent that an array of molecules can, by this optical feedback and ultimate oscillation, become a relatively powerful little broadcasting station and of stable frequency, it becomes relatively simple to establish its identity.

To switch for a moment to the region of much lower frequencies—i.e., in the microwave region—the ammonia maser (microwave amplification by stimulated emission of radiation) can be excited by injecting microwaves into a resonant cavity filled with ammonia. At a frequency of 24,000 megahertz, amplification at very low noise levels occurs because this corresponds to an excitation level of the ammonia molecule. The corresponding wavelength is 1.25 cm or wavenumber $v = 0.8$ cm^{-1}. At higher amplification, the initiating signal can be turned off because sustained oscillations occur. The maser is then a constant frequency source or "ammonia clock." An accuracy of 1 part in 10^{10} is characteristic of the ammonia maser. The hydrogen maser has a stability estimated to be 1 part in 10^{13}, but for ruggedness, relative simplicity, and wide experience in its use, the cesium beam clock with stability of 1 in 10^{10}

would supercede both in practical applications. Atomic clocks have attained acceptance as international standards, replacing celestial measurements, since the October 1967 conference in Paris. The cesium beam clock is the basis of the collision avoidance system (CAS) expected to be operational in the United States for all commercial aircraft in late 1970. (*Reader's Digest*, July 1969, pp 106–110).

This digression on masers or related high stability oscillators has significance in another sense. With stability of this order, would one feel any uncertainty that one was dealing exclusively with ammonia molecules, hydrogen molecules, or cesium atoms in the respective examples? In the case of laser action in a Raman oscillator or dye oscillator with comparable frequency, stability could afford quick identification by relatively simple means.

These considerations lead us to suggest that it may be profitable to study the analytical possibilities of inducing laser action in samples as a direct approach to identification and possible quantitative estimation.

In the general problem of laser development, there is an extensive background of information on physical optics, much of it of only passing interest today in other respects. In mixtures of gases, much is known about the influence of inert gases on the population in metastable states. For example, mercury vapor excited to the first resonance level at 4.86 volts (2537 A) is easily thrown into the adjacent metastable state by the admission of nitrogen with an increase of lifetime by a factor of 10^4. The early studies of R.W. Wood on the step-wise excitation of the mercury spectrum by successive irradiation with lines from another mercury arc were completely in accord with the known energy levels in this system, but beyond that, afforded several neat tricks to artifically enhance certain transitions. As he once remarked—"It's all very simple, gentlemen, it's done with mirrors."

Ultracentrifuge

Some Analytical and Preparative
Uses in Biomedical Research

Ronald J. Casciato
Director, Research Applications, International Equipment Co.,
A Division of Damon, Needham Heights, Mass.

The ultracentrifuge is an increasingly important instrument in biomedical research. Separations of cellular and subcellular species are exceedingly important to a better understanding of human disease mechanisms

TODAY'S SCIENTIFIC RESEARCH community is experiencing a revolution within its ranks. Fewer than 20 years ago not only were basic research and medical research widely separated regarding the level of activity and direction, but within each group, subdivisions of specific subgroups were well defined. Research was carried on almost in competition with other disciplines in the same major group. As the level of technology and research activity increased, it became clear that this division was not always in the best interest of research, and growing interrelation between groups has evolved. Dr. N. G. Anderson, chairman of the MAN program at the Oak Ridge National Laboratories, has aptly stated, "Each science, as it matures is concerned with defining those basic units which in turn are definable by the next subjacent discipline."

The effect is seen in many of today's important scientific achievements, both technological and theoretical. To make the system work effectively, each discipline must work toward diminishing its boundaries to achieve a more complete understanding of today's research problems. This interdisciplinary approach has provided many new and exciting approaches to old problems.

Separation Necessary in Cell Research

It is now generally accepted that the majority of human diseases are ultimately to be understood at the molecular level, and it must be concluded that separation and cataloging the molecular constituents of human cells are necessary groundwork for attempts to describe human pathological states in molecular terms. It is clearly this attempt to separate, purify, isolate, and ultimately characterize the molecular constituents of a cell that occupies much of the research attention today.

Separations are made by utilizing one or more of the characteristic properties of the specific entity, whether it is an organelle, such as a mitochondrion, or soluble serum protein. Some useful properties include surface potential, ionic charge, membrane permeability, enzymatic specificity, solubility and/or insolubility, color, absorptiveness in light, morphological appearance, size, shape, density, and sedimentation coefficients. It is these latter properties that almost totally govern the effectiveness of the ultracentrifuge in biomedicine.

Many cellular components, including macromolecules, exhibit characteristic size, shape, and density phenomena. In combination, these parameters coupled with a specific technique of sedimentation result in a "sedimentation-coefficient" which is in itself, characteristic of the component. By utilizing centrifuge technology along with sedimentation coefficients, a separation or fractionation of a mixture of cellular components may be effected.

Pertinent to this discussion is the development of some principles of centrifugal force and technology of design, as well as discussion of the application to biomedical research problems. Centrifugal force, simply stated, is an enhancement of gravitational force. The concept of using gravitational force for separation of particles is easily envisioned when one considers a cylindrical glass tube of water into which is dropped a mixture of sand and gravel. As the mixture sediments, the gravel stones are only slightly retarded in the water and rapidly drop to the bottom. The sand, however, settles more slowly and after a short time a visible separa-

tion is evident with the sand layered on top of the gravel and the liquid above is again clear. If the process had been stopped after the gravel had reached bottom and before the sand had fully sedimented, a separation would have been accomplished. This separation is the total result of various forces (including gravity) on the size, shape, and density of the suspended material. This example is directly related to the separation of cellular particles or organelles and components, where the differences are magnitudes of the size, shape, and density of the parameters; in cellular particles they are thousands of times smaller than those of the example above. The ability to separate cellular entities depends on the ability to increase the effective force of separation, in this case gravity. The most desirable means from an engineering point of view, is to increase the force by rotation. Not only are space requirements lessened, but resulting instrumentation facilitates easy sample handling.

By rotating a cylinder about an axis at one end it is evident that the force generated at the opposite end is increased. By increasing the force, the time of separation can be lessened and/or the sizes of the particles to be separated may be much smaller. If increased g force is to be achieved, thus permitting the separation of cellular components, certain design requirements must be satisfied. By briefly considering existing commercially available equipment, it becomes evident that the ultimate ultracentrifuge is a very complex interrelated laboratory unit. A variable-speed drive system is required. Currently available units have speed ranges to 60,000 rpm (approximately six times that of a jet aircraft turbine engine) and equal to 405,-000 × g. To maintain the physiological state of the samples, refrigeration systems must be considered; sample temperatures should be held constant at a range from 0–20 °C. Rotation at controlled high rpm alone is an arduous task. Add to this the load forces of a rotor and sample material, and the task becomes impossible. It is therefore necessary to construct the instrumentation so that the rotation systems operate in a vacuum, thus eliminating the windage forces acting upon the rotor. High vacuum is also essential for ultimate temperature control. Current vacuum levels approach 0.1μ. This provides essentially a cabinet housing a vacuum system, refrigeration, controls, and drive system, all operative to a chamber into which the rotor and samples will be placed.

Rotor Types

The samples themselves are self-contained in a variety of "rotors" in either tubes or bottles in which the actual sedimentation takes place. The design of centrifuge rotors is very complex. Rotors have historically taken only a few configurations: fixed angle, horizontal, or swinging bucket, and more recently, zonal.

The fixed angle rotor is one in which the cavity that holds the tube is set at an angle to the central axis (the angle is determined by the types of pelleting material desired). It is useful for pelleting materials at all speeds, and if the sample suspension is homogeneous, a concentration of material is obtained by decanting the supernatant. In contrast, the horizontal or swinging bucket rotor is one in which the centrifuge tube hangs in a vertical position at rest, and during acceleration the tube swings out to a horizontal position. In this rotor the material is rarely pelleted but separations are based on liquid density gradients.

The zonal rotor is basically a large-volume cylindrical pressure vessel rotating on an axis. The design of zonal rotors permits larger volumes to be used at existing speeds, using materials such as titanium and aluminum. In all cases it is important that the design of the rotating system be sufficient to withstand the repeated stress of cycling a rotor to high rpm's and g levels.

In practice, the ultracentrifuge is utilized for pelleting or separation of subcellular components or particulate material. In fixed-angle rotors the operation is straightforward—homogeneous suspensions of sample material, whole tissue homogenates, or cellular components are contained within a sealed tube at a fixed angle to the rotating axis. The fluid within the tube is affected by convective forces that set up a swirling motion from the top to bottom of the tube. As the heavier material in the suspension nears the tip of the tube, increasing g force causes pelleting as the suspension fluid is centrifuged for longer periods of time; smaller fragments are subsequently pelleted as a result of high g force for an extended period of time. This process removes suspended particulate material from a homogeneous solution. It is, however, of little value in the separation of subgroups within the sample, since pelleting causes overlapping of the material.

Types of Centrifugation

Sample fractionation may be accomplished by the technique of differential centrifugation. In this method a given sample suspension is spun in a horizontal tube, and since the particles in the suspension are distributed uniformly, the material at the bottom of the tube experiences the highest g force and sediments (pellets) rapidly. This initial pellet captures with it some smaller material and soluble nonsedimenting material. Continuing centrifugation causes smaller particles to layer over the initial material. The process is continued, but contamination of each layer with its subsequent layer is approximately proportional to the differences in sedimentation rates. The net result is that only the smallest particles may be obtained in a pure state in a single centrifuge run. Since many cellular components have a rather wide range of sedimentation rates, it has been, in many cases, sufficient to use differential centrifugation for preparation. This can easily be seen when whole cells, organelles, and soluble proteins are centrifuged in the same run. It is increasingly difficult, if not impossible, to satisfactorily use this method for the subsequent separation of particles with a narrow sedimentation-rate range. Particle-sizing studies are also cumbersome using this method. It is difficult, from an experimental point of view, to consider successive stages of differential centrifugation, such as a resuspending and recentrifuging, until additional separations are achieved.

An excellent method for sophisticated separations using liquid density gradients is, however, in widespread use. Liquid-density centrifugation was initially employed to provide a means for selectively separating families of particles during a single run. The liquid density gradient is simply made by placing a lighter liquid over a heavier liquid so that two layers are formed. Diffusion at the interface gradually lessens the sharpness of the boundary between layers and after some time a more less-continuous transition in density from the top to bottom can be seen. In this gradient it is possible to effect separations based on either size and/or density of the suspended particles. The density of the gradient is centrifugally stable, and the sample is layered on top of the gradient. During centrifugation, material is separated into bands of material spread proportional to the sedimentation rates.

Two types of separations are possible. *Isopycnic,* or equilibrium centrifugation is based on the principle that particles sediment to their own density in a liquid density gradient whose density range encompasses extremes within which the density of the particles lie. As the centrifugal force acts upon the particles, their immediate location may be either above or below their specific gravity. They will be forced in an axial direction until the density of the surrounding fluid matches the specific gravity of the particles. *Rate-zonal* centrifugation uses the relative size of the particles as well as density to effect

separations. If a gradient is constructed over a narrow density range, a shallow gradient, and several size particles are centrifuged, the resulting separation will be based on the rate at which each particle travels against the viscous forces of the medium. This type of separation is usually more efficient. Both of these methods offer advantages for increased separation capability. This "resolution" is the narrowest distance that separates two different particles in a specific system.

Since resolving power is oftentimes the desired result of any cellular separation, rate-zonal centrifugation has become the method of choice for research use. This is due mainly to the use of an "isokinetic gradient" developed by Dr. Hans Noll (1). This gradient is shaped to compensate for the increasing velocity of sedimentation as the particle enters an increasing g field during the run. The classic example of this technique is seen in the experimental separation of ribosomal aggregates, called polysomes.

Polysomes are groups of ribosomal particles connected by a strand of messenger RNA. These aggregates are seen within the cytoplasm of cells and under certain conditions may be isolated intact. Since the number of attached ribosomes is a measure of the position of the messenger RNA in the system of protein synthesis, it would be helpful to achieve separations whose length showed sequential variation.

By use of 14.5 × 96-mm thin-walled tubes in a swinging bucket rotor rotating at 41,000 rpm (283,000 g), an analysis of consecutive polysome bands shows separations of up to 10 ribosomal subunits, demonstrating that the instrument is capable of resolving two macromolecular objects of equal density and similar shape whose mass differs by as little as 10%. Techniques are currently available to increase even this measure of resolution.

There are, however, serious limitations to either of these methods of separation. They stem more from the physical conditions imposed on the system than from limitations in the application of the theory. For example, any zone centrifugation was originally performed in a tube inserted into the rotor. Since sedimentation occurs radially from the axis of rotation, it is clearly seen that the tubes provide an impedance to the centrifugal path of all particles with the sample. The area encompassed by a group of particles at the top of a tube increases directly with an increase in radius of the tube. It follows directly then, that if the tip of a centrifuge tube is twice the radius of the meniscus of the fluid at the top of a tube, approximately 25% of the particles will impact the wall and give rise to incomplete separation. Also, construction of a static density gradient, on top of which a sample of material is to be placed, has proved to be a cumbersome and an experimentally limiting condition. Berman (2) shows that the theoretical capacity—mass of the sample to be separated on swinging bucket gradients—is never approached even when optimum conditions are maintained. This reaction attributed to the uneasy stability of the gradient and excessive manipulation in handling

and sample loading. Added to this difficulty is the additional handling of the gradients in the recovery phase of the run. Another problem is the decrease in usable sample volumes in rotors when higher rpm and g force are required. At high rpm's the volumes usable for separations become increasingly small. This seriously limits the amount of material that can be used in any separation requiring high speed and g forces. It is not difficult to see why this technique has been used more in analytical than in preparative-type systems.

Zonal Rotor System

In 1964, however, a major breakthrough in the field of centrifugal technology was presented by Dr. N. G. Anderson at Oak Ridge. This was the development of the zonal system of rotors for use in ultracentrifugation. The invention of a zonal rotor system and subsequent commercial production by ultracentrifuge manufacturers has provided a state-of-the-art approach to the solution of the inherent problems with earlier gradient systems. As discussed previously, the zonal rotor is a cylindrical pressure vessel—the most recent design can be seen in Figure 1. More than 75 rotor designs have been considered. This bowl-shaped rotor rotates about is central axis. Inside the rotor are compartments interconnected through fluid channels in the vane-septa. This design eliminates several of the aforementioned problems (such as wall effects). The septa form essentially sector-shaped compartments which do not allow a large percentage of the sedimenting particles to impact the wall. Rotor capacity is increased at operating speeds many times that of existing rotor volumes. Gradient materials may be introduced into the rotor while it is spinning at low speed through fluid channels and an effluent feed seal system, causing the gradient to be less sensitive to manipulative operations since the gravitational field stabilizes the fluid. Increased gradient

A DIV. OF DAMON

COURTESY INTERNATIONAL EQUIPMENT CO.

Figure 1. A commercial rotor constructed of titanium. Edge unloading is facilitated through the channels at the rim of the septa. The tapered wall of the bowl facilitates better resolution

283

capacity is also facilitated by this system. Since fluids may be interchanged without stopping the rotor, the recovery phase becomes simply a pumping operation with a flow monitor and a collection device.

An added advantage of this system of rotors is the flexibility in gradient shape *vs.* rotor volume. Preliminary experiments by Spragg *et al.* (*3*) indicate that the zonal rotor may be even more quantitative as an analytical tool than other systems.

In operation (Figure 2), the seal is attached to the rotor at the beginning of the run and the rotor is accelerated to a loading speed of approximately 2000 rpm. The gradient, sample, and overlay are then pumped into the rotor through the feed and effluent seal, the seal is removed, and the rotor is capped and accelerated to its maximum operational speed. After the desired separation has been accomplished, the rotor is decelerated to its unload speed, approximately 2000 rpm. The cap is removed, and the effluent and feed seal are reattached. Heavy gradient material is then pumped to the periphery of the rotor to displace the fractionated contents to the core portion of the rotor, and through the feed and effluent seal for fraction collection and further analyses.

For many specific applications, the commercial availability of zonal rotors has provided a means for the researcher to obtain large quantities of the separation need while at the same time improving purity.

Three basic types of zonal rotors are shown in Figure 3. The "A" series, low speed rotors (up to 4600 rpm) are designed for the isolation and purification of large subcellular organelles such as nuclei mitochondria, and lysosomes. The intermediate speed rotor, Z-series, allows centrifugation up to 8000 rpm and purification of subcellular components such as chloroplasts, mitochondria, and large viruses such as tobacco mosaic virus. Both the A series and Z series rotors are constructed of alumi-

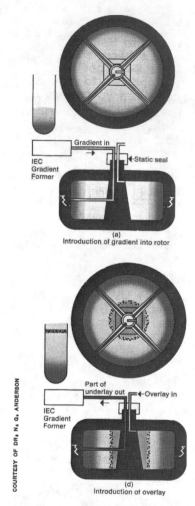

Figure 2.　Fractionation by zonal ultracentrifugation

num and plexiglass allowing visual observation and measurement of band migration during operation.

The "B" series are high speed rotors —up to 50,000 rpm—that allow isolation of subcellular particles down to the ribosomal, ribosomal subunit, and macromolecule size. Particle studies with sedimentation coefficients to 10s such as

284

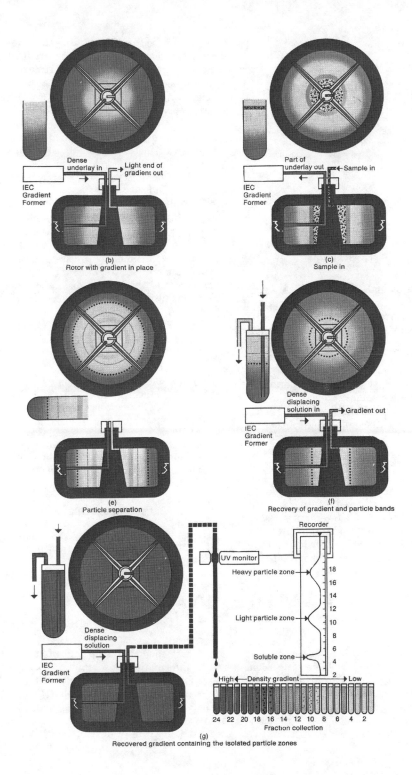

(b)
Rotor with gradient in place

Dense underlay in

Light end of gradient out

IEC Gradient Former

(c)
Sample in

Part of underlay out

Sample in

IEC Gradient Former

(e)
Particle separation

(f)
Recovery of gradient and particle bands

Dense displacing solution in

Gradient out

IEC Gradient Former

(g)
Recovered gradient containing the isolated particle zones

Dense displacing solution

IEC Gradient Former

Recorder

UV monitor

Heavy particle zone

Light particle zone

Soluble zone

High ◄—— Density gradient ——► Low

24 22 20 18 16 14 12 10 8 6 4 2
Fraction collection

18
16
14
12
10
8
6
4
2

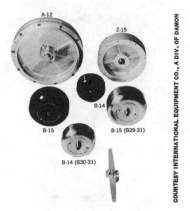

Figure 3. Several different types of commercial zonal rotors

large proteins, bacterial RNA, serum macroglobulin, and smaller viruses, such as polio, are ideally suited for use in these rotors. The B series rotors are constructed of titanium and afford the maximum versatility needed to perform any zonal run.

The B-29 and B-30 incorporate the latest design innovations of the Oak Ridge National Laboratories. These rotors allow the unloading sequence to proceed through the rim line instead of the core line facilitating separations where path length is too short in standard rotors. In practice, the "edge-loading" has the advantage of allowing removal of faster sedimenting particles while lighter and slower particles are still sedimenting. This, in effect, increases the available path for sedimentation to infinite dimensions. Rate sep-

arations based on flotation of the sample from the dense end of the gradient are now possible also. As a result of this design, exotic gradient materials, often prohibitively expensive, do not need a piston or cushion fluid of greater density to unload the rotor, since either air pressure or water supplied to the core is satisfactory. Zonal systems offer biomedical research a new outlook for the isolation and study of cellular and subcellular systems.

The ultracentrifuge is a sophisticated research instrument whose capabilities are ever increasing to meet the demands of research. If a complete understanding of human disease mechanisms and their treatment is to be achieved and the level of disease mechanisms is ultimately at the macromolecular level, then separation of cellular and subcellular species is of the utmost importance.

References

(1) Hans Noll, *Nature*, 215, 360, 1967.
(2) A. S. Berman, "Theory of Centrifugation: Miscellaneous Studies," National Cancer Institute Monograph 21, p 41, 1966.
(3) S. P. Spragg, R. S. Harrod, and C. T. Rankin, Jr., "The Optimization of Density Gradients for Zonal Centrifugation (1969)," in press.

Suggested Reading

National Cancer Institute Monograph No. 21, 1966. The Development of the Zonal Centrifuge.

COMMENTARY

by Ralph H. Müller

DR. CASCIATO's discussion emphasizes the importance of large-scale zonal ultracentrifugation permitting, as it does, the separation of cellular and subcellular species. In his opinion, this is the ultimate role of the ultracentrifuge in biomedicine. In a preparative

sense, this might well be true, but so much information is obtained during the course of ultracentrifugation that we wonder if biochemists would agree with the finality of this significant achievement.

In the 46 years of development since

The Svedberg's pioneer work, enormous advances have been made. Few methods or techniques in science borrow so heavily on related disciplines. The ultracentrifuge has never been a simple device. Even the mere task of spinning a relatively heavy rotor smoothly at speeds approaching 100,000 rpm is an engineering feat of the first order.

Even the earliest investigations of Svedberg could have been published as outstanding examples of engineering, applied optics, or physical chemistry, entirely aside from the main problem of ultracentrifugation. Most experts seem to agree that the 1940 summary of the field by Svedberg and Pedersen (5) is much more than a historical review; it is a classic which embodies practices and concepts still accepted and only greatly improved upon with the passage of time.

The earliest studies coincided with our graduate studies at Columbia University and they were required reading particularly since our research was concerned with the optical and electrical properties of colloidal dispersions of gold. All students were urged to study these papers as models of skilled research, interpretation, and documentation.

Detailed mechanical studies with the original oil turbine–driven centrifuge revealed interesting effects. Ball bearings proved to be unsatisfactory (loss of sphericity at high speed?). Rotors with definite ellipticity rather than exact cylindrical shape were capable of higher speed before fracture occurred. More recent observations have shown a decrease in temperature of the rotor upon acceleration from rest to 60,000 rpm. This was attributed to stretching of the rotor and a consequent adiabatic change, causing a temperature drop of about −1.0°C. This was contested by Hiatt (2) but completely confirmed by Biancheria and Kegeles (1) by melting point techniques and by Pickles (3) by direct measurement with a thermistor mounted in the rotor. It has been shown that the temperature decrease is linear with the square of the rotor speed, and the cooling for acceleration to 60,000 rpm is very close to 0.8°C. According to Schachman (4) much of the data in the literature must be modified because of neglect of this factor. Svedberg made extensive studies of frictional losses of all kinds. Windage losses were examined in detail. By operation in hydrogen at reduced pressure, the loss was strictly proportional to the 3/2 power of hydrogen pressure. Today, operation in a good vacuum is preferred.

Speed control and speed measurement have been a continuing problem. The earliest control method involved a differential gear arrangement in which one input shaft was driven by the main motor or turbine, the other input shaft by a synchronous motor of known speed. The output shaft of the differential gear drove a rheostat controlling the main motor. Rotational speed was measured stroboscopically.

Today, the measurement and control of speed can be as precise and elegant as one can afford. For example, a single reflecting spot on the rotor can be scanned photoelectrically, yielding an output pulse for each revolution. By substituting an encoding disk for the single spot, several hundred pulses per revolution can be obtained. The pulses can be counted directly or after preliminary electronic pulse shaping. A scaler can indicate the total number of counts for a precisely defined time interval. In the limit, an 8-decade scaler with crystal controlled counting of time (operation at 100 MHz) can measure speed to 1 part in 10^8. This is quite feasible but not necessarily sensible economically.

Speed control can be realized indirectly by other means. Thus, in the magnetically levitated rotor, designed by J. W. Beams, the rotor is brought up to speed by an air turbine. After drive cutoff, the rotor is operating in a high vacuum and its decrease in speed can be as low as 0.3 rps per day! The researches of Jesse Beams at the University of Virginia over a period of more than 30 years provide another classic in the field. Numerous key references are to be found in Ref. (4).

The resources of physical optics have been adequate since the very beginning of ultracentrifuging techniques. Some of them, notably light absorption, have been vastly improved by modern methods of photoelectric photometry. The Jamin, Mach-Zender, and Rayleigh interferometers provide elegant means of following boundaries during ultracentrifugation. Since one can measure to within 0.04 of an interference fringe, with the sodium line equal to 5893Å, this amounts to a distance of about 4 $\times$ 10^{-7} in.

The original Lamm-scale method yielded excellent results but was exceedingly tedious and time consuming. The schlieren technique of Töpler and Thovert as modified by Philpot, Svensson, and Longsworth is widely used for viewing and analyzing refractive index gradients. Excellent examples of the patterns given by the various methods are shown in the frontispiece of Ref. (4). Absorption methods are being revived for two reasons. The first is the great advance in photoelectric photometry and the second arises from the use of much shorter wave lengths in the ultraviolet for which the absorbance by proteins and related substances is very high. This has permitted operation in the μg/ml range.

We refrain from a discussion of the biochemical applications of the ultracentrifuge because our ignorance of microbiology and biochemistry is so great. These things have been going on for 45 years and many of the details are fully comprehended only by experts. We are enthusiastic about the field because it is so refreshing to note how good engineering and inspired instrumentation have contributed to the solution of such vital problems. Many years ago, Wolf-gang Ostwald wrote an intriguing little book entitled, "The World of Neglected Dimensions," referring, of course, to colloidal dispersions. Modern developments and refined concepts have differentiated between ill-defined lumpy aggregates and true, but gigantic molecules. That these things are at once of interest to the polymer chemist, the biochemist, and the clinician is obvious. Even the casual reader can profit from a reading of Refs. (4-6). Dr. Schachman's book (4) gives an excellent balance between technique and theoretical interpretation and, as the table implies, deals with biochemistry. Beyond this, he has contributed heavily to the field. Ref. (6) covers a symposium, somewhat more recent, by experts lending their varied experiences in the field. The editor, John Warren Williams, is one of the pioneers in ultracentrifugation. All reviews, monographs, and summaries are out of date by the time they are published, particularly if the field is an active one. This does not diminish their value for those of us who hope to get caught up in a strange field. In this one, the periodic bulletins, entitled "Fractions" published by the Spinco Division of Bechman Instruments, Inc., of Palo Alto, Calif., is a useful source of reports on current developments.

Literature Cited

(1) Biancheria, A., and Kegeles, G., J. Am. Chem. Soc., 76, 3737 (1954).

(2) Hiatt, C. W., Rev. Sci. Insts., 24, 182 (1953).

(3) Pickels, E. G. (1953). See Ref. (4).

(4) Schachman, Howard K., "Ultracentrifugation in Biochemistry," Academic Press, New York and London, 1959.

(5) Svedberg, The, and Pedersen, Kai, "The Ultracentrifuge," 1940, Oxford, Clarendon Press.

(6) "Ultracentrifugal Analysis," 1963, J. W. Williams, Ed., Academic Press, New York and London.

Modern Aspects of
Air Pollution Monitoring

R. K. Stevens[1] and A. E. O'Keeffe

National Air Pollution Control Administration
U. S. Department of Health, Education, and Welfare
Cincinnati, Ohio 45226

Instrumentation with improved sensitivity and specificity
for air pollution monitoring will appear in the years ahead.
The advancement from wet chemical methods to directly
responding sensors requires much research,
engineering, and testing

Today's continuous measurements of air pollutants are being performed by what we have come to call the first generation of air pollution instruments. The all-inclusive characteristic of first-generation instruments is that they were adapted from instrumentation in use in the chemical process industry at the time—mid 1950's—when the need for monitoring of air pollutants was first recognized. Typically, but not exclusively, they operate on a principle of solution in an aqueous medium of the gas to be determined, reaction with a color-forming reagent in that medium, and measurement by photoelectric means of the color formed.

Such analyzers are being used to measure atmospheric concentrations of sulfur dioxide, nitrogen dioxide, nitric oxide, and total oxidants. These in-

struments require considerable attention and, except for one analyzer employing the modified West-Gaeke procedure for measuring sulfur dioxide, their responses are affected by interfering substances likely to be found in polluted air. For example, sulfur dioxide drastically interferes with the neutral potassium iodide procedure that measures the "oxidant" in air.

Some first-generation instruments operate on physical rather than chemical principles. An example is the nondispersive infrared instrument used to measure carbon monoxide. Even this, however, suffers from serious interferences from water vapor and, to a lesser extent, from hydrocarbons and carbon dioxide. Too, it is unable to give accurate measurements at concentrations below about 2 ppm.

In attempts to instill the desired single-substance specificity into first-generation instruments, some manufacturers have incorporated scrubbers to remove interfering substances. Un-

[1] Present address: National Air Pollution Control Administration, 3820 Merton Drive, Raleigh, N. C. 27609.

fortunately, most scrubbers also tend to remove some of the pollutant of interest. Often, the efficiency of such scrubbers is temperature-dependent and short-lived, especially in high humidity.

The National Air Pollution Control Administration (NAPCA), aware of the deficiencies of currently available analyzers, is devoting some effort to developing, both in-house and by contract, a second generation of air pollution monitors. These devices have improved sensitivity, specificity, and speed of response when compared to their first-generation progenitors. They tend to operate either on a principle of direct observation of a physical property characteristic of the gas to be measured, or on a principle of measurement of a physical change (such as the release of light energy) upon reaction of the gas of interest with another gas or with a reagent supported on a surface upon which the measured gas impinges.

Some of these devices are (1) an ozone analyzer based on the chemiluminescent reaction of O_3 with an organic dye, (2) two versions of an SO_2 analyzer based on the chemiluminescent reaction of SO_2 in a hydrogen flame, and (3) a CO and CH_4 analyzer based on gas chromatographic principles.

This paper describes these analytical developments and points out their advantages over the current instrumentation.

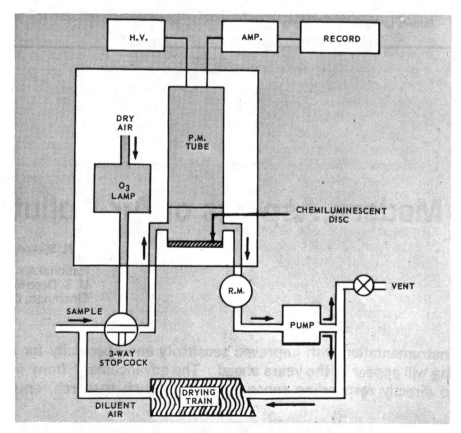

Figure 1. Chemiluminescent ozone analyzer

Chemiluminescent Ozone Sensor

A highly sensitive and specific procedure for determining ambient levels of ozone was described by Regener (1) in a report of his experiments with ozone sondes in the upper atmosphere. His procedure is based on the chemiluminescent reaction between ozone and a disk coated with Rhodamine B adsorbed on silica gel. The resultant emission is detected by a phototube, and the current generated is directly related to the mass of ozone per unit time flowing over the dye.

The Rhodamine B:silica gel disks used by Regener and others suffer a serious decrease in sensitivity on exposure to ozone, light, or water vapor. Recent experiments by Hodgeson, Stevens, and Krost (2), with disks cut from thin-layer chromatographic sheets and coated with Rhodamine B combined with a silicone resin, produced a surface that is moisture-insensitive; thus the life of the disk is extended, and continuous monitoring of ozone is possible.

Previous Regener designs required complicated plumbing to prevent high concentrations of moisture from reaching the disk. Figure 1 is a diagram of the ozone instrument used by Hodgeson et al. (2). The system was designed to blend clean, ozone-free air with ambient air, passing the blend directly over the surface of the disk. Continuous monitoring of ambient air in Cincinnati for several weeks has shown excellent qualitative agreement between results obtained with the chemiluminescent sensor and with a Mast ozone monitor. In almost all instances the Regener ozone meter records values 30 to 400% higher than those given by the Mast meter and 20 to 50% higher than those given by an analyzer based on the neutral KI colorimetric method. The differences in values may have been caused by negative (SO_2) interference with responses of the Mast and the colorimetric KI instrument.

Six automated chemiluminescent ozone analyzers fabricated under contract to NAPCA are being field-tested. If these devices perform satisfactorily, NAPCA probably will consider the chemiluminescent method as a standard method for measuring atmospheric concentrations of ozone.

Flame Photometric Detector (FPD) for Sulfur Dioxide

Crider (3) in 1965 and Brody and Chaney (4) in 1966 described a flame photometric detector designed to produce a semispecific response to volatile phosphorus and sulfur compounds. The flame detector designed by Brody and Chaney was arranged with a photomultiplier tube viewing a region above the flame through narrow-band optical filters. Sulfur compounds, when introduced into the hydrogen-rich flame, produce strong luminescent emissions between 300 and 423 mμ. By use of a narrow-band optical filter that permits transmission at 394 mμ $\pm$ 5 mμ a specificity ratio of sulfur to nonsulfur compounds between 10,000 and 30,000:1 was achieved.

The response characteristics of the flame photometric detector (FPD) and its applicability to the measurement of ambient concentrations of sulfur dioxide have been reported (5). Figure 2 is a diagram of the flow scheme used to calibrate the FPD.

Figure 3 represents the response of the FPD as a function of concentration between 5 ppb and 10 ppm for SO_2, CS_2, H_2S, and CH_3SH. The detector response produces a straight-line relationship between 5 ppb and about 0.9 ppm on a log-log scale for compounds whose sulfur contents exceed 50% by weight.

The calibrated FPD was arranged to draw atmospheric air through a glass microfiber filter (to remove particulates) into a Teflon tube leading to the inlet of the detector. Figure 4 shows a

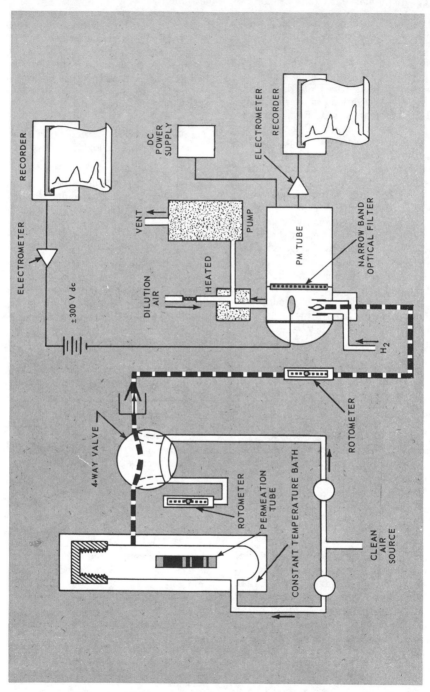

Figure 2.   Flame photometric detector for sulfur compounds

continuous recording with the FPD of the air outside the fourth floor of a NAPCA laboratory in Cincinnati. Response of the flame photometric detector was compared for several weeks with that of an instrument incorporating a (modified) West-Gaeke reagent at the NAPCA Continuous Air Monitoring Program (CAMP) station in Cincinnati. The responses agreed both qualitatively and quantitatively, suggesting that the gaseous atmospheric sulfur concentrations in the vicinity of the monitoring station consist largely of SO_2.

Gas Chromatographic System Using Flame Photometric Detector to Measure Ambient Concentrations of SO_2

Reactive sulfur compounds, SO_2 and H_2S in particular, have proved resistant in many attempts to separate them chromatographically at levels below 1 ppm. Because of tenacious sorption on column walls or solid supports, or irreversible reaction with column, support, stationary liquid phase, or carrier gas, these important compounds have failed to appear in eluates from chromatographs after samples of low con-

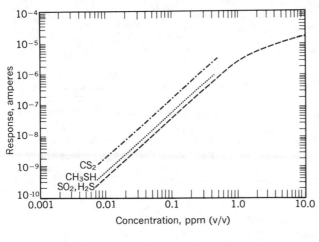

Figure 3. Calibration of FPD showing response as function of concentration of several sulfur compounds

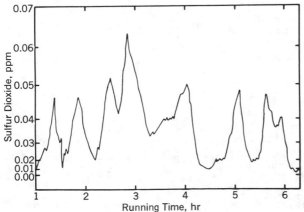

Figure 4. Continuous recording of the response of FPD to air outside NAPCA Laboratory in Cincinnati, Ohio

293

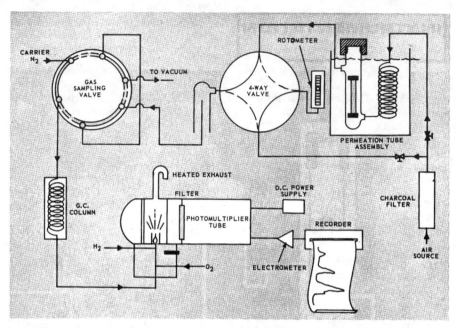

Figure 5. Automated gas chromatographic-FPD atmospheric SO₂ analyzer

centration were injected.

Our initial work with the FPD total-sulfur analyzer demonstrated that we must acquire the ability to discriminate among the several volatile sulfur compounds that conceivably could produce responses from this instrument. One route we pursued was gas chromatography. Figure 5 diagrams the automated analytical gas chromatographic system that was successfully developed. It can quantitatively measure low-ppb levels of SO_2, H_2S, CH_3SH, and CH_3CH_2SH in ambient air. The analyzer consists of (1) a Perkin-Elmer six-port gas sampling valve; (2) a gas chromatographic oven; (3) a flame photometric detector; (4) 34 ft of 0.085-in.-id fluorinated ethylene-propylene copolymer (FEP Teflon) tubing packed with polyphenyl ether five-ringed polymer containing phosphoric acid on 40- to 60-mesh Teflon (column is packed first with 40- to 60-mesh Teflon, then coated by pushing through, at 1 ml/min, 100 ml of acetone containing 12 g polyphenyl ether and

500 mg phosphoric acid); (5) 1-mv recorder; and (6) an industrial cam timer to actuate the gas-sampling valve. For the studies reported here the flame photometric detector was equipped with a 394 mμ $\pm$ 5 mμ optical filter.

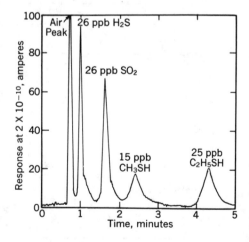

Figure 6. Typical chromatograms of mixtures of SO₂, H₂S, CH₃SH, and C₂H₅SH in air

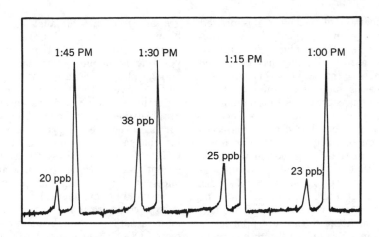

Figure 7. Typical chromatograms of ambient air showing SO₂ concentrations

The detector was maintained at 120°C, and the column temperature was 50°C.

The detector can respond reproducibly to concentrations of volatile sulfur compounds as low as 2 ppb. Although CH_3CH_2SH, CH_3SH, SO_2, H_2S theoretically should all produce the same response at equal concentrations, this was true only for SO_2 and H_2S; the responses for CH_3SH and CH_3CH_2SH were somewhat lower. Figure 6 is a chromatogram of a mixture of SO_2, H_2S, CH_3SH, and CH_3CH_2SH.

This gas chromatograph was used to measure ambient concentrations of SO_2 outside the NAPCA laboratory in Cincinnati almost continuously between December 1968 and April 1969. Figure 7 shows typical chromatograms of ambient air automatically assayed during this period. Practically no differences have been observed between the SO_2 concentrations measured gas chromatographically and the values obtained with a flame photometric detector designed to measure continuously total gaseous sulfur in air. Figure 8 compares responses of the gas chromatographic system and the FPD analyzer to samples of ambient air. This response study indicates that total gaseous sulfur in the Cincinnati atmosphere probably is synonymous with

SO_2. The device will be used to survey the atmosphere in several major urban areas and in areas with kraft mill activities to determine the ratios of SO_2 to total atmospheric gaseous sulfur. Studies to date show the potential of this analyzer for replacing the modified West-Gaeke colorimetric procedure as the referee method for measuring ambient concentrations of SO_2. In addition, this analyzer can be used to measure low-ppb levels of H_2S in areas where this gas could occur naturally or from commercial activities. Development of a commercial version of this chromatograph would be an important contribution to agencies concerned with limiting the emissions of this gas.

Gas Chromatographic Approach for Measuring Atmospheric Carbon Monoxide and Methane

Although some commercially available nondispersive infrared carbon monoxide (CO) analyzers can measure concentrations of 1 or 2 ppm, their responses are generally nonlinear and are affected by carbon dioxide, water vapor, and hydrocarbons. For this reason NAPCA is interested in developing a more sensitive and reliable CO monitor.

295

One method of achieving this desired sensitivity and specificity is by applying gas chromatographic techniques.

Gas chromatography has been used for years to measure low concentrations of gaseous materials; however, it has not been accepted as a routine monitor for certain air pollutants. Development has been slow because materials common to polluted air tend to interfere with most gas chromatographic measurements. Recently, several investigators (6, 7) have coupled gas chromatographic techniques with the catalytic conversion of carbon monoxide to methane to measure ppm levels of CO in air samples. These investigators, however, did not mention the quantitative aspects of the CO to CH_4 conversion and restricted their techniques to manual introduction of samples. NAPCA scientists (8) have developed an automated gas chromato-

graphic procedure for analysis of ambient concentrations of CO and CH_4 between 0.010 and 200 ppm by incorporating a unique pre-column system and a catalytic surface that quantitatively converts CO to CH_4. Figure 9 diagrams the analytical gas chromatographic CO–CH_4 analyzer. This monitor has been used almost continuously for the past year to measure CO and CH_4. The unit requires little attention except for changing the pre-column once every 3 months and replacing air and carrier gas cylinders monthly.

Two commercial gas chromatograph manufacturers are fabricating automated CO–CH_4 analyzers based on design criteria developed by NAPCA. Both manufacturers have developed procedures to store the signal from the detector until it can be interrogated by a data acquisition system, an arrangement that simplifies data reduction.

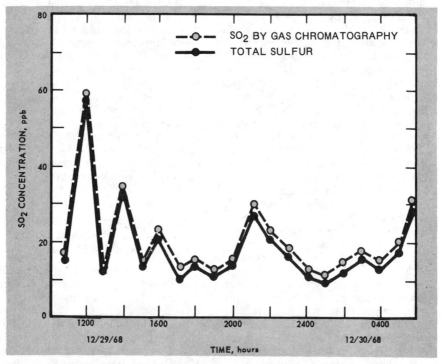

Figure 8. Comparison of responses of gas chromatographic system and the flame photometric detector to ambient air containing SO₂

296

Future Activities

The second-generation instruments described here are not as inexpensive and uncomplicated as would be desired to satisfy the growing need for pollution monitors. For example, an instrument designed to measure one pollutant is unlikely to bear much resemblance to one designed for another pollutant. For these reasons, second-generation instruments will tend to be as expensive as those of the first generation, both in terms of original acquisition and in terms of the skills required for operation and maintenance. Nevertheless, the urgent demand for working, reliable instruments dictates that development of further second-generation instruments will proceed along the lines typified by the examples given, at least for the immediate future.

At some point in that future, the authors hope to have a part in the development of still more advanced air pollution instruments. Members of that generation of instruments will be marked by two characteristics: simplicity of principle and commonality of components.

By simplicity of principle we mean that the instruments will operate through a change, chemical or physical, that takes place when the gas to be measured impinges on another gas, or on a liquid or solid surface. By commonality of components we mean that the subsystem comprising sensor and signal-processing hardware will bear a close resemblance to their counterparts responding to other pollutants. As an example of what advanced-generation instruments might be like, consider the semiconductor devices so widely used

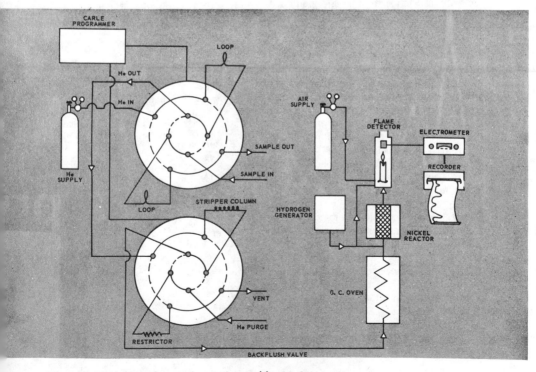

Figure 9. Automated gas chromatographic CO-CH₄ analyzer

today in the electronics industry. These devices are universally protected by an impervious coating whose purpose is to prevent unwanted changes in the electrical properties of the semiconductor that might be imposed by atmospheric contaminants. It is readily evident that the omission of this impervious coating offers an opportunity for using semiconductor devices as sensors for these contaminants. Indeed, much work has already been done in attempts to capitalize on this fact.

This work has largely been abandoned because of the difficulty of finding a sufficient number of semiconductors, each of which will respond specifically to a single contaminant. It is the authors' belief that the development of a totally specific solid-state sensor may never be fully realized. It is their hope, however, that a family of quasispecific sensors can be found. In an array of such sensors, each member will be primarily sensitive to a single pollutant and secondarily sensitive to each of a number of other pollutants. Given this situation, it will require a fairly simple computer to accept the signals from an array of such quasispecific sensors and to solve the matrix of equations relating their responses to the total composition of the atmosphere observed.

Recent developments in microwave absorbance spectrometry indicate high potential as a technique for measuring certain air pollutants without prior separation of atmospheric samples. The unique spectral lines occurring at high microwave frequencies provide very high specificity for component identification and measurement. Recent developments in solid-state microwave components may well lead to inexpensive, compact microwave analyzers capable of monitoring some of the most important air pollutants.

The third generation of analyzers is still some years from reality. Meanwhile we must concern ourselves with meeting immediate needs for reliable air pollution monitors. NAPCA has made considerable progress, but more research and development are required. Each analyzer must be tested in the field to provide the analytical air pollution chemist with data to give him confidence in its performance.

References

(1) H. V. Regener, *J. of Geophys. Res.,* **69,** 3795 (1964).

(2) J. Hodgeson, R. K. Stevens, and K. J. Krost, "Chemiluminescent Ozone Sensor," presented at the 156th National American Chemical Society Meeting, Atlantic City, N. J., 1968.

(3) W. L. Crider, ANAL. CHEM., **37,** 1770 (1965).

(4) S. S. Brody and J. E. Chaney, *J. Gas Chromatog.,* **4,** 42 (1966).

(5) R. K. Stevens, A. E. O'Keeffe, and G. C. Ortman, *Environ. Sci. Technol.,* **3,** 652 (1969).

(6) L. Dubois, A. Zdrojewski, and J. L. Monkman, *J. Air Pollut. Contr. Ass.,* **16,** 135 (1966).

(7) (a) A. P. Altshuller, and I. R. Cohen, *Int. J. Air Water Pollut.,* **8,** 611 (1964). (b) A. P. Altshuller, S. L. Kopczynski, W. A. Lonneman, T. L. Becker, and R. Slater, *Environ. Sci. Technol.* **1,** 899 (1967).

(8) R. K. Stevens, A. E. O'Keeffe, and G. C. Ortman, "A Gas Chromatographic Approach to the Semi-Continuous Monitoring of Atmospheric Carbon Monoxide and Methane," presented at the 156th National American Chemical Society Meeting, Atlantic City, N. J., 1968.

Mention of a commercial product does not constitute endorsement by the National Air Pollution Control Administration.

COMMENTARY

by Ralph H. Müller

THIS REPORT by Drs. Stevens and O'Keeffe is an informative and encouraging view of what may be expected of second-generation air pollution monitors. We were particularly intrigued by Hodgeson, Stevens, and Krost's improvement of the Regener ozone detector. For many years, the Regener ozone radio sonde has been sent aloft in all corners of the earth. In ignorance of much which has been done with the device, we wonder if the analyst has been too much concerned with the mere chemical aspects of this detector. To what extent have the many collateral aspects of the radio sonde been utilized? Modern radio telemetry is a vast and highly developed subject. It is entirely possible, in addition to the recording and transmission of ozone concentration, to transmit azimuth, altitude, slant range, temperature, pressure, wind direction and velocity, as well as the intensity of solar radiation. It would seem that a detailed vertical profile of ozone concentration could be recorded which would provide important information on the formation of ozonides, peroxides, and other obnoxious and lachrymatory products. The full resources of computer analysis of the data could furnish far more information than the mere "spot" determination of ozone concentration. Much of this is done routinely in meteorology and weather forecasting, but, in the latter, the chemical factors are of negligible interest.

The other chemiluminescent reactions, such as SO_2 in a hydrogen flame, are good examples of the second-generation approach, and it is to be hoped that the NAPCA in-house scientists and their out-of-house scientists under contract, will continue and redouble their efforts in this direction. This is important because all too much work is being done by conventional instrumental methods, and the state of the art is limited by the capabilities of commercially available instruments. We wonder what might be done with abridged mass spectrometers which could be sequentially set for selected mass numbers, says a parent peak or one or two prominent fragment peaks for a given substance. The helium leak detector is a portable single mass spectrometer, and its extension to a limited number of other masses might be reasonably economical. It would, of course, be quite impractical for CO whose mass differs from nitrogen only in the second or third decimal place.

Another approach which might merit serious attention is a phenomenon long known to spectroscopists. This is the effect of foreign gases on the spectrum of the rare gases. Helium admitted to a high-vacuum system can be excited to luminescence and, even if unresolved by a spectroscope, yields a brilliant yellow-pink glow. If the region around a suspected crack or leak is exposed with a tiny unignited jet of illuminating gas, the discharge will at once attain a dull slate color due to the presence of CO. This is about as sensitive as the modern helium leak detector (mass spectrograph) except that in the latter, helium gas has a much higher diffusion coefficient. Band spectra of the foreign gas are excited partly by direct electron bombardment but principally by collisions of the second kind with helium ions. The ionization potential of He (24 eV) is the highest of any known gas and will, therefore, excite them to luminescence. Indeed, at high concentrations of foreign gas, the bulk of the current is carried by the contaminant. It would seem a simple matter to photograph the spectrum of the pure He and then, after photographic intensification of the spectrum plate, to use it as a mask in the image plane, thus obscuring most of the helium radiation. Refocusing all other regions into a slit in front of a phototube would record the radiation of the contaminant. The arrangement could be simple and inexpensive. However, it is not likely that the information would be simple to interpret. Much empirical calibration would be required with mixtures of contaminants. With sufficient data, it might be possible to select a series of readings in different spectral regions. Once more, we emphasize the sensitivity of the method. It was many years before the correct values of the ionization potentials could be obtained by direct electron bombardment (true values are known from spectroscopic data). It was a question of rare gas purity. The rare gases even affect each other.

Certain aspects of electron spectroscopy, discussed in detail in the January 1970 issue of ANALYTICAL CHEMISTRY (pages 20A, 43A), would also appear to have a role in future air pollution studies.

TEMPERATURE COMPENSATION USING THERMISTOR NETWORKS

Ray Harruff and Charles Kimball
Yellow Springs Instrument Co., Inc.,
P.O. Box 279, Yellow Springs, Ohio 45387

In temperature compensation, it is usually possible to tailor a thermistor's response to match a needed correction. The complex interrelationships involved make a trial-and-error procedure often the simplest. Examples are given to show just how these networks can be designed

MOST PHYSICAL and chemical processes exhibit variations in their behavior as a function of their temperature. Those processes involved in creating electrical "signals" may be amenable to using electrical corrections for their temperature artifacts. Thermistors offer several advantages over other temperature transducers for such applications.

Some of the thermistor characteristics that create these advantages are speed of response, negative temperature coefficient of resistance, small size, high sensitivity, wide range of resistance values, rugged construction, and interchangeability if needed. In many situations any one of these characteristics may be surpassed by another transducer, but often the thermistor offers the best compromise among them.

The following are the most common difficulties in applying thermistors: lack of knowledge about specific data, how to design for the extremely non-linear (approximately exponential) slope characteristic, and if more than one is needed, how to achieve repeatable results from unit to unit. The rest of this article will try to deal with these problems.

Thermistors have a temperature $vs.$ resistance characteristic that can be grossly approximated by the equation $R_t = R_0 (e^{B/T})$ where R = resistance at desired temperature, R_0 = resistance at the temperature for which B is evaluated (usually 25°C, B is presumed constant for any particular material and is a slope or sensitivity term), and T = absolute temperature (degrees Kelvin) at which the resistance is desired. Often a more useful number is alpha (α) which is the slope of the resistance $vs.$ temperature curve at the temperature specified.

A better equation for approximating thermistor behavior is $R_t = R_0 e^{(B/T + C^2/T^2)}$. B has about the same meaning as above and C can be

Table I. Specific Data: Typical Range of Values

	B & tol., % @ 25°C	α, %/°C @ 25°C	R (ohm) vs. T (°C)						Resis. tol., % @ 25°C	TC, sec.	P_d, MW/°C	Max. temp., °C
			−25	0	25	50	75	100				
Small beads	2800 ± 8	−3.2	68K	2.4K	1K	480	250	140	5 to 20%	0.5 to 1	0.1	300 best
	4300 ± 8	−4.8	10M	3M	1M	350K	130K	54K				600 max
Large beads	2800 ± 10	−3.2	680	240	100	48	25	14	5 to 20%	2 beads	1	300 best
	3900 ± 10	−4.4	103K	32K	10K	3700	1500	700		25 prbs.		600 max
Glass probes	5100 ± 10	−5.7	438M	70M	17.5M	3.9M	1.1M	350K				
Close tolerance disks	2750 ± 1.0	−3.1	627.9	232.7	100	48.5	25.9	15.0	1%	10	1	100 best
	3300 ± 1.0	−3.7	8489	2691	1K	423.9	200.4	103.6	1%			150 max
	3640 ± 0.5	−4.1	102.9K	29.49K	10K	389.3	1700	816.8	1/2%			
	4600 ± 1.0	−5.2	...	3.966M	1M	295.9K	100.3K	38.2K	1%			
Wide tolerance disks	3400 ± 3 to 5	−3.8	45	12	5	2.1	1	0.5	5 to 20%	150	25	100 to
	3900 ± 3 to 5	−4.4	103K	31K	10K	3.7K	1.5K	700		10	3.5	200
Special mounts	Any above	...	...	...	...	...	...	...	...	Up to 30 min.	Up to 10 X above	...

thought of as a correction. This improved equation becomes important when tight tolerance thermistors are used. B and C are not usually published because it is normally easier for the user to work with resistance *vs.* temperature tables directly.

The power that can be dissipated is a function of many things: size and style of thermistor, surrounding environment including mounting, "thermal runaway" prevention, and the loss of accuracy that is tolerable. The standard industry technique for comparing one thermistor to another is to measure the temperature rise above surrounding still air when a known power is dissipated in the thermistor while it is suspended by its leads. Since this is the most extreme (except vacuum) case, it is often useful to have another situation described, and some manufacturers use the same measurement with the thermistor immersed in a liquid. Since every use is different, the vendor can only provide these guidelines, and the specific use will have to be evaluated by the user. The values in Table I are improved about 10 times with well-stirred liquids for unmounted thermistors, and from 4 to 20 times for various mounting configurations. If the current is limited so that the maximum allowable temperature is never exceeded, only the error due to temperature rise need be considered. The test for current limiting is as follows:

$$T_{amb} + I^2 R P_0 < T_{max}$$

T_{amb} = maximum ambient temperature around thermistor

I = current through thermistor if thermistor were at maximum allowable temperature

R = resistance of thermistor at maximum allowable temperature

T_{max} = maximum allowable temperature

Time constants and dissipation constants are both an indication of the ability to effect a heat transfer between the medium and the thermistor. Thus, again the user must evaluate his system and the vendor can only provide some guidelines. One factor to consider with the thermistor time constant is the time constant of the element being compensated. With the thermistor in its final position, they should either match or a sufficient waiting period should be allowed for both to come to equilibrium.

Most thermistors are available with resistance *vs.* temperature accuracy of ±10 to $\pm20\%$ at 25°C. Additionally the sensitivity values are ±5 to $\pm10\%$. Beads are available with special matching to tighter tolerances. Small disks are readily available in accuracies to $\pm^1/_2\%$ over the range of 0 to 75°C, gradually changing to $\pm5\%$ at −80°C, and $\pm2\%$ at 150°C. This corresponds to an equivalent temperature error of ±0.05°C at 25°C.

With all these different facets and such wide ranges of values within each, designing to use the thermistor is somewhat complex. Also, there is enough versatility to cover a wide range of problems. There are some guidelines to help overcome the complexities.

A simple compensation problem for thermistors is a resistor with a positive temperature coefficient needing to be compensated to yield a constant resistance. One example of this is a meter being used to measure a voltage in a direct reading Wheatstone bridge. The current through the meter (for any particular value of R_1) is a function of the resistance in series with the meter and the resistance of the meter itself (Figure 1). Many Wheatstone bridges use microammeters as detectors to maximize the sensitivity. These meters have significantly large resistance. For example, one 50 μA $\pm$ 2% meter has a 1712 ohm $\pm$ 10% at 25°C coil resistance. Because this is a copper wire coil, the resistance at different temperatures is calculated by $R_T = R_{25}(1 + 0.00385\,T)$.

To compensate for this effect some additional data are required. What

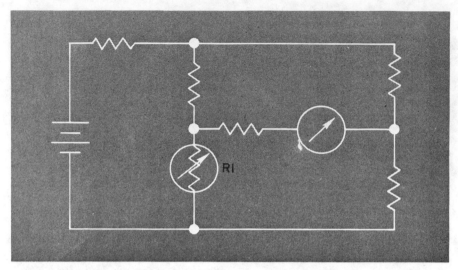

Figure 1. A resistor with a positive temperature coefficient needs compensation to yield a constant resistance. Example is a meter used to measure a voltage in a direct reading Wheatstone bridge

temperature range is important, how much error is allowable, what can the worst-case voltage sensitivity be, and what compensation component values are available? For this example, some typical values are: 0 to 50°C ± 2% error due to imperfect temperature compensation and no more than 130 mV required from the bridge output. Compensation components will be chosen from standard values after the first trial designs.

The maximum voltage required by the meter is that needed by the meter with the maximum current and the maximum coil resistance. Thus 51 μA × 1883 ohms = mV. The maximum voltage available in this example is 130 mV: 130 mV − 96 mV = 34 mV that can be dropped in the compensation network and 34 mV/51 μA = 666 ohms maximum series resistance.

The nominal rate of change of resistance is calculated by multiplying the meter resistance by its coefficient. Thus, 1712 ohms × 0.00385 ohms/ohm/°C = 6.6 ohms/°C.

If a thermistor and a fixed value resistor are paralleled and are placed in series with the meter, a circuit with the tendencies shown in Figure 2 is produced. As the temperature increases, the meter resistance increases, the thermistor resistance decreases

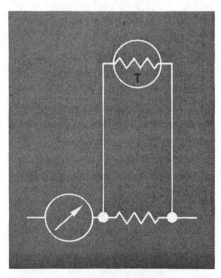

Figure 2. A thermistor and fixed value resistor are paralleled and placed in series with the meter

greatly, and the shunt resistor diminishes the thermistor decrease. Since the network must have about 6.6 ohms/°C change and since the fixed resistor is going to reduce the effect of the thermistor, the thermistor must have a change of more than 6.6 ohms/°C at the highest temperature to be compensated. To keep the total added resistance as low as possible and to avoid the very large change per degree of the thermistor at low temperatures, the lowest value thermistor having about twice the needed change at high temperatures should be chosen.

It can be seen in Table I that the 1 K disk shows a change of 576 ohms from 25 to 50 and 224 ohms from 50 to 75. This is an average of 400 ohms for 25°C change or 16 ohms/°C at 50°C. The 1 K bead is 520 ohms and 230 ohms or 15 ohms/°C. To satisfy the sensitivity requirement either would be acceptable for the first try. The allowable temperature compensation error of 2% dictates the tighter tolerance one.

Since the meter changes 6.6 ohms/°C and the range of interest is 0 to 50°C, the total meter change is 330 ohms.

The compensation network should also change 330 ohms. When the thermistor to try has been selected, the approximate value of shunt resistor can be calculated. The resistance of the thermistor and resistor in parallel at 0°C must be 330 ohms larger than at 50°C.

$$\frac{2691\,R}{2691 + R} = 330 + \frac{424\,R}{424 + R}$$

$$R = \text{(See Figure 2)}$$

$$1937\,R_2 - 1.03 \times 10^6\,R -$$
$$0.377 \times 10^9 = 0$$

$$R = 783$$

Trying this value with actual thermistor data shows the worst error, for a nominal meter of 1712 ohms, to be 0.21%. Trial and error attempts with standard value 1% resistors show 768 to yield the worst case of 0.13%. To use the same components for meters with ±10% from 1712 causes a compensation error with the 783 to be 0.59 to 0.75%.

In addition to these errors are the errors associated with the 1% resistor

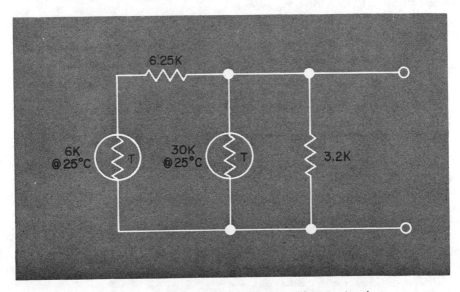

Figure 3. Circuit diagram for a Thermilinear network

tolerance and the thermistor tolerances. With tight tolerance thermistors, it is feasible to compensate meter circuits to a few per cent over wide temperature ranges. At 25°C, the thermistor is 1000 ohms and the resistor is 768. This yields 434 ohms as the compensation resistance.

If no manual resetting is used each time the temperature changes, the maximum voltage required would occur at 50°C with a maximum resistance and current meter. The meter resistance is 1712 ohms + 10% × 1712 + 6.6 ohms/°C × 25°C change = 2048 ohms; 2048 ohms × 51 μA = 105 mV. The maximum voltage required with the automatic compensation is the same 105 mV plus the voltage drop across the compensation network. With both the thermistor at 50°C and the 768-ohm resistor at their +1% tolerance with 51 μA, this is 14.1 mV + 105 mV = 119.1 mV. This meets the original requirements for a maximum of 130 mV and shows that automatic compensation cost 14.1 mV or a 13% reduction in sensitivity compared to ideal.

The resistance variation of the thermistor can also be used to correct other variations than resistance. Some transducers generate a current as the desired signal. If this current is passed through a resistor to develop a voltage, then the resistor can include a thermistor.

One example of this is the membrane-covered polarographic electrode. The membrane permeability is temperature sensitive. The coefficient of TFE Teflon membranes is approximately +4%/°C, while the thermistor is approximately −4%/°C. Thus, just using a thermistor of the right value to derive the desired voltage will give a "first-order" correction. Resistor and thermistor combinations, as in the meter coil problem, can be applied to modify the fit.

Remember that either series or shunt resistance with the thermistor will reduce its temperature coefficient. Similarly a thermistor can be used to vari-

ably shunt a current around a circuit when currents are being used or derived from voltages.

The ability to compensate with thermistors can be improved by the use of more than one thermistor. At Yellow Springs Instrument Co. (YSI), we have exploited this concept and produced a commercial product, Thermilinear Components, consisting of two thermistors and two resistors which will produce linear current and voltage or resistance information for specific temperature ranges. In addition, it is possible to produce other curve shapes by proper manipulation of the two resistors. This concept then is a powerful tool for compensating a number of temperature phenomena.

For example, assume a test instrument with an ordinary panel meter in a Wheatstone bridge circuit with the meter exposed to a temperature variation of 0 to 100°C. Furthermore the coil resistance of that meter is 3520 ohms at 25°C and the coil is made of copper. Then the coil resistance will change from 3190 ohms at 0°C to 4510 ohms at 100°C. This could cause an error in the meter indicated value of as much as 30%.

A solution to this problem can be found by first returning to the instrument to determine if an additional compensating resistance can be tolerated. For this example, assume a total of 6600 ohms ± 1% of meter coil; compensation resistance is the goal. For compensation, then, a network whose resistance is 3410 ohms at 0°C and 2090 ohms at 100°C at a rate of −13.2 ohms/°C is needed.

An investigation of what is available to satisfy the network needs discloses that a Thermilinear Component for 0 to 100°C with a sensitivity of −17.115 ohms/°C and a resistance level of 2768 ohms at 0°C is the closest thing. The circuit diagram for the Thermilinear network is shown in Figure 3. The difficult part of the problem is to decrease the sensitivity of the network of Figure 3. The resistance level can be ad-

306

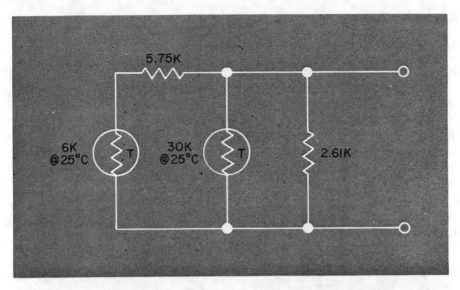

Figure 4. Revised network of Figure 3 produced by computer calculation

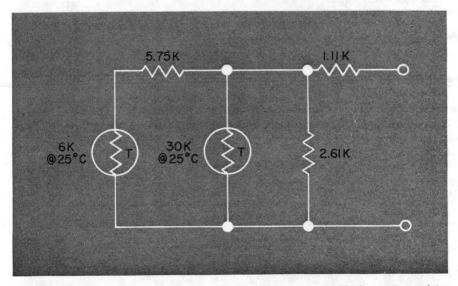

Figure 5. Revised network of Figure 4 with the addition of a 1110-ohm series resistor

justed with series resistance.

Since sensitivity is to be decreased, the 3200-ohm resistor must be decreased, but if the resistance change with temperature is to remain reasonably linear, the 6250-ohm resistor must also be altered. The best way to find these changes is an "educated cut and try" method.

We have developed a computer program which cycles the thermistors through their range of values while allowing an operator to change the fixed resistor values at the beginning of each cycle. The revised network developed by this method is shown in Figure 4. This network will produce a resistance of 2309 ohms at 0°C, and 972 ohms at 100°C. When an 1110-ohm series resistor is added, Figure 5, the total at 0°C is 3419 ohms, at 100°C 2082 ohms. Table II gives results every 5° for the entire 0–100°C range.

Another fairly common compensation area is the feedback loop of an amplifier—for instance, the circuit shown in block diagram Figure 6. It is quite common for each element in Figure 6 to have a temperature coefficient. A composite coefficient may be determined by exposing the entire circuit of Figure 6 to the expected temperature variations. At the same time, by varying the feedback it is normally possible to compensate for sensitivity changes in the circuit. If this feedback circuit requires decreasing resistance with increasing temperature, a thermistor network is a natural.

The circuit of Figure 7 is a common operational amplifier used in an inverting mode.

Suppose that a linear-with-temperature gain change (the ratio of R_f to R_1)

Table II. Results of Figure 5 Network

°C	Desired resistance, ohms	Actual resistance, ohms
0	3410	3419
5	3344	3361
10	3278	3299
15	3212	3235
20	3146	3170
25	3080	3103
30	3014	3037
35	2948	2971
40	2882	2905
45	2816	2840
50	2750	2775
55	2684	2708
60	2618	2642
65	2552	2574
70	2486	2505
75	2420	2435
80	2354	2364
85	2288	2293
90	2222	2222
95	2156	2152
100	2090	2082

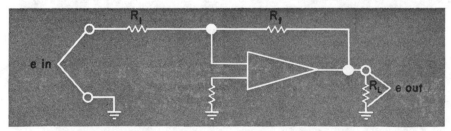

Figure 6. Block diagram of feedback loop of an amplifier

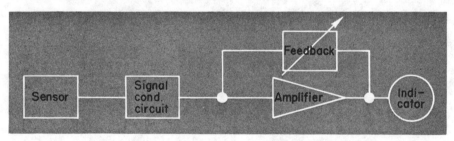

Figure 7. Circuit of an operational amplifier used in an inverting mode

of -10% in going from 0 to 50°C is needed. A YSI Thermilinear Component with a resistance of 12,175 ohms at 0°C and 5820 ohms at 50°C at -127.096 ohms/°C is available as a standard part. This component was chosen merely for the example. Other available components might be chosen for a more specific problem:

Let

$$R_f = R_T + R$$

R_T = thermilinear network resistance

Then

$$\frac{(5820 + R)}{R_1} = 0.9 \frac{(12,175 + R)}{R_1}$$

$$R = \text{required unknown}$$

$$R = 51,375$$

Therefore:

$$R_f \ @ \ 0° = 63,550$$

$$R_f \ @ \ 50° = 57,195$$

R_1 is then chosen to produce the initial gain required. Likewise if the gain change needed was say -5% for 0 to 50°C, then:

$$\frac{(5820 + R)}{R_1} = 0.95 \frac{(12,175 + R)}{R_1}$$

$$R = 114,920 \text{ ohms}$$

Therefore:

$$R_f = 127,095$$

$$R_f \ @ \ 50° = 120,740$$

Again R_1 is chosen to produce the initial gain required.

Shown above are some of the possible techniques of temperature compensation using thermistors. A thermistor has a large negative temperature coefficient of resistance. Since this coefficient is larger than most other devices' coefficients, it is usually possible to tailor the thermistor's response to match the needed correction. Because of the complex nature of thermistor's temperature relationship, as well as the relationships of the devices to be compensated, a trial-and-error procedure is usually the simplest. If the volume of use is large, then spending the initial time on a long mathematical analysis might be warranted. Because it takes many words to describe, the trial and error methods shown sound laborious. However, the design of these networks goes quickly after a little practice; especially if the best compensation is not needed or if the temperature range is narrow.

Addendum

Since this article was written, another equation (1) form has been discovered which is yet a better approximation. One form of this equation is:

$$\frac{1}{T} = A + B \ln R + C (\ln R)^3$$

This can be rewritten to solve for R in the form:

$$R = e^{[(-y+z)^{1/3} - (y+z)^{1/3}]}$$

which expands to:

where e = natural logarithm base, and $X = T + K$. (K is a constant *required* to translate temperature to absolute temperature. For example, when T is °C, K is 273.15.) A, B, and C are constants of the particular thermistor obtained by measuring the thermistor at three known temperatures and solving the three simultaneous equations that are then available.

Typical results are interpolation errors of 0.007°C or less for temperature spans on the order of 100°C.

$$R = e^{\left(\left\{ \left\{ \frac{-AX+1}{2\,CX} + \left[\frac{4\,B^3 + 9\,C\left(A - \frac{1}{X}\right)^2}{36\,C^3} \right]^{1/2} \right\}^{1/3} - \left\{ \frac{AX-1}{2\,CX} + \left[\frac{4\,B^3 + 9\,C\left(A - \frac{1}{X}\right)^2}{36\,C^3} \right]^{1/2} \right\}^{1/3} \right) }$$

Reference

(1) J. S. Steinhart and S. R. Hart, *Deep Sea Res.*, **15**, 497 (1968).

Semiconductor Light-Emitting Diodes

Richard A. Chapman

Texas Instruments Incorporated, Dallas, Tex. 75222

A wide range of light-emitting semiconductor diodes is now available for applications in the visible and infrared. Devices with specific wavelengths of direct interest to analytical chemists can be developed

SEMICONDUCTOR p-n junction diodes which emit infrared or visible light have attracted considerable attention during the last year. (In this article, light emitter will be used as the generic term for devices which emit radiation in the ultraviolet, visible, or infrared.) The noncoherent devices go under various names such as light-emitting diode (LED) and solid-state lamp (SSL). With proper device design, these diodes can be fabricated as lasers or used to optically pump other solid-state lasers. The noncoherent emitters and lasers emit radiation in a narrow wavelength band. Devices have been constructed which operate in the range of wavelengths from 4800 Å to 30 μm depending on the semiconductor used. Commercial interest has centered on visible and near infrared emitters and lasers, but there are also some extremely interesting projects using far infrared lasers.

Interest in semiconductor light emitters goes back many decades. Research shifted to semiconductor diodes in the mid-1950's. The discovery in 1962, that efficient emitters (both lasers and noncoherent emitters) could be constructed from GaAs, caused considerable R&D activity. Recent advances in materials and in device fabrication have now brought this technology to a position which should interest analytical chemists in the application of these devices in laboratory measurements or in process and pollution control.

A part of the interest in semiconductor light emitters stems from their unique features in comparison with those of other light sources, such as incandescent, fluorescent, and gas discharge light sources. For instance, the semiconductor emitters are extremely small (typically a mm or less in cross section) and they require small voltages (typically 1 to 2 V). Pulsed semiconductor lasers have extremely large spectral radiant emittance. Parameters which vary widely with device type are operating temperatures (4.2 to 300°K), bias currents (millamps dc to 75 A pulsed), linewidths from 1 to 500 Å for GaAs devices), and requirements for dc $vs.$ pulsed operation.

High-power devices available commercially include pulsed GaAs lasers with 1 to 20 W peak power emission at 9050 Å and dc-operated GaAs noncoherent emitters with 200 mW emission at 9350 Å. Visible noncoherent red, green, and yellow light emitters are commercially available. Other than the InAs emitter which emits at 3.2 μm, most of the longer wavelength

311

emitters and lasers are not yet commercially available.

Noncoherent light emitters should be useful in applications requiring single wavelength-band monitors of reflectance, transmittance, and absorbance. Semiconductor lasers should be particularly useful as pulsed sources for luminescence and light scattering or in applications requiring more narrow linewidths than available with noncoherent emitters.

In this article, the author will attempt to acquaint analytical chemists with the characteristics of available sources. A discussion of theory and device technology necessary for proper application is included. An attempt is made to stimulate the interest of the readers so that they may in turn direct device development into areas of particular importance to analytical chemistry.

Semiconductor Theory and Materials

Before proceeding to a discussion of the devices, it is useful to review terminology for those not active in the field. In a semiconductor, the atomic wavefunctions have combined in such a manner that a region in energy devoid of states is formed in a perfect crystal. The region is called the forbidden gap; this gap separates the valence band from the conduction band. In Figure 1, the lower energy limit of the conduction band, E_C, and the upper limit of the valence band, E_V, define the energy gap $(E_C - E_V)$. Impurities and other defects in the crystal cause impurity states such as donors (energy E_D) and acceptors (energy E_A). By adjusting the impurity content, the semiconductor can be made n-type (conductivity dominated by electrons in the conduction band) or p-type (conductivity dominated by holes in the valence band). Generally, p-n junctions are formed in semiconductors by impurity diffusion or by changes in doping during crystal growth.

Absorption of radiation with energy greater than the band gap $(E_C - E_V)$

causes the generation of excess electrons and holes. The excess electrons and holes generated may recombine directly as is shown on the left in Figure 1. The electrons and/or holes may be trapped first on donors or acceptors, respectively, and then recombine. Two examples of such recombinations are also shown in Figure 1. There are many other important recombination mechanisms. An excellent review of recombination processes has been given by P. J. Dean (1). If the recombination process results in the generation of a photon (with energy $E_C - E_V$, $E_C - E_A$, $E_D - E_A$, etc.), the process of absorption and recombination results in photoluminescence. If the excess holes and electrons have been created by an incident beam of high-energy electrons, the process is called cathodoluminescence.

The unique advantages of a semiconductor luminescent source are only ob-

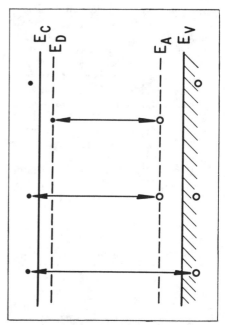

Figure 1. Energy band diagram showing several possible radiative recombination mechanisms between electrons (dots) and holes (open circles)

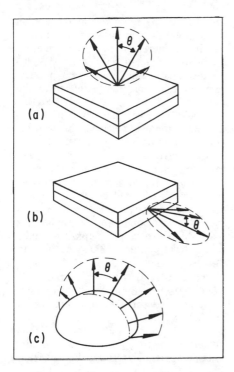

Figure 2. Three types of noncoherent light emitter structures: (a) flat source, (b) flat using edge emission (laser geometry), and (c) dome emitter. The line through the source is the *p-n* junction in the flat sources

tained when *p-n* junctions are used. When forward bias is applied to the *p-n* junction, excess electrons are injected into the *p*-type region and excess holes into the *n*-type region. Radiative recombination of these excess carriers in their respective regions causes the emission of light. Often only the *p*-type side is an efficient source of luminescence.

Semiconductors of major importance as light emitters and detectors include compounds of chemical groups IIIA and VA (the III–V compound semiconductors such as GaAs, etc.), the IVA–VIA compounds (such as PbTe, etc.), and in some cases the IVA elemental semiconductors (alloys such

as SiC). The II–VI compounds have not been utilized because *p-n* junctions are difficult or impossible to form in these materials.

Alloys of semiconductors are of great technological importance. Each semiconductor has a given energy gap which in large part sets the wavelength of the radiation emitted by the diode. Alloying two similar semiconductors such as GaAs and GaP allows the fabrication of a device with a band gap intermediate between that of the constituents.

Light-emitting injection diodes in the blue and ultraviolet are not yet available. The development of such devices is a formidable task. Large band gap semiconductors typically have high melting temperatures and poor electrical characteristics, and tend to be more ionic. Blue and ultraviolet semiconductor lasers have been constructed using cathodoluminescence. In a different approach, infrared emitting diodes have been used to pump green and blue phosphors by the multiple absorption of two or three infrared photons.

Noncoherent Emitters and Lasers

Both semiconductor diode lasers with coherent emission and semiconductor diode emitters with noncoherent emission are built. In this section, the characteristics of these two types of LED will be compared.

A wide variety of noncoherent emitter designs are available. In many applications, an emitter and detector are placed in close proximity without additional optics; in this type of application the radiant intensity in watts per steradian is most important. Some available GaAs devices include a parabolic reflector or lens to increase radiant intensity. In other applications, maximum radiant flux or power output (watts) is desired. To meet this goal, the emitter must be designed to make the external quantum efficiency (photons emitted per carrier crossing the device) approach the internal quantum efficiency (photons *created* per carrier crossing the device). A discussion of the parameters determining external quantum efficiency helps to explain the variety of noncoherent emitter designs and the relation to laser design.

There are three major loss mechanisms decreasing external quantum efficiency: (1) internal absorption of the created photons can be a large effect when the photon energy is close to the forbidden gap energy; (2) a fraction of the radiation is internally reflected if the device is not antireflection coated; and in any case, (3) all rays striking the exit surface with angles to the normal greater than the critical angle $\phi_C = \sin^{-1}(1/n)$ will be totally reflected ($\phi_C \sim 16^0$ for GaAs), where n is the index of refraction in the semiconductor and the index of refraction of air is taken as equal to 1. A thorough discussion of figure of merit for noncoherent emitters is available (*2*).

Figure 2 shows three types of device design. The length of the arrows represents the intensity of the emitted radiation as a function of the external angle θ, ($\sin \theta = n \sin \phi$). Figure 2a shows an emitter design which uses the emission exiting from the face parallel to the junction. The angular distribution of radiation intensity is almost Lambertian (intensity proportional to $\cos \theta$). Figure 2b shows a design which utilizes the light exiting along the p-n junction periphery. The emitter designs of Figure 2a and 2b are often packaged in a miniature mount which has a parabolic mirror to collect the radiation and increase the radiant intensity. Figure 2c shows a hemispherical emitter (the p-n junction is near the center of the hemisphere). This dome greatly reduces the internal total reflection loss and results in a larger external radiant flux and external quantum efficiency. The highest power output noncoherent emitters use the hemispherical design.

Semiconductor diode lasers are constructed using the geometry of Figure 2b with two narrow sides made parallel by cleaving the sample. As the current density is increased in a p-n junction diode, the injected carrier density increases until it finally creates a population inversion in the region near the junction. For currents above this threshold current, the absorption coefficient becomes negative in the inverted region near the junction. Radiation reflected back and forth between the two cleaved faces can increase in intensity, and if the gain is greater than the losses, lasing will occur with emission from the cleaved faces. The rays leave the semiconductor at angles less than the critical angle helping to make the external quantum efficiency high in the laser structure.

The angular distribution of radiation from a laser is limited by diffraction. Unlike other lasers, the inverted region in a semiconductor diode laser is quite narrow (a few micrometers), thus causing a relatively large far-field diffraction angle. Typically, the radiation from a GaAs laser is contained in a beam about 5° wide at half-intensity in the plane perpendicular to the junction and 15° wide in the plane of the junction. This angular distribution is more narrow than that for noncoherent semiconductor emitters which do not have a built-in parabolic mirror, but the 5° by 15° beam is considerably larger than obtained with gas lasers and solid-state lasers such as ruby. The far-field diffraction angle is directly proportional to wavelength, making the diffraction angle larger for the far infrared semiconductor lasers.

Because of the small emitting area, semiconductor lasers have tremendous radiance (watts/steradian/mm² of emitting area). Pulsed GaAs lasers have radiances in the range 10^5 to 10^6 peak W/steradian/mm² with a 0.1% duty cycle (pulse duration divided by pulse repetition period). Radiance is the important factor if the experiment involves focusing the radiation onto a sample to obtain maximum illumination. Noncoherent GaAs hemispherical emitters are available with 20 mW/steradian/mm² radiance. A radiance of this value is sufficient for many experiments, and dc operation is often extremely important.

Lasing is obtained near the maximum of the noncoherent emission line, and the luminescence from a laser can have a linewidth much smaller than that of a noncoherent emitter. If the laser is operated with current densities only slightly above threshold, some semiconductor lasers can be operated in one cavity mode which is less than 1 Å in linewidth for GaAs. More often, however, the device is driven above threshold to obtain maximum power output; under these circumstances, many lower gain cavity modes lase such that numerous lines are observed in a bandwidth in the order of 20 to 40 Å. Detailed information on cavity modes active in GaAs lasers is available (3). GaAs noncoherent emitters operated at room temperature have bandwidths ranging from 250 to 500 Å depending on device type and dopant used.

Noncoherent emitters may be operated in a dc mode or in a modulated or pulsed mode. Response times the order of a few nsec for standard GaAs can be obtained and the order of 0.2 μsec for GaAs doped with silicon. When pulsed, a GaAs laser can have a radiant output response time of less than a nsec.

Light output degradation with time has been observed to occur in light emitters. At least for GaAs, the degradation rate is directly proportional to current density and is, therefore, much more rapid with lasers (which must be driven past a certain threshold current density). The required lasing threshold current density increases with increasing temperature. This is one of the reasons that cw room temperature semiconductor lasers have not been developed. Typically, a room temperature GaAs laser must be operated with a duty cycle of 0.1% or less and a pulse width of 200 nsec.

Noncoherent emitters can also be used to pump other solid-state laser materials (4) such as neodymium-doped yttrium aluminum garnet (YAlG:Nd). The high efficiency silicon-doped GaAs emitters are particularly useful in this application. This system has the advantage of better laser optics (smaller diffraction angle and fewer modes), and the prospect of room temperature cw operation since room temperature dc noncoherent emitters are available.

Devices and Materials

Table I shows a partial list of semiconductor materials from which lasers or noncoherent emitters have been made. Some entries are or have been commercially available. Of the commercially available devices, GaAs, Ga[As,P], and InAs are all available as either lasers (including Ga[As,P] at 8600 Å) or noncoherent emitters (Ga[As,P] at 6600 Å). The III–V alloys cover a wide range of wavelengths. Ga[As,P] alloys can be made which emit in the wavelength region between GaAs (9000 Å) and GaP (5600 Å). The quantum efficiency decreases for the shorter wavelength alloys. The alloy system [Ga, Al]As covers about the same wavelength region as Ga[As,P].

Because the band gap, E_G, changes with temperature, the wavelength, λ, of the emission depends on temperature. For the III–V semiconductors, $\Delta E_G/(\Delta T)$ is about $-4 \times 10^{-4} \times$ eV/°C. Since $\Delta\lambda/(\Delta T) = (\lambda/E_G) \times (\Delta E_G/\Delta T)$, the fractional change in wavelength with temperature is larger for the smaller band gap semiconductors. The entries for InAs and InSb in Table I show the variation of wavelength with temperature from 300 K to 4.2 K. A change of 3 Å/°C is usual for GaAs. The PbSe laser has been pressure-tuned over 8 to 22 μm using pressures to 15 atmospheres. The alloys [Pb,Sn]Te and [Pb,Sn]Se have a range of wavelengths available, de-

Table I. Lasers and Noncoherent Emitters

Material	Wavelength	Noncoherent emitter response time, sec
GaAs:Si–phosphor-coated (Blue)	4700 Å, 6500 Å, 8000 Å	...
GaAs:Si–phosphor-coated (Green)[a]	5400 Å, 9400 Å	2×10^{-3}
GaP (Green)[a]	5600 Å	10^{-9}
SiC (Orange)[a]	5900 Å	$<10^{-7}$
Ga[As,P] (Amber)[a]	6100 Å	10^{-9}
GaP (Red)[a]	7000 Å	5×10^{-7}
Ga[As,P] (Red)[a]	6600 Å	10^{-9}
Ga[As,P][a]	8600 Å	10^{-9}
GaAs[a]	9000 Å	10^{-9}
GaAs:Si[a]	9200 Å to 9500 Å	3×10^{-7}
InAs[a]	3.1–3.6 μm	10^{-7}
InSb	5.2–6.2 μm	...
PbSe	8.0–22 μm	...
[Pb,Sn]Te, [Pb,Sn]Se	9.5–30 μm	...

[a] These devices are commercially available.

pending on alloy composition. The wavelength of any emitter can be slightly tuned by diode current (a heating effect).

The first two entries in Table I are phosphor-coated noncoherent emitters. The wavelength is determined by the phosphor. Several different wavelengths are obtained simultaneously from each phosphor; these are listed for each phosphor-coated device. The linewidth of the green emission is about 30 Å. Visibly bright emission can be obtained if these devices are driven with high intensity infrared emission from the high efficiency GaAs:Si emitters.

The last column in Table I lists the fall-time of commercially available emitters. It should be emphasized that the response time can be larger than stated if the light emitter is not properly packaged for fast response. The response times of all injection lasers will be 1 nsec or less. The fall-times of the blue and green phosphor-coated devices are similar and are set by the phosphor response.

The maximum dc power output is obtained with silicon-doped GaAs noncoherent hemispherical emitters which emit approximately 200 mW radiant flux at 9300 Å with a 350 Å linewidth. This represents a total radiant emittance of 7.5 W/cm² of emitting area and a spectral peak emittance of about 210 W/cm²/μm. This total emittance could be obtained with an incandescent tungsten source, but a tungsten filament would have to be operated near its melting point to obtain the same peak spectral radiant emittance at 9300 Å.

Room temperature GaAs lasers are available with peak radiant flux of 1 to 20 W peak at 9050 Å using 10 to 75 A peak drive current with a 0.1% to 0.02% duty cycle. The radiance and spectral radiance of these sources should surpass all types of conventional discharge lamps (5). Applications requiring high illuminance with single-mode dc operation would presently require the use of gas lasers such as the argon-ion laser.

The output of visible light emitters is usually given in terms of photometric units which express the response of the human eye to the output emission. For instance, lumens can be correlated with watts, candela with watts per steradian, and candela/cm² with watts/steradian/cm². The reader is referred to the literature on photometric measures (6).

Information on infrared detectors to be used to detect the emission from semiconductor emitters can be obtained from an earlier article by Levinstein (7) in this column. Both emitters and intrinsic infrared detectors must be operated at reduced temperatures. The long wavelength [Pb,Sn]Te and InSb lasers are usually operated at temperatures approaching 4.2 K. InAs is usually operated at 77 K as a detector and incoherent emitter. GaAs and materials with shorter wavelength emission are operated at room temperature and use detectors such as photomultipliers and silicon photodiode detectors which are also operated at room temperature.

Applications

One of the major scientific applications of noncoherent emitters and lasers has been the use of these devices as sources for optically pumping other materials. The use of silicon-doped GaAs noncoherent emitters to pump solid-state laser materials has already been mentioned. Laser radiation from GaAs diode lasers has been used optically to pump other semiconductor materials such as InSb. GaAsP lasers have been used to stimulate lasing in GaAs. Radiation from GaAs lasers has also been used to study light scattering in semiconductors (8); the many modes present in radiation from high-intensity GaAs lasers can make difficult the interpretation of quantum-loss light scattering experiments.

Radiation from either noncoherent emitters or diode lasers can be used to study response times of infrared detectors. Pulsed semiconductor lasers are particularly useful in measuring response times of detectors designed for high-frequency response infrared detectors. A number of investigators, including the author, have used GaAs and [Pb,Sn]Te lasers and InAs emitters and lasers in the study of detector response times.

Noncoherent-light-emitter silicon-photodiode-detector pairs can be used to monitor transmittance. The availability of Ga[As,P] alloy emitters should permit the monitoring of absorptance of materials with absorptance bands in the near infrared and visible. These devices can be operated at room temperature without sophisticated electronic pulse equipment.

The narrow linewidth of semiconductor lasers and the ability to slightly tune the emission wavelength permits ir spectroscopy research of much improved resolution. For instance, a [Pb,Sn]Te laser operating at 10.6 μm can be tuned over 0.05 μm (4 cm⁻¹ wavenumbers) by varying the diode current from 0.7 A to 1.4 A. A current tuned [Pb,Sn]Te laser has been used to study the ammonia absorption line at 941.75 cm⁻¹ wavenumber and sulfur hexafluoride absorption lines near 10.6 μm wavelength (9). A Doppler broadened linewidth of 0.0027 cm⁻¹ was measured for the 941.75 cm⁻¹ line. This line cannot be resolved with a grating spectrometer having a 0.3 cm⁻¹ resolution at that wavelength. Hundreds of previously unresolved lines are observed in the study of sulfur hexafluoride. Both direct detection and heterodyne detection using the difference signal between a carbon dioxide laser and a [Pb,Sn]Te laser have been used.

Acknowledgments

The author wishes to thank those manufacturers who supplied requested information on semiconductor lasers and emitters. Product information is available from American Electronics Laboratories, Electronuclear Laboratories, General Electric Miniature Lamp Department, Hewlett Packard, Monsanto Electronic Special Products, RCA Electronic Components Department, Raytheon, Spectronics Incorporated, and Texas Instruments Incorporated—Optoelectronics Department.

Special thanks are also due to David Hinkley of Lincoln Laboratories for preprints on infrared spectroscopy experiments using [Pb,Sn]Te lasers and for permission to mention results on experiments in progress.

References

(1) P. J. Dean, *Trans. Metal. Soc. AIME*, 242, 384 (1968).

(2) W. N. Carr, *Infrared Phys.* 6, 1 (1966).

(3) T. H. Zachos, *IEEE J. Quantum Electron.*, QE5, 29 (1969); T. L. Paoli, J. E. Ripper, and T. H. Zachos, *ibid.*, QE5, 271 (1969).

(4) R. B. Allen and S. J. Scalise, *Appl. Phys. Lett.*, 14, 188 (1969).

(5) L. Levi, *Appl. Opt.*, John Wiley & Sons, Inc., New York, N. Y., 1968, Chap. 1 and Tables 3 and 4.

(6) L. Levi, *ibid.*, Chap. 5.

(7) H. Levinstein, ANAL. CHEM., 41, 81 A (1969).

(8) A. Mooradian, "Advances in Solid State Physics," *Festkörperprobleme IX*, edited by O. E. Madelung, Pergamon Press, New York, N. Y., 1969, pp 73–98.

(9) E. D. Hinkley, MIT Lincoln Lab. Opt. Res. Rept., 1969:2 p 3.

Mülheim Computer System for Analytical Instrumentation

Engelbert Ziegler, Dieter Henneberg, and Gerhard Schomburg

Max-Planck-Institut fur Kohlenforschung,
Mulheim/Ruhr, Germany

The close connection of an analytical on-line system to a large computational system will certainly stimulate the application of advanced analytical methods

DURING THE PAST few years a number of different approaches in the computerization of analytical instruments have been described. Most commonly the instruments are connected to small "dedicated" computers. Another approach consists of a medium-sized computer system (*e.g.*, IBM 1800) servicing several not too dissimilar instruments simultaneously (*1, 2*).

A different solution has been developed at the Max-Planck-Institut für Kohlenforschung in Mülheim, Germany (*3, 4*). This institute is a research institute for organometallic chemistry and radiation chemistry with extended analytical, physical, and physicochemical laboratories. A large time-sharing computer (PDP 10) has been installed to meet the computational requirements for both general off-line computations and real-time acquisition and analysis of data from a variety of on-line analytical instruments.

This paper describes the hardware configuration of the Mülheim computer system and the basic structure of the analytical on-line system incorporated into the interactive, conversational PDP 10 time-sharing mode.

Computational Requirements

There are two distinct areas of computer applications at the Mülheim institute:

(1) Off-line computations for
X-ray diffractometry
Molecular orbital calculations
Chemical kinetics
Spectra simulations and iterations
Documentation and search of spectra
Documentation of literature
Administration

(2) Real-time data acquisition and
analysis for
  (a) "Slow" instruments (data
      rates ≤ 20 Hz)
      20-30 Gas chromatographs
      1 Mass spectrometer
      2-3 Nmr instruments
      1 Ir-spectrometer
      1 Raman spectrometer
      1 Esr instrument
      1 Spectropolarimeter
  (b) "Fast" instruments (data
      rates between 1.25 and 20
      kHz)
      2 Low-resolution, fast-scan
      mass spectrometers
      1 Pulsed nmr instrument

Hardware Configuration (Figure 1)

The hardware of the PDP 10
central processing unit includes relocation and protection registers,
floating-point and byte-handling
instructions and seven levels of pri-
ority-nested interrupts. Core
memory consists of 32K words of
36 bits (1 μsec cycle time) and is
connected through a fast data
channel to a fixed-head disk with
17 msec of average access time, 13
μsec transfer time per word and a
capacity of 500K words. User programs are temporarily transferred
from core memory to this disk
(swapping), whenever core space
becomes too small for all of the
concurrently running programs.
But part of the disk is also used
for data storage and for system
programs such as compilers, as-
semblers, and text editors.

The present set of peripheral devices includes a line printer, card
reader, plotter, magnetic tape units,
paper tape units, and 14 teletype
terminals.

Several special devices are interfaced to the PDP 10 for the real-
time data acquisition. There is one

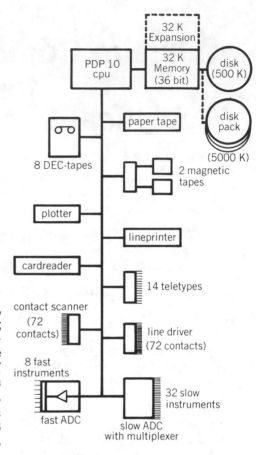

**Figure 1. Hardware configuration of the
Mülheim Computer System**

Dashed lines indicate the main expan-
sions scheduled for 1970

A/D converter (analog-to-digital)
to service up to 32 slow, simultaneously running instruments connected through a multiplexer. This
ADC provides 13 bytes of resolution
within a useful dynamic range of
about 10^6 subdivided into 11 different gain ranges. The maximum
over-all data rate of this device is
8 kHz with programmed selection
of gain ranges and 3.3 kHz with
automatic range selection. Auto-

matic (hardware) selection is slower because the range for the full conversion is determined by an additional preceding conversion cycle. For fast instruments, fast-scan mass spectrometers, another A/D converter is used, capable of data rates between 1.25 and 20 kHz. This ADC provides 10 bytes of resolution within a dynamic range of 2.5×10^5 subdivided into three gain ranges. A multiplexer connects eight data lines to this ADC. But, to avoid conflicts with the general time-sharing system, only one fast instrument is allowed to transmit data at any given time. Because of the very short time periods—typically 2 or 3 sec— when such a fast instrument will be active, this restriction is not a serious one.

The timing for the real-time data acquisition is provided by a real-time clock, operating with 20 kHz pulses and a preset counter for each A/D converter, loaded by program.

A contact scanner and a line driver are able to control 72 external relay contacts each. With these devices, the operating system services start and stop requests, as well as certain instrument control functions. The capability of the contact scanner to interrupt the processor allows discontinuously scanning instruments, such as infrared spectrometers, to request an A/D conversion by closing the associated external relay.

Time-Sharing Organization

The present system allows up to 15 concurrently running jobs (each job may run one program), which are initiated and controlled by the various teletype terminals distributed all over the institute.

The hardware protection of the central processor provides the possibility of keeping several jobs in core separated, without interference, by allowing only "legal" memory references. Jobs may be shuffled in core or swapped out to the disk and located into different core areas when brought in again, because memory references are displaced by adding the contents of the relocation register during execution.

Commands from the teletype terminals are interpreted by the operating system "monitor." The monitor also determines the job which is actually run by the central processor, and controls the allocation of the system resources to the various jobs. Whenever a running job becomes unrunnable—for instance, when starting input or output activities *via* one of the seven interrupt levels—or after certain time slices, the monitor looks for another job to run.

Because of this time-sharing capability, the system is flexible enough to handle simultaneously many analytical instruments requiring different data acquisition and different data analysis methods.

Software Organization

The analytical instrumentation software is of course different for the slow and the fast instruments. These two software packages are each subdivided into

(1) Code within the monitor
(2) Data acquisition programs
(3) Basic-analysis programs
(4) Refined-analysis programs
(5) Miscellaneous auxiliary programs (*e.g.*, data plotting and testing programs)

Software for Slow Instruments

Within the monitor a special routine looks at start or stop requests of any instruments every 20 msec. Every 50 msec (thus limiting the maximum data rate for a single instrument to 20/sec), a list of slow instruments is created, for which a conversion has to be started. When a conversion is completed, the data word from the ADC is brought in on interrupt level and stored into buffers of the data acquisition job.

This data job (GCDIR) is started at system-setup time and put into a deactivated status. Whenever one of the data buffers has been filled with ADC data, this program is activated by the monitor to empty the accumulated data into files on the disk. It is important that no reduction or analysis of ADC data is performed at this stage. Because ADC data are generated continuously, the data job may not be swapped but has to stay resident in core. In this aspect, this job differs from normal time-sharing jobs.

There is only one data job to service all of the slow instruments, but various different analysis programs may be started by the analyst in the laboratories at any time. The basic-analysis programs process the files of unreduced ADC data after an instrument run has been stopped. These programs may be initiated as to check at certain time periods automatically for any finished runs and they may control a large number of instruments which are using the same analysis methods, such as a group of gas chromatographs.

The flexibility of this kind of software organization can be illustrated by gas chromatography. But the procedure with spectroscopic methods is quite similar.

If a technician wants to run a gas chromatogram, he first starts a parameter - type - in program (PARAM) from the teletype in his laboratory. If no parameters are specified, those of the previous chromatogram on the same channel are taken for the new run. The conversational parameter program asks him for the channel number, the data rate, the run time, and for report headings. Then he may inject his sample and start the acquisition of data by pressing the start button of the proper channel. The run is stopped either by the technician's pressing the corresponding stop button or if the specified run time has elapsed.

If the channel is under automatic control of a basic analysis program (GCAN), then this program will print an analysis report on the teletype and on to the disk. This report consists of a list of peaks with peak positions, intensities, normalized area percentages, peak half widths, and information about peak overlapping and shoulders.

If the analyst is not satisfied with this basic analysis, he may request a more sophisticated one by starting one of the so-called refined analysis programs. For instance, he may wish to apply any special standardization, such as a determination of Kovats indices (INDEX), or he may want to discard several regions of the chromatogram, or he may want to start the analysis on the tail of a large peak to determine the areas of small peaks on the tail with better accuracy. All this refined analysis is performed without requiring a rerun of the chromatogram, because all the unreduced ADC data, as well as the report of

the basic-analysis program, remain on the disk until the end of the next run on the same channel.

This type of operation (analyzing the different chromatograms individually with a set of programs dependent on the result of the basic analysis) is especially suited for a laboratory in a research and development environment with a variety of different problems. In the case of industry-type routine chromatography, some standard steps of the refined analysis would be incorporated into the basic-analysis program.

A detailed description of the gas chromatography system and its performance and results will be given in a further paper.

Software for Fast Instruments

The organization of the software for fast instruments differs in some aspects from that described for the slow instruments. For each fast instrument on-line an extra data-acquisition program is used, whereas all slow lines are handled by a single data job. As soon as a run of a fast instrument, such as a fast-scan mass spectrometer, is requested by closing one of the contacts controlled by the contact scanner, the data job associated with the requesting instrument is brought into core to accept all the data from the ADC and stays in core as long as the instrument runs—just for a few seconds. During this period, no other start requests from fast instruments are accepted, because only one fast line is allowed to be active at any given moment. The data acquisition of the simultaneously running slow instruments, of course, continues as well as time-sharing response for other jobs and service of computational programs are still guaranteed. This is possible because all input and output activities occur on different hardware interrupt levels and are accomplished within short periods of time. The normal data rate for fast-scan mass-spectrometer runs is 5 kHz. Thus, every 200 μsec, a data word from the ADC is transferred. But the transfer itself takes only 6 μsec, leaving the remaining 194 μsec available for other tasks.

Up to now, one fast-scan, low-resolution mass spectrometer (MAT CH4) and a spin-echo nmr setup are connected on-line. For the latter, the data job just transfers the ADC data without any reduction to the disk. A time-averaging program or a Fourier transform program then reads the data from there.

For the mass spectrometer, the associated data job performs a data reduction and writes these reduced data on the disk. Similar to the organization of the slow instruments, the analysis program is running in parallel to the data job waiting for any finished runs.

A considerable amount of work has been put into the development of programs for the fast-scan low-resolution mass spectrometer.

1. Data reduction program (MSDAT)

   By preliminary peak-search, the 10,000 samples of a low resolution mass spectrum are reduced to 50–1000 intensities and positions of potential peaks, depending on intensity and mass range of the spectrum

2. Calibration of mass scale (MASCAL)

   Out of the reduced data of a polyfluoro kerosine mass spectrum, a list of positions of masses 1–1000 is created automatically

3. Evaluation of spectra out of reduced data (MACO)

   Masses and intensities of sharp peaks and metastables are determined peak intensities with a dynamic range of 3.10^4

4. Spectra are printed out in a normalized form as table on the line printer (REMAS)

   Presently a Varian-Statos recorder is interfaced to the system to provide fast output of spectra in the form of bargraphs in two different intensity scales simultaneously

5. A set of programs has been developed to test and to analyze unreduced ADC data and to check the performance of analysis algorithms.

The operational procedures and the algorithms used for analysis and calibration purposes will be described later.

The most difficult application incorporated into the Mülheim System is the automatic operation of a GC/MS combination. In this setup, high-resolution capillary-gas chromatography is applied, causing fast narrow peaks within the chromatogram. Therefore the system response time is a very critical parameter for this application. When a start of the mass spectrometer is necessary, the corresponding data job to receive the ADC data from the spectrometer normally resides on the swapping area of the disk. Some time is lost until this job is brought back into core memory. To achieve the required response times of less than 1 sec, special modifications of the time-sharing monitor were necessary.

In the present implementation, the monitor checks at certain time intervals for beginning and end of peaks within the signal from the total ion current (representing the chromatogram) which is recorded *via* one line of the slow ADC system. Between the beginning and end of a peak, the monitor starts successive mass spectra. Up to now, this application is the only case in which, not only a real-time data acquisition, but also a real-time data analysis is performed. However, this mode of operation will be changed as soon as the core memory of the system is expanded. Then during the entire experiment, successive mass spectra will be started every 5 sec with cycling magnetic field (2-sec duration of the active data transfer), but only valuable spectra are transferred to the disk by the data acquisition and reduction program (MSDAT).

During the remaining 3 sec, another mass spectrometer or fast instrument may be started.

The GC/MS combination is not yet running in a routine fashion, because some work on the analysis methods, especially for the intensity corrections of mass peaks with the total ion current, has still to be done.

Present System Performance

As the data acquisition is independent of the spectroscopic method, any instrument may be connected to the system, which delivers voltage signals between -10 and $+10$ V and runs either with data rates of 20 Hz and less or with rates between 1.25 and 20 kHz. Up to now, one fast-scan, low-resolution mass spectrometer and one pulsed-nmr instrument are con-

Status of Real. 7XA at 11:51:2∅ on 2∅-Feb- 7∅
Uptime 17:17:2∅, 93%. Null Time

Job	Who	Where	What	Size	State	Run time
1	**, **	DET	GCDIR	2K	SL	∅∅:∅∅:27
2	3∅, 5∅	TTY∅	PIP	4K	↑C	∅∅:∅2:15
3	**, **	DET	TRADAT	3K	SL	∅∅:∅∅:∅1
4	11, 3∅	TTY15	INDEX	5K	TT	∅∅:∅∅:23
5	**, **	DET	DSKILL	1K	IO	∅∅:∅∅:14
6	**, **	DET	MSDAT	8K	SL	∅∅:∅∅:35
7	15, 35	TTY12	REMAS	6K	↑C	∅∅:∅7:5∅
8	5∅, 1∅1	TTY14	TECO	2K	TT	∅∅:∅4:52
9	14, 34	TTY3	TECO	2K	↑C	∅∅:∅∅:19
1∅	**, **	CTY	SYSTAT	1K	RN	∅∅:∅∅:∅∅
11	52, 1∅2	TTY1	PIP	4K	TT	∅∅:∅2:25
12	5, 11	TTY6	PIP	4K	↑C	∅∅:∅∅:17
13	13, 33	TTY5	GCAN	13K	TT	∅∅:∅∅:37
14	22, 42	TTY11	PARAM	1K	TT	∅∅:∅∅:∅1

Figure 2. "System snapshot" illustrating the variety of simultaneously active time-sharing jobs: GCDIR (job 1) is the data acquisition program for "slow" on-line instruments; TRADAT (job 3) transfers data from the fast ADC to the disk (used for pulsed nmr); MSDAT (job 6) transfers reduced mass spectra to the disk; GCAN (job 13) and REMAS (job 7) are analysis programs for gas chromatography and mass spectrometry; INDEX (job 4) is a refined analysis program for gc; PARAM (job 14) a parameter-specification program for runs of slow instruments

nected to the "fast" ADC, whereas 10 gas chromatographs (the electronic cleanup of the other gas chromatographs is not yet accomplished), 1-nmr, 1-ir spectrometer, and 1 spectropolarimeter are on-line *via* the slow ADC.

The reliability of the system is improving after some difficulties in the beginning (mostly hardware failures of the special real-time devices and a few, but nevertheless, disturbing monitor malfunctions). As an average the system has one hardware breakdown every three weeks and about three software "crashes" every week.

As shown by the null time in the system snapshot (Figure 2), the central processor is far from being used to its full capacity. But nevertheless there are two bottlenecks in the present system causing difficulties under heavy system load:

(1) As the installation is not only used for the real-time applications, the present core memory is too small for the number of simultaneous users. The time-sharing monitor including the real-time code resides in 16K of core. To allow up to 32 simultaneously running slow on-line instruments, the associated data job needs 3K of core permanently. Temporary restriction to 20 simultaneously active lines saves 1K. But the remaining 14K of core is not sufficient for 15 concurrent jobs (even, if larger computational programs are run during the nights with a very small monitor), because too much processor time is wasted by swapping actions, and time-sharing response is decreased considerably, especially for programs with many input/output actions (*e.g.*, Fortran compilations). Therefore

the system is now being expanded by 32K of core. As swapping does not need CPU time, even during swap periods, the processor is available to other programs which are in core. But if core memory is too limited no other programs will fit into the remaining core when large programs are being swapped.

(2) The capacity (500K words) of the disk is too small for extensive data storage for all the users and especially for extensive storage of real-time data. Furthermore the future development of documentation systems for molecular spectra will need a large bulk of backup storage. Therefore the system will be expanded by the addition of disk packs.

Discussion of System Layout

When we consider the system economics, we should keep in mind that the Mülheim system had been designed to handle both off-line and on-line applications in theoretical and analytical chemistry. The purchase price of the system without the special devices for the real-time applications totals $450,000. This system would have been necessary for the off-line applications anyhow. The costs of the additional real-time hardware are about $65,000.

The availability of resources such as a line printer, plotter, magnetic tapes, and disk, as well as the high computing power of a large computer system, proved to be extremely useful in the development of the analytical instrumentation software. The use of high-level languages, such as Fortran, text editors, and debugging aids facilitate the writing, testing, and modifying of programs. To find suitable algorithms for the analysis programs, a close look at the unreduced ADC data is necessary. This can be accomplished by especially developed test programs. The fact that all this program development could be done within the interactive time-sharing mode, thus avoiding long turn-around times of batch systems or tedious loading and testing procedures on small computers, has, without doubt, accelerated the development of programs.

It seems to be obvious that there will be an increasing need in the future for large computational programs in the field of analytical chemistry (*e.g.*, extensive spectra iterations, spectra identification by documentation systems, "artificial intelligence" programs), which will use the output data from on-line systems for input. The close connection of the analytical on-line system to a large computational system will certainly stimulate the application of advanced analytical methods.

As far as the special hardware for the real-time data acquisition is concerned, some improvements to the Mülheim system are possible. In the present layout, all ADC data are handled by the CPU of the PDP 10, which has to perform all the bookkeeping and sorting of these data. Connecting the slow ADC and its multiplexer to a small satellite computer as a data concentrator would decrease the load on the PDP 10 processor. The satellite could overtake most of this bookkeeping and data formatting and transfer the ADC data blockwise to the PDP 10 (*5*). The Mül-

heim system will most certainly be modified in this way, if by increasing computational requirements the CPU time of the PDP 10 should be used to a much higher degree than it is now.

Small satellites could also become desirable if the need for more real-time experimental control arises. It is difficult to incorporate extensive real-time functions into the time-sharing monitor, and modifications to this code are tedious. Therefore small satellites are suited more for this type of application.

References

(1) Sederholm, C. H., Friedl, P. J., and Lusbrink, T. R., Pittsburgh Conference on Analytical Chemistry and Applied Spectroscopy, March 1968.
(2) Sederholm, C. H., 21st ACS Annual Summer Symposium on Analytical Chemistry, Penn State University, June 18–21, 1968.
(3) Ziegler, E., Pittsburgh Conference on Analytical Chemistry and Applied Spectroscopy, March 1969.
(4) Ziegler, E., Henneberg, D., and Schomburg, G., Pittsburgh Conference on Analytical Chemistry and Applied Spectroscopy, March 1970.
(5) For a more general discussion: J. W. Frazer, Pittsburgh Conference on Analytical Chemistry and Applied Spectroscopy, March 1970.

Additional References

The system described in this article in 1970 has since been expanded and improved. For further details the following publications are recommended:

Six articles in *Angew. Chem., Int. Ed.*, 11 (5) (1972).

J. Chromatogr. Sci., 9, 735 (1971).

Chromatographia, 5, 96 (1972).

E. Ziegler in "Computers in Chemical and Biochemical Research," Vol 1, Academic Press, New York, N. Y., 1972.

A Modular Approach to Chemical Instrumentation

Richard G. McKee

McKee-Pedersen Instruments (MPI), P.O. Box 322,
Danville, Calif. 94526

Analog and digital circuit modules offering a wide variety of electronic functions are now available. These powerful and versatile tools are reliable and easy to use. By combining appropriate modules with a few accessories, the analytical chemist can design new instruments

THE ANALYTICAL CHEMIST often must develop a specialized or new instrument for his work. In the past, this required a significant knowledge of electronics or the assistance of an electronic engineer. The advent of modern transistorized and integrated circuits has changed this. Now complex electronic functions are easily obtained using integrated circuit or discrete component operational amplifiers and digital integrated circuits as modular building blocks. A detailed knowledge of electronics is not required to use these modules so that instrumentation design is simplified.

The recent article by Springer (1) is a good reference on the use of integrated circuits, both linear and digital. This article will not repeat the material he covered. Instead, it attempts to go one step further and show you how to go about designing an analytical instrument. The principles of operational amplifiers (op amps) will not be covered as they are familiar to most readers. [If you are an exception, the book by Morgenthaler (2) is an excellent primer.] However, the limits of op amps and the other factors which should be considered when designing an instrument will be discussed.

The modular approach to instrument design involves a simple analysis of requirements. First you must put down your desired specifications. Then you analyze your instrument in terms of the functions the different parts must perform. Finally you select electronic circuits and modules to perform the various functions. In this step, the selection depends on the original specifications. Some of the important considerations in selecting modules are covered below.

What Is an Instrument

To best understand the functional approach to instrument design, first review what an instrument is. Simply

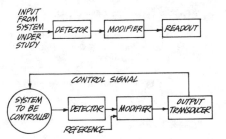

Figure 1. Top, instrument. Bottom, control system

stated, all instruments have three parts: input transducer, signal transformation modules, and output transducer. The input transducer, or detector, responds predictably to the physical property to be measured. The signal transformation modules perform necessary and/or desirable operations on the electrical output from the detector. The output transducer converts the final electrical signals back to physical ones we can read and interpret. With the addition of a reference signal and a feedback loop, an instrument becomes a control system (Figure 1). The signal from the detector is compared with the reference. The modifier output is then used to operate some device which will make a change in the physical system and reduce the difference to zero. An example is the pION-stat discussed at the end of this article.

Essentially every effect which can occur in an electrical circuit has been used in some type of input transducer. Table I gives a few familiar examples. Note that only those devices which produce a voltage may be used without an external power supply.

Most input transducers are inherently analog devices, as they are measuring continuous physical properties. The detectors respond steadily to their environment and give continuous electrical outputs. Where events occur in a digital fashion, the detector is designed to give pulse outputs. For example, a lithium-drifted germanium radiation detector gives a current pulse whenever struck by a gamma ray.

The performance of an instrument is usually limited by the quality and capabilities of the input transducer. Detector design is difficult. Fortunately, a large variety is available, and you can probably avoid designing your own. But, you cannot always avoid the design of the associated equipment—*e.g.*, such items as sample holders, detector housings, shielding, light sources, and optics. Some of these items are sold as individual modules by various manufacturers. However, many will be specific for your application and will have to be built to order.

When the detector requires a power supply, you should obtain one that is well regulated. Otherwise, your instru-

Table I. Some Examples of Input Transducers

Detector	Independent Variable	Dependent Variable
Specific ion electrode/ reference electrode	Activity of some ion in solution	Voltage
Thermocouple	Temperature	Voltage
Photomultiplier tube	Light level	Current
Polarographic cell	Concentration of electroactive specie	Current
Thermistor	Temperature	Resistance
Photodiode	Light level	Resistance
Hall effect device	Magnetic field	Resistance

Table II.   Some Signal Transformation Operations

Amplification
(multiplication by a constant >1)
Attenuation
(multiplication by a constant <1)
Integration
Differentiation
Addition
Subtraction
Taking logarithms
Taking antilogarithms
Arbitrary function generation
Squaring
Square root
Absolute value
Comparison

Analog to digital conversion
Digital to analog conversion
Track-and-hold
Isolation
Current-to-voltage conversion
Logical AND
Logical OR
Logical NOT
Logical NAND
Logical NOR
Counting
Filtering
Multiply variables
Divide variables
Rectification

ment may not give reproducible results. For example, if resistance is the dependent variable, it might be measured by applying a constant current and observing the resulting voltage drop (Ohm's Law: $R = E/I$). If the current is not constant, there will be a varying output even when the detector resistance is constant. An excellent way to provide constant currents and voltages for detectors is by means of operational amplifier circuits (3–6). Two examples are shown in Figure 2. A stable reference voltage is required for both. The resistors used should be metal film ± 0.1% tolerance or better with a low temperature coefficient.

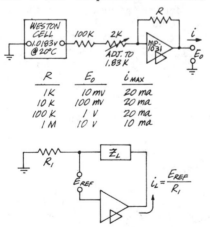

R	E_0	i_{MAX}
1 K	10 mv	20 ma
10 K	100 mv	20 ma
100 K	1 V	20 ma
1 M	10 V	10 ma

Figure 2. Top, precision voltage source. Bottom, constant current source

Many detectors require voltages or currents that are beyond the capability of operational amplifiers. If you are using one of these, a special power supply will have to be obtained. You should make certain that the supply may be operated floating—i.e., the common is not connected to the chassis. Then the common may be tied to the same ground point used by your other instrument building blocks. This helps to eliminate ground loops which can cause serious electronic noise.

An op-amp circuit can often be used when the detector requires excitation with some waveform. Very accurate and stable waves (e.g., sine, square, triangular, sawtooth, ramp) are easily generated (7, 8). An advantage of the op amp circuits is their ability to generate very slow waves. The disadvantage of many general-purpose op amps is their inability to produce waveforms at frequencies much above 5–10 kHz. For higher frequencies, you can go either to high-frequency op amps such as the Fairchild type 715 or Analog Devices types 148 and 149, or to a commercial waveform generator. Again, you should make certain that it has a floating output.

Signal Transformation Modules

The signal transformation modules operate on the signal representing the detector output. Depending on the type of detector, some of the operations may be necessary rather than optional. With high-impedance voltage sources, for example, it is impossible to draw any current without changing the voltage because of the resulting ir drop. The first job that must be done is isolation of the source with a circuit that draws a negligible current. Voltage amplification may or may not be required. With a current source, the objective is to measure the short-circuit current. Here a very low impedance device is needed. Since the signal current may be very small, the device must again draw a negligible current. The commonly used circuit is called a current-to-voltage converter (Figure 3).

Table II lists some of the many operations performed by signal transformation modules in instruments and control systems. These are extremely diverse. If you tried to buy separate modules to do all these tasks, you would wind up with a high price and probably have serious ground loop and compatibility problems. Fortunately, most of these operations can be carried out by op amps and digital logic circuits. For background on the use of these devices, see references (1–5, 9–11). With a little care in providing proper power supplies and grounding, you should get good results without a lot of effort.

Once you have determined the operations you wish to perform on signals, you have the problem of physically implementing the design. Much assistance is available from the applications literature noted before. Most of the basic circuits have been worked out. Only rarely will you have to design a totally new signal transformation module. To assemble your circuit, you can make use of the op amp manifolds offered by a number of manufacturers (Philbrick, Burr Brown, Analog Devices, MPI). These include the necessary power supplies. Also available are a number of systems for breadboarding IC op amps and logic elements (Correlated Educ. Serv., DEC, EDTEC

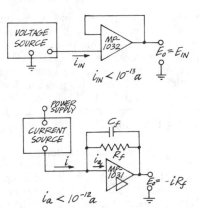

Figure 3. Top, isolation using the voltage follower circuit. Bottom, current-to-voltage converter

Assoc., EL Insts., Heath, MPI).

The op-amp manifolds and IC breadboard systems are only partially helpful, however, as many additional building blocks are usually needed to implement all of the desired signal transformations. Examples are voltage references, timers, relays, integration networks, switches, and precision resistors. Also amplifier–voltage balance and input–current compensation circuits are essential and must usually be supplied by the user. When building a specialized device to be housed in a box and used as a finished instrument, you have no choice but to wire up the unit with individual components. You have to go to the trouble of locating the various parts and assembling them. For methods development and short-term projects, one system is available which offers the amplifiers ready to use and also includes a wide variety of accessory units (MPI System 1000).

Output Transducers

The instrument readout device may be as simple as a meter or complex as a high-speed teletypewriter. Some of the commonly used output transducers include: meter, strip chart recorder, X-Y recorder, oscilloscope, photographic film, potentiometer and null

331

meter, digital printer, and alpha-numeric display. Occasionally ear phones are used to detect a null point.

Most readouts can be purchased as ready-to-use modules. Again, you should avoid ground loops when low-level signals are fed to the device. It is best to buy equipment which can be operated floating.

Practical Considerations

Practical considerations in the design of an instrument include drift, accuracy, stability, and noise. An understanding of the factors affecting each of these items is necessary, if you hope to design an instrument that will meet your desired specifications. These factors are discussed below.

Drift can occur in each of the parts making up an instrument, but as often as not, the detector gives more trouble than the others. For example, we all are familiar with the glass electrode and its unstable performance if it is not properly soaked before use. Since each detector has its own peculiarities, it is best to carefully follow the manufacturer's instructions.

The need for well-regulated power supplies, for those detectors which use them, has already been mentioned. Power supplies can drift too. The modern units have temperature-compensated zener diode references, which help considerably in reducing drift caused by temperature changes. Mercury cells are still often used for reference voltage sources in instruments. These cells are reasonably stable when handled properly. Short-term drift can be as low as 0.005% of the half-scale range. Power supply variations can seriously affect the drift of some cheaper op amps. An important specification for amplifiers is the power supply rejection ratio. A reasonable value for a quality unit is 10^5 (100 db).

Drift in solid state amplifiers is largely caused by temperature differentials. Integrated circuit amplifiers can have quite low voltage and current drifts, since the devices are so small and all the circuit components are close together. In fact, they are on a single silicon chip. Typically, the offset voltage drift of IC op amps runs from 3–30 μV/° C. The input bias current drift is about 0.6–2 nA/° C. Well-designed differential units built from discrete components can do about as well. Amplifiers employing field-effect transistor (FET) input stages have voltage drifts from 3–50 μV/° C. The input current doubles for every 10° C temperature rise. This may seem like a large drift. However, the bias current is so low that this change is usually unimportant.

Chopper stabilization can be used to reduce drift to below 1 μV/° C. Some amplifiers are available with temperature drift as low as 0.1 μV/° C. Most chopper amplifiers have very low-input current drift. Chopper amplifiers draw about 10–100 pA at 25° C. Some can be adjusted to draw less than 1 pA. Bias current drift varies from 0.03–100 pA/° C for these devices.

The effect of drift on performance can best be seen by examining the general equation for an op-amp circuit in which the positive input is referred to ground. See Figure 4. [Note that a resistor is often placed in series with the positive input. Its value is equal to the resistive component of the sum of the impedance connected to the inverting input: $R = R_i R_f / (R_i + R_f)$.

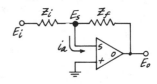

$$E_o = -E_i \frac{Z_f}{Z_i} + E_s \left[1 + \frac{Z_f}{Z_i} \right] + i_a Z_f \quad (1)$$

Figure 4. General inverting amplifier network

332

In this way the input currents will tend to remain balanced during temperature changes.]

Let us start by assuming that the $i_a Z_f$ term is negligible. Now if E_i is constant, E_s will be constant at $-E_0/A$ volts. A is the open loop dc gain of the amplifier. If E_s then varies a little because of temperature changes, the output voltage will change by ΔE_s $[1 + Z_f/Z_i]$. For example, assume that E_i is -1 V; A is 10^5; Z_f/Z_i is 10; and the offset voltage drift is -10 μV/° C. For a 10° C temperature change, the output voltage will change from 9.9989 V to 9.9988 V. If i_a is 10^{-9} A and drifts at 0.1 nA/° C, we can figure in its contribution. First Z_f must be specified. Its value depends on the type of amplifier used. Obviously, if i_a is -1 nA, you should not use too large a Z_f. Otherwise, the product $i_a Z_f$ would not be negligible. Let us assume Z_f is 10^5 Ω. Then we find E_0 is really 9.9988 V at the first temperature and 9.9986 V at the second. In this example, drift is not serious. The output changed less than 0.002% over a 10° C temperature range. However, if Z_f/Z_i is increased to 100 by increasing Z_f to 10^6 Ω, the effect of drift is much larger. Assume E_i is now -0.1 V; then at the lower temperature, E_0 will be 9.9889 V, and it drops to 9.9869 V at the higher temperature. The change amounts to 0.02%.

Although amplifier drift affects precision, the overall precision of an instrument may be limited by the readout device. Meters, for example, are normally good to $\pm 2\%$ of full scale. If you buy an expensive one, you may be able to get $\pm 1/2\%$. An oscilloscope is usually readable to only $\pm 2\%$ of full scale. A good chart recorder (Y-T or X-Y) may be accurate to ± 0.1–0.25% of full scale. If you use a null meter and potentiometer for readout, the precision will depend on the linearity of the potentiometer and whether you bother to calibrate the unit against a lab standard. Most 10-turn potentiometers run ± 0.5 to $\pm 0.1\%$ of full scale. Be careful when you buy a potentiometer, as there are several different commonly used ways of specifying linearity. Digital volt meters are available with precision of $\pm 0.1\%$ of full-scale ± 1 digit. In some readout devices, drift or susceptibility to line-frequency hum pickup may also be factors affecting precision. It pays to carefully read the manufacturers' specifications. Good-quality chart recorders, for example, usually have built-in hum-rejection filters tuned to the line frequency.

To obtain the desired results from the modifier portion of your instrument, you must consider the following: precision and temperature coefficient of circuit components, amplifier open-loop gain and frequency response, input current, source impedance, common mode rejection, and noise. These factors are discussed below.

At low frequencies (dc to 100 Hz), the open-loop gain of an op amp is normally high enough so that accuracy of a circuit is solely determined by how well you know the values of the components. The output of the circuit is given by the general equation

$$E_0 = -E_i Z_f/Z_i +$$
$$E_+ (1 + Z_f/Z_i) \quad (2)$$

When the positive input is not used— i.e., it is grounded—the output is given by Equation 3:

$$E_0 = -E_i Z_f/Z_i = -i_i Z_f \quad (3)$$

The circuit components Z_f and Z_i may be composed of resistors, zener diodes, capacitances, etc. Be sure to choose parts that are adequate for your application. Resistors, for example, are commercially available in $\pm 0.05\%$ tolerance and with temperature coefficients as low as $\pm 0.0002\%$/° C. Precision capacitors are more difficult to obtain.

Polystyrene units are available with $\pm 0.1\%$ tolerance and temperature coefficient as low as -0.008%/° C. Zener diodes are available with temperature coefficients of voltage down to 0.0005%/° C. When you are working at higher frequencies, the component values will not be what you have at dc. A good-quality resistor, for example, can be made with very low inductance. Still, there is some inductance, and it will change the impedance of the resistor at high frequency. These points are mentioned here only to make you aware of them. In some applications, they may be important. In most chemical instrumentation, low frequencies or essentially dc signals are encountered, and you need not worry about such items as stray inductance and capacitance. Just remember to use the proper values for your components at the frequency at which you are working, since they may be different from the dc values.

The simple equations above are only usable if amplifier open-loop gain, A, is very high. Ideally, A is infinite. As signal frequency increases, however, gain decreases. This is a necessary requirement for stability. The frequency-gain relationship is easily plotted from two specifications given by the manufacturer. These are dc open-loop gain and the unity-gain frequency. Most amplifiers are frequency compensated so that A decreases by a factor of 10 for each decade increase in frequency. Simply make a log-log plot. Draw a line with the proper negative slope intersecting the frequency axis at the unity-gain frequency. Draw a second line parallel to the frequency axis intersecting the gain axis at the dc gain. The curve may be used to determine A at any frequency (Figure 5). The error resulting from finite gain is then given by Equation 4:

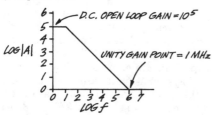

Figure 5. Gain-frequency curve for a typical operational amplifier

$$\text{Error} = \epsilon = 1 -$$

$$\left[1 + \left(\frac{1 + Z_f/Z_i}{A} \right) \right]^{-1} \times$$

$$\frac{Z_i + Z_f}{Z_f + Z_i(1 + A)} \quad (4)$$

A more complete discussion of factors affecting amplifier accuracy has been presented (4, p II-4).

The current drawn by an op amp is one of the most important sources of error in instrumentation applications. Equation 1 above is general. It shows that the output of an amplifier circuit includes a contribution equal to $i_a Z_f$. You would like to be able to neglect this term. However, if Z_f is large and/or i_a is large, the product may be very significant. The rule of thumb is that the signal current, E_i/Z_i, must be large compared to i_a if you wish to make meaningful measurements. For 0.1% precision, i_a should be three orders of magnitude smaller than the signal current. Remember that both inputs of an op amp draw current. When you use the positive (noninverting) input, i_a is drawn from the signal source. If it has a high impedance, your effective signal voltage will be reduced by the product $i_a Z_{\text{source}}$. These errors may be avoided by paying attention to the amplifier specifications and properly selecting the unit for the task. The better op amps include controls so that input current can be adjusted to low values. A good general-purpose unit will draw less than 10^{-8} or 10^{-9} A at either input. Some chopper stabilized units are available with input currents adjustable to as low as 10^{-12} A. FET amplifiers are available with input currents from 10^{-10} to 10^{-13} A. One unit uses a MOSFET input stage and draws less than 10^{-15} A. You should avoid overly large feedback impedances to minimize the $i_a Z_f$ term. This is a good rule for resistors, since large feedback impedances are thermal noise generators. Incidentally, amplifier input current is very troublesome when you are carrying out long integrations, as it is integrated along with your signal.

The impedance of your signal source may lead to error if you are not careful. As mentioned above, current drawn from the source will change its voltage. At the positive input, E_+ is really $E_{source} - i_a Z_{source}$. When you use the inverting amplifier, remember that Z_i is really $Z_i + Z_{source}$ as far as the amplifier is concerned. Hence $E_0 = -E_i Z_f/(Z_i + Z_{source})$. Fortunately, the output impedance of an op-amp circuit is usually less than 1 Ω. [See (4, p II-6) for the equations used to calculate the exact value.] This is low enough so that it normally will not cause any error in subsequent sections of your instrument.

In those applications where you use the noninverting input of an op amp, you must make certain that you do not exceed the common mode limit, and that the common mode rejection ratio (CMRR) is adequately high for your desired accuracy. The maximum voltage that may be applied to an amplifier input is called the common mode limit. This is always specified by the manufacturer. If the limit is exceeded, the output will not be the expected function of the input. Also, the amplifier may possibly be damaged. With inverting circuits, the negative input (or S, the summing point) is held virtually at ground potential. The signal voltage is applied to an impedance connected to S. It is very rare that the potential at S gets high enough to damage the amplifier. However, when the positive input is used, both it and S are raised to whatever potential is applied to the positive input. Common mode error results, because the gain of the amplifier is not the same at both inputs. (Ideally, A is the same at both inputs, and the equations used to calculate output make this assumption.) The CMRR is a measure of the error. If you need to use setups involving the positive input, be sure to select a unit having a high CMRR. Otherwise, accuracy will be poor. Most good-quality differential op amps have a CMRR of 10^4 (80 db) or higher.

Noise is another source of error in most instrumentation. Any spurious output which is not contained in the input signal is called noise. Amplifier drift, for example, is simply a very low frequency noise. Voltage noise at the amplifier inverting input will appear at the output multiplied by $(1 + Z_f/Z_i)$. Current noise injected at S is multiplied by Z_f. Any noise voltages induced in Z_i are multiplied by Z_f/Z_i. Noise voltages induced in Z_f are unamplified. See (2, p 25) for a discussion of the various types of noise.

Noise pickup can generally be kept small by adequate shielding. Shielded cables can eliminate pickup of the electrostatic component of electromagnetic radiation. To reduce the magnetic component, keep the area included within signal-carrying loops to a minimum. In other words, keep your leads short! Shielded cables connected to the inverting input of an op amp can cause stability problems, because they introduce a capacitance between S and ground. If the network uses a large feedback resistor, use a parallel feedback capacitor to ensure stability. (Be sure to calculate the impedance of the feedback capacitor as a function of frequency to determine the gain errors.)

Random noise is generated in all circuit components. It can be minimized by restricting the bandwidth of your amplifier network and by using the lowest impedances possible. Random noise generated in resistors (and the resistive component of any impedance) is called thermal noise. It is proportional to the square root of the absolute temperature, resistance, and bandwidth. Remember that the rate of information transfer is proportional to the bandwidth. If you take your date more slowly, you can use a feedback capacitor to narrow the bandwidth of your setup and reduce thermal noise.

If digital logic is required in the instrument design, it is necessary to select an integrated circuit family. At present, the most common logic elements are TTL (Transistor-Transistor Logic) that operate between 0 and +5 V. This logic is compatible with modern small computers, and its low impedance design minimizes the effect of switching transients. When using this logic, it is advisable to use a low-impedance capacitor between the supply voltage and signal ground for every five integrated circuits. Typical capacitance values are 0.01 μfd. In addition, the power pin on every operational amplifier should be decoupled to ground through a similar type of capacitor.

Circuit stability is another factor to consider when using op amps and/or building a control system. Keep in mind that any positive feedback at a gain greater than one will make a system unstable—*i.e.*, oscillate. An op amp is designed to be stable over a wide range of feedback conditions. This requires that the open-loop gain decrease as input signal frequency increases so that the signal transformation modules and the overall system are stable. A description of operational amplifier stability is given (*5*). For a more detailed explanation, see (*12*).

The two major sources of instability in op-amp networks are input capacitance to ground and load capacitance. The input capacitance can come from a coax cable connected to the inverting input. It will easily introduce a 90° phase shift (lag) when Z_f is a resistor. This is corrected by placing a small capacitor in parallel with your feedback impedance. The capacitor introduces a compensating phase lead. A load capacitance forms an RC network with the output resistance of the amplifier

and can thus give you a 90° phase shift. The best solution is to isolate the load as shown in Figure 6. R is equal to the specified output impedance of the amplifier. In electrochemical applications, it is very common to have a large amount of capacitance associated with the cell. When driving such a capacitive load, you should examine the output of your amplifier with an oscilloscope to make certain it is not oscillating. If it is, it may be damaged. The above suggestions should prevent such instability.

Some of the additional factors which should be considered when using digital logic circuits are fanout and delay. These have been discussed in the recent article by Springer (*1*).

To summarize, instruments can easily be assembled from functional building blocks, but good performance results only if you heed the practical points discussed above. It pays to use modules designed to work together. Then some of the burden is shifted to the manufacturer. It also pays to read the specifications and instructions!

A Digital pION-stat

The design of a digital pION-stat will serve as an example of the modular approach to instrument design and assembly. The first step is to set down your desired specifications. For this example, let us assume that pION is to be controlled to within ±0.01 unit. The instrument will be used for kinetics studies. Therefore we need a readout of volume of reagent added to maintain constant pION as a function of time.

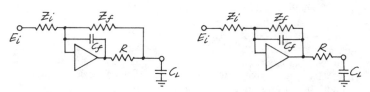

Figure 6.   Two methods for isolating a capacitive load

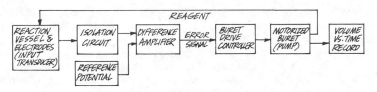

Figure 7.   Block diagram of a pION-stat

(If the pION-stat were to be used for synthesis at constant pION, this record could be omitted.) In this application, the reaction produces an ion which must be removed to keep pION constant. A potentiometric cell will be used to monitor pION. The indicating electrode will be assumed to have a dc impedance of 100 MΩ. A digital approach will be used. Small aliquots of reagent will be added to bring the pION back to the set point. The aliquot size must be chosen so that it will change the pION by no more than the allowable error of 0.01 unit. Finally, we must specify how much the temperature of the instruments will change. Perhaps the lab is not air conditioned, and it is summer. When the experiment is started, the lab might be at 25° C and could reach 35° C by the end of the run. (For simplicity, we will assume that a separate system will be used to control cell temperature. Hence this discussion will not include the design of the thermostat.)

The next step is to draw a block diagram. Each function you think will be needed should be included. A potentiometric pION-stat requires the functions shown in Figure 7.

The potentiometric cell consists of a reference and an indicating electrode. We assumed 100 MΩ as the impedance of the latter. (This is a reasonable value. The glass electrode, for example, typically has an impedance of 10 to 500 MΩ.) Cell impedance will be essentially the same, as most reference electrodes have relatively low impedance. To select an op amp for the isolation circuit, we must first calculate the voltage corresponding to our allow- able error in pION. The cell potential is 2.303 RT/nF for 1 pION. At 25° C, the voltage is 59.15 mV/pION for monovalent ions. To control to ±0.01 pH, for example, the instrumentation must respond to ±0.59 mV. For divalent ions, instrument sensitivity should be half this figure. The isolation circuit will have to use a high-quality FET amplifier setup as a voltage follower. The basic circuit was shown in Figure 3. The output of the op amp will be the input voltage minus any ir drop caused by the input bias current.

To get an idea as to the maximum allowable input current, we could assume that the allowable error of 0.01 pION was all caused by this current. For divalent ions, this comes to 0.59 mV/ (100 MΩ) or 2.95 pA. Since the bias current doubles for every 10° C temperature rise, the input current at 25° C must be less than half this amount, or about 1.5 pA. Voltage drift will also add to the error. In practice, the op amp should contribute less error so that there is room for the errors caused by other parts of the system. It would be wise to look for an amplifier drawing about one-tenth the above current. For example, MPI offers an FET op amp that draws less than 0.1 pA at 25° C and has voltage offset drift less than 20 μV/° C (4, p I-35). At 25° C, the output voltage of this amplifier would be in error by only 0.00034 pION for a divalent ion. At 35° C, the output voltage error will be 200 μV + 0.2 pA × 100 MΩ. This corresponds to 0.0075 pION for a divalent ion. This amplifier would certainly be suitable for this application.

The next portion of the instrument which must be considered is the reference voltage. It has to be very stable. After all, any change in it will affect the system the same as a change in cell potential. A suitable reference source could be a mercury cell. It would be most convenient to build a reference module directly settable in pH units or millivolts and including controls for temperature compensation and calibration with buffer solutions. The circuit for such a module is given in (2, p I-3). The reference voltage will be most stable if negligible current is drawn from the mercury cell. One way to accomplish this is to use the reference to buck against the test cell. Then our FET voltage follower doubles as the difference amplifier. When the reference and test cells are at the same potential, the net input voltage is zero. The output of the isolation circuit is zero, and may be observed with a null meter. This is a very convenient circuit. In effect, it is a null balance pH meter. You can readily calibrate it for pH work by placing buffer in the test cell, setting the buffer pH on the reference module, and properly setting the temperature on the compensation dial. Then by adjusting the pH standardize control to bring the meter to null, you complete the procedure. When the test solution is placed in the test cell, the desired control potential is then set on the reference module.

The amplifier output will be the error signal used to operate the rest of the pION-stat.

One way to get response to a very small error signal is to integrate it with respect to time. The integral builds up until it reaches a preset level where it triggers the addition of an aliquot of reagent and also a reset circuit. The addition of reagent should then bring the pION back to the desired level. If the reaction in the test cell runs very slowly, aliquots will not be added very often. Integration errors could then be serious. Once more, a very low-input current amplifier should be used. A chopper-stabilized unit would be a good choice.

The remainder of the buret drive controller could be built with a switch amplifier, relay, another reference voltage, and some electronic components. The reference potential determines how large an output from the integrator is required before an aliquot will be added. The polarity of this reference and direction of the diodes and of the electrolytic capacitor (Figure 8) depend on the chemical system. The figure shows the setup for experiments in which cell potential tries to go less positive and a reagent is added to prevent it. For a pH-stat, this represents addition of base to hold pH constant while the reaction generates acid. When a cell potential tries to go more positive, you must reverse the diodes, capacitor, and switch-

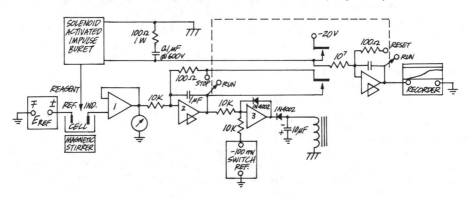

Figure 8.   Digital pION-stat

reference polarity. The diode and capacitor in the output of amplifier 3 are used to provide a time delay. Without this delay, the relay could reset so fast that the impulse buret may fail to operate. The relay must be selected so that it can operate on the output wattage of amplifier 3. If necessary, a booster amplifier can be used in conjunction with amplifier 3 to increase the output power. An ordinary general purpose op amp may be used for the switch circuit, since low-input current or high stability is not required.

To record volume of reagent added as a function of time, an integrator can be used to sum a constant input each time the relay closes. The output consists of a series of steps displayed by a strip-chart recorder. It is better to integrate a signal from another source rather than use the output of amplifier 3. This is a rather messy signal and may not give constant steps. The figure shows a setup in which the impulse buret is operated by a 20-V dc supply. This same supply provides the input for the integrator. Note the addition of a resistor and capacitor across the relay contacts to suppress arcing.

The complete pION-stat is shown in Figure 8. The op-amp power supplies are omitted for clarity. As discussed above, they should be fairly well regulated. Note that the relay and impulse buret are grounded to the instrument chassis rather than the high-quality signal ground. This is to keep these large currents from mixing with the low-level signal currents and causing ir drops along the signal ground buss.

The digital pION-stat may seem like a fairly complex instrument. However, it was built up from only a few basic building blocks. By analyzing other instrument tasks the same way we approached this one, you, too, can make use of functional electronics to fill your instrumentation needs.

References

(1) John Springer, ANAL. CHEM., 42, No. 8, 22A (1970).

(2) L. P. Morgenthaler, "Basic Operational Amplifier Circuits for Analytical Chemical Instrumentation," 2nd Ed., McKee-Pedersen Instruments, Danville, Calif., 1968.

(3) "Applications Manual for Computing Amplifiers," Philbrick Researches, Inc., Dedham, Mass., 1966.

(4) "MP-System 1000, Operation and Applications," 3rd Ed., McKee-Pedersen Instruments, Danville, Calif., 1968.

(5) "Handbook of Operational Amplifier Applications," Burr-Brown Research Corp., Tucson, Ariz., 1963.

(6) MPI Applications Notes, 4, 17–20 (1969).

(7) Ibid., 1, 5–8 (1966).

(8) Ibid., 4, 1–4 (1969).

(9) Ibid., McKee-Pedersen Instruments, Danville, Calif., Bi-monthly publication, illustrating the use of operational amplifiers in chemical instrumentation.

(10) H. V. Malmstadt, "Digital Electronics for Scientists," Benjamin, New York, N. Y., 1969.

(11) P. M. Kintner, "Electronic Digital Techniques," McGraw-Hill, New York, N. Y., 1968.

(12) H. N. Bode, "Network Analysis and Feedback Amplifier Design," D. Van Nostrand, Princeton, N. J., 1951.

Image–Analyzing Microscopes

Philip G. Stein

National Bureau of Standards, Washington, D. C. 20234

Image-processing technology, through expanded data gathering and storage capabilities, now has the potential to solve difficult problems in metallurgy, polymer and fiber chemistry, and cell biology. Many studies involving the counting and sizing of particles or fibers can be automated through the use of image-processing techniques

M ANY PROBLEMS in metallurgy, polymer and fiber chemistry, and cell biology are being tackled today using image processing techniques. A typical measurement of the area of voids in a metallic specimen, for example, would require making a photomicrograph, printing it, and measuring the resulting photo. Some laboratories use electromechanical scanners to make these measurements on the print, or directly from the negative. Elementary tests may be done with logic in the scanner. More sophisticated problems are attacked by making a digital tape recording of the print and processing it with a general-purpose computer.

A different approach to this type of problem has recently been embodied in several instruments, known as image-analyzing microscopes. At the very least, they can eliminate the time-consuming and inaccurate photographic steps by scanning the specimen directly in the image plane of the microscope. Further sophistication could lead to improvement of the data gathering process with on-line computer control of the scanning and other microscope functions.

Each of these instruments contains a scanning microphotometer, which divides the image plane of the microscope into a number of resolution elements and measures the luminous flux transmitted through (or reflected from) each element. Data are then processed by special or general purpose computers connected directly to the microscope. Some instruments include variable-wave-length illumination and a motor-driven specimen stage which allows positioning of more than one specimen field under the microscope objective.

All of these systems share one basic consideration that colors their design: the amount of data is immense. In the SPECTRE II system at the National Institutes of Health, a single scan at one wavelength produces 2^{19} information bits. A single scan in this case, resolving at or near the limits of the optical microscope, covers only a 50- $\times$ 50-μm square on the specimen. A 10- $\times$ 10-mm specimen would require 40,000 such scans at each wavelength of interest to acquire all of the data.

For this reason, most of the instruments do not store the raw data. They may, in fact, have provisions for recording or transmitting the data to a computer, but in normal operation they rely on the specimen itself to act as a data storage medium. By repeated

340

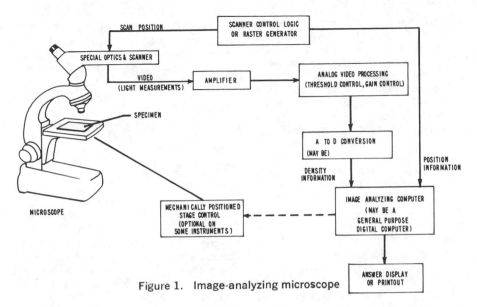

Figure 1. Image-analyzing microscope

reference to the actual image, very little storage of the basic density-position data is required.

Gathering the Data

The heart of any image-analyzing microscope is the scanner. It partitions the image into a number of equal areas and measures the light through each. The size of the area measured is a function both of the microscope objective and the scanning aperture. An aperture of 10-μm diam at the image plane, for example, corresponds to a 0.25-μm diam spot on the specimen (with a 40× objective). To take full advantage of the resolution implied by this aperture-objective combination, one must use large aperture, oil-immersion objectives and condensers, and be extremely careful with the incident illumination. All image-analyzing microscopes require the operator to set up the condenser system manually, and the resultant measurements are somewhat dependent on how well this is done. For less critical problems, smaller magnifications and apertures obviate many of these problems. The user must also make some trade-off between resolution

and contrast. At high resolution, the image contrast is reduced, and more amplification is required in the scanner light-measuring circuit. This, in turn, results in increased noise. Again, lower-resolution studies are relatively easy.

Types of Instruments

Figure 1 is a block diagram of an image-analyzing microscope. Clearly, the technical details of the equipment following the scanner depend on the scanner design. Breaking these into broad categories, they are TV scanners (raster), electronic scanners (using a random-scan tube), and mechanical scanners.

The TV scanner is most amenable to use with special purpose computers. The data are scanned in a raster fashion, completing one line of X for every increment in Y until the entire field is covered. This is done at a fixed rate. For vidicon-type camera tubes, this is important, as the tube photocathode integrates all of the light reaching each position. When a given spot is interrogated by the beam, the information is erased and a new integration period begins. For the data to be uniformly noisy (or quiet), the integration time at each point must be the same. Therefore, use of a vidicon for random scans is not recommended.

Since each line is being scanned at a fixed rate, an object in the field will be transformed into a time series of voltages representing the transmission at each point. If we set a threshold (Figure 2) which says "anything darker than this level is an object," we then get a pulse whose width is determined by the object size, the scanning rate, and the threshold setting. This is the basic principle behind the TV-type special computers.

Analysis-TV Microscopes

From this point, there are many ingenious ways to process the pulse data using logic circuits. Some of the possibilities are:

(1) Using a TV display of the image, intensify or otherwise identify all points above the threshold enabling the operator easily to find a proper threshold setting.

(2) Similarly intensify only the leading or trailing edge of an object.

(3) Restrict the image processing to a rectangular subportion of the screen, and indicate the area to the operator. The operator may continuously adjust this "window."

(4) Reject pulses shorter than a certain minimum length, thereby ignoring much dirt and some artifacts of the specimen as being too small.

(5) Save the previous scan line in a memory and compare it with the current one. If an object was present on the last line and is still present, do not count it again.

With these and other processing techniques, it is possible to automate many problems involving counting and sizing

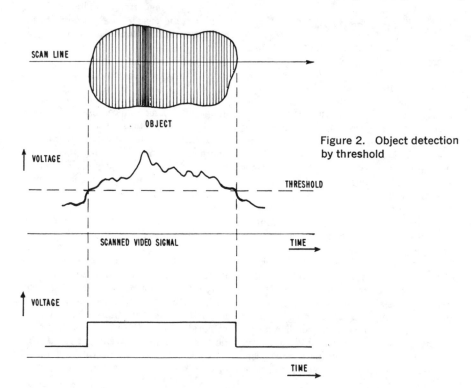

Figure 2. Object detection by threshold

of particles or fibers. By rotating the image 90°, either mechanically or optically, measures of out-of-roundness or form factor are also easily generated. The data may be transferred by an automatic sequencing programmer to a desktop calculator; statistics developed about the field, and measurements which are numerically derived from two or more pulse variables can be calculated. In addition, the data may be transferred to a general purpose computer through an appropriate interface, and further computations done.

Why Not?

The most serious problem with instruments of this type lies in their assumption of data uniformity. The threshold control must be set manually by an operator, and so must the illumination. This is indeed true of any image-gathering device of this type, but the TV system relies very heavily on this setting for the repeatability of the data. In addition, the logic must depend on an implicit requirement of spatial uniformity of both specimen and illumination. The threshold remains uniform over the field, even though the data may not be. In some machines, a threshold may be locally defined by measuring the change in level across a boundary and "splitting the difference." This may help in a few cases. Another problem is that, even with the most sophisticated logic now available, it is not possible to recognize all objects unambiguously. Some regions of the field will be blind to the presence of objects owing to "paralysis" of the logic by other objects. Whether this is of any consequence, of course, depends on the problem. It would be possible to connect a computer in such a way that it could adjust the threshold and interpret the data at each different setting, thus generating "local" thresholds in the image. This is cumbersome, though. It is better to approach those problems requiring this degree of sophistication with a different instrument.

Other Ways to Do It

Random-access scanners may use a moving specimen stage, vibrating mirrors in the light path from objective to image plane, or an image-dissector scanning tube. Of these, only the image dissector shares with the vidicon the problem of photocathode nonuniformity, but it has a great speed advantage over the mechanical scan methods. In all cases though, the output of the scan is transmitted directly to a general-purpose digital computer. In some instruments, this computer can directly control the scan pattern. It can also be connected to make large motions of the specimen stage (such as from field to field), change the color, intensity, and bandwidth of the illumination, and change the focus. None of these systems has a storage photocathode, so there is no need to scan every point in the field at a fixed rate. The transmitted or reflected intensity data are gathered directly by the computer, and the image processing may be done either directly or by transmitting the data to a large, high-speed machine.

This approach is far more complicated and expensive than the purchase and operation of a simple TV microscope. It is also much more flexible, permits a great deal of investigation into the interaction of the machine and the properties of the data, and allows unlimited sophistication of the image-processing techniques employed. These might involve, even on the most basic level, continuously variable local thresholds, depending on the data themselves, thresholds of sharpness (rate of change of light) rather than simple intensity, and programs that follow boundaries with the scan to resolve ambiguities about shapes owing to paralysis.

Since the data are available in their raw form, the computer might select a proper wavelength for a study on the basis of contrast, or it might adjust the integration time at each point in the image to improve statistics while wasting little or no time on background areas.

The computer also might instruct the scanner to look at a field with very low resolution (perhaps every tenth line), and locate all of the large objects. The scan would then return to those areas where data were found to scan them in detail.

Artificial Intelligence

It is also possible, in some special cases, to use the information about geometry and spectral transmission or reflection to identify and classify isolated specimens. This hopefully would lead to the ability to distinguish normal and abnormal specimens automatically. This development depends strongly on processing the image in a programmable fashion, making it easy to change classification techniques. A very general facility as in Stein *et al.* (*1*) could be used to simulate the operation of any number of specialized image-gathering and image-analyzing microscopes. It could be used, therefore, as a design tool for single-purpose instruments.

Finally, the computer can produce a display for the operator which shows not only the raw data plus a threshold, but also the result of the data processing by any algorithm available to the programmer. In this way, the image-analyzing techniques themselves may be iteratively improved.

Literature Cited

(1) P. G. Stein, L. E. Lipkin, H. M. Shapiro, *Science* **166**, 328–33 (1969).

Suggested Source Material

(1) R. C. Bostrom, H. S. Sawyer, W. E. Tolles, *Proc. IEEE* **47**, 1895 (1959); P. H. Neurath, B. L. Bablouzian, T. H. Warms, R. Serbagi, A. Falek, *Ann. N. Y. Acad. Sci.* **128**, 1013 (1966).

(2) M. L. Mendelsohn, B. H. Mayall, J. M. S. Prewitt, R. C. Bostrom, W. G. Holcomb, *Advan. Opt. Electron Microsc.* **20**, 77 (1968).

(3) M. Ingram, P. E. Norgren, K. Preston, Jr., *Ann. N. Y. Acad. Sci.* **157**, 275 (1969).

(4) G. L. Wied, P. H. Bartels, G. F. Bahr, D. G. Oldfield, *Acta Cytol.* **12**, 180 (1968).

(5) W. Prensky, *J. Cell Biol.* **39**, 157a (1969).

(6) K. Preston, Jr., P. E. Norgren, *Ann. N. Y. Acad. Sci.* **157**, 393 (1969).

(7) L. E. Lipkin, W. C. Watt, R. A. Kirsch, *Ann. N. Y. Acad. Sci.* **128**, 984 (1966).

(8) L. E. Mawdesley-Thomas, P. Healey, *Science* **163**, 1200 (1969).

(9) S. A. Rosenberg, K. S. Ledeen, T. Kline, *Science* **163**, 1065 (1969).

Data Domains—An Analysis of Digital and Analog Instrumentation Systems and Components

Data domains concepts offer a means of effectively utilizing new electronic devices which requires only an understanding of basic measurement processes. These concepts can be used to great advantage in designing or modifying systems and in assessing and minimizing the sources of measurement errors.

C. G. ENKE

Department of Chemistry,
Michigan State University,
E. Lansing, Mich. 48823

SCIENTIFIC INSTRUMENTATION is being revolutionized by the availability of an ever-increasing array of electronic devices which increase measurement speed, accuracy, and convenience while decreasing instrument size and power requirements. Integrated circuits and hybrid circuits have made many measurement techniques, which were previously only theoretically possible, a reality. The continual decrease in the cost of digital and linear circuits has made many sophisticated devices such as frequency meters, digital pH meters, signal averagers, and minicomputers practical for most laboratories. As electronic technology continues to advance, we can expect more and more of the sampling, control, and data analysis of scientific measurements to be performed by the instrument itself.

Digital instrumentation has been the scene of much development and interest because of its inherent accuracy capability, convenient numerical output, and potential digital computer compatibility. However, just "digitizing" an instrument does not insure these advantages. There are literally hundreds of data handling and digitizing devices available today and an unwise combination of units can actually degrade the output accuracy. Also, many digitized measurement systems, while providing excellent accuracy and convenience, are unnecessarily complicated.

Data Domains Concepts

The convenience and power of the amazing new electronic devices are irresistible to almost all scientists, but few are in a position to understand these new tools in detail. A means of applying new devices efficiently and effectively, which requires only an understanding of the basic measurement concepts, is needed for most. The data domains concepts described here are very useful in analyzing, describing, modifying, and designing analog, digital, and analog/digital measurement systems and devices and in assessing and minimizing the sources of measurement errors (1). In addition, a much better understanding of the instrumental data handling process is gained as a result of the study and application of the data domains concept. The first four concepts of data domains analysis are given below:

(1) Measurement data are represented in an instrument at any instant

by a physical quantity, a chemical quantity, or some property of an electrical signal. The characteristics or properties used to represent the measurement data can be categorized in groups called "data domains."

(2) As the data proceed through the instrument, a change in the characteristic or property used to represent the measured data is called a "data domain conversion."

(3) All electronic measurement systems can be described as a sequence of two or more data domain converters, each of which can be analyzed separately.

(4) Methods of using electrical signals to represent measurement data fall into three major categories or domains: analog, time, and digital.

Since there are only three data domains for electrical signals, the electronic sections of complex measurement systems can be easily analyzed (or designed) as combinations of only a few basic interdomain converters. Also the hundreds of data handling devices available can be shown to be simply various methods of accomplishing the basic interdomain conversions.

Electronics-Aided Measurement

In an electronics-aided measurement, the quantity to be measured is converted into an electrical signal and then amplified or otherwise modified to operate a device which visually displays the value of the measured quantity. This process is illustrated for a typical case in the block diagram of Figure 1. An input transducer such as a photodetector, thermistor, glass pH electrode, or strain gage is used to convert the quantity to be measured into an electrical signal. The current or voltage amplitude of this signal is related in some known way to the quantity to be measured. The electrical signal from the input transducer is then modified by an electronic circuit to make it suitable to operate a readout device. The electronic circuit is most frequently an

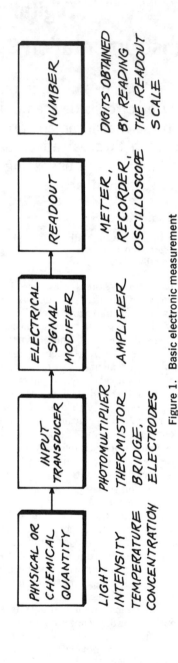

Figure 1. Basic electronic measurement

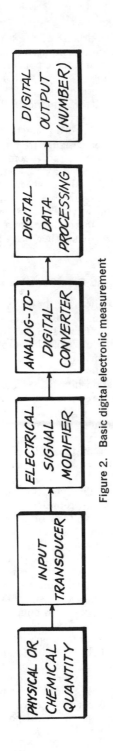

Figure 2. Basic digital electronic measurement

amplifier with the appropriate adjustable parameters (zero, standardization, position, etc.) and sometimes with automatic compensation for nonlinearities, temperature variation, etc. of the transducer. The output is a readout device from which a number can be obtained, generally by observing the position of a marker against a numbered scale.

Using the data domains concepts, the basic electronic measurement of Figure 1 is described as follows: The measurement data exist first as the physical or chemical quantity to be measured. At the output of the input transducer, the measurement data are represented by an electrical signal and are thus in one of the three "electrical" domains. The input transducer is thus a device which converts quantities or translates information from a physical or chemical domain into an electrical domain. The measurement data remain in an electrical domain through the electrical signal modifier. However, the output device converts the electrical signal into some readable form such as the relative positions of a marker and a scale— *i.e.*, a nonelectrical domain. Thus the entire measurement can be described in terms of conversions between domains and modifications within domains.

In the basic electronic measurement, at least two converters are required; one to transfer into an electrical domain and one to transfer out. The characteristics of each interdomain converter and each signal modifier affect the quality of the measurement. To take advantage of special input transducers, particular readouts, and available signal processing techniques, an instrument may involve many data domain conversions and signal modifiers. The data domains concept allows each step to be blocked out and analyzed separately. This will be shown to be particularly desirable in assessing sources of errors and the relative advantages of various digitizing or interfacing possibilities.

Digital Measurement. A common form of digital measurement system is shown in block form in Figure 2. At some point after the measurement data have been converted into electrical amplitudes, an analog-to-digital converter is used (2). This is an electronic circuit which converts an analog electronic signal (where the measurement data are represented by the signal amplitude) to a digital electronic signal (which represents integer numbers unambiguously by coded binary-level signals). If the digitization was performed to take advantage of the great accuracy, power, and versatility of digital data processing, that will be done next. Finally, the numerical binary-level signal is decoded into a number which is displayed, printed, and/or punched.

Because so many advantages are claimed for digital techniques, many techniques have claimed to be "digital." In fact, any type of device which has dial settings or outputs which are numerals in a row is likely to be called digital. By that standard a decade resistance box is a digital instrument. Since the end result of any measurement is a number, all instruments could be called digital, but the meaning of the word in that sense becomes trivial. Some confine the use of the words "digital instrument" to those instruments which contain binary-level

electronic logic circuits of the type developed for digital computers. As will be shown later, it is common for measurement data to be represented by a binary-level electronic signal and still not be "digitized" or numerical. Therefore, in this paper, a digital instrument will be defined as one that uses a digital electronic signal to represent the measurement data somewhere within the instrument. The analysis and design of digital measurement systems necessarily involves an understanding of the ways electrical signals can represent data and how conversions from one form to another are accomplished.

Electronic Data Domains

There are only three basic ways by which measurement data are represented by an electrical signal: Analog, symbolized E_A, in which the amplitude of the signal current or voltage is related to the data; time, $E_{\Delta t}$, in which the time relationship between signal-level changes is related to the measurement data; and digital, E_D, in which an integer number is represented by binary-level signals. The characteristic signals in each of these domains and examples of their use are described in this section.

Analog, E_A. The measurement data in this domain are represented by the magnitude of a voltage or a current. The analog domain signal is continuously variable in amplitude. Also, the analog amplitude can be measured continuously with time or at any instant in time. Most input transducers used today convert the measurement data from the physical and chemical domains (P) to the E_A domain. Examples of P-to-E_A converters are: photodetectors which convert light intensity to an electrical current, a thermistor bridge which converts temperature difference to an electrical potential, a combination pH electrode which converts solution acidity into an electrical potential, and a flame ionization detector which

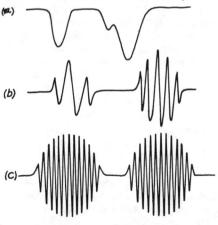

Figure 3.   Analog (E_A) domain signals

converts the concentration of ionizable molecules in a gas into an electrical current. Figure 3 shows some typical E_A signals.

At each instant in time, the measured quantity is represented by a signal amplitude. The variations in the signal amplitude may be plotted against time, wavelength, magnetic field strength, temperature, or other experimental parameters as shown in Figure 3. From such plots, additional information can often be obtained from a correlation of amplitudes measured at different times. Such information would include simple observations like peak height, peak position, number of peaks, or more complex correlations such as peak area, peak separation, signal averaging, and Fourier transformation. The techniques of correlating data taken at different times must be distinguished from the techniques of converting the data taken at each instant into a usable form. It is the latter problem that this paper is primarily concerned with. The former problem is handled by data processing techniques, once the required instantaneous data points have been converted to a useful form and stored.

Signals in the E_A domain are susceptible to electrical noise sources contained within or induced upon the circuits and connections. The resulting signal amplitude at any instant is the sum of the data signals and the noise signals.

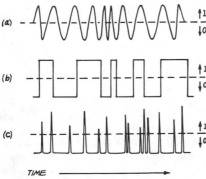

TIME

Figure 4.   Time ($E_{\Delta t}$) domain signals

Time, $E_{\Delta t}$. In this domain, the measurement data are contained in the time relationship of signal variations, not the amplitude of the variations. Typical $E_{\Delta t}$ domain signals are shown in Figure 4. The most common $E_{\Delta t}$ domain signals represent the data as the frequency of a periodic waveform (a), the time duration of a pulse (b), or as the time or average rate of pulses (c). These are logic-level signals—*i.e.*, their signal amplitude is either in the HI or *1* logic-level region or the LO or *0* logic-level region. The data are contained in the time relationship between the logic-level transitions. The greater the slope (dE/dt) of the signal through the logic-level threshold region, the more precisely the transition time can be defined. Because the data in an $E_{\Delta t}$ domain signal are less amplitude-dependent than in an E_A domain signal, they are less affected by electrical noise sources. A common example of this is the FM radio signal ($E_{\Delta t}$ domain) *vs.* the more noise susceptible AM radio signal (E_A domain). The greater the difference between the average *0* or *1* signal-level amplitude and the logic-level threshold, the less susceptible the signal will be to noise-induced error. In these respects, the signal shown in Figure 4b is better than those of Figures 4a and 4c. The logic-level transitions of signals like Figures 4a and 4c are generally sharpened to those like 4b before the significant time relationship is measured. This is accomplished by a *comparator* or *Schmitt trigger* circuit. Examples of converters producing $E_{\Delta t}$ domain signals from physical domain quantities are: a crystal oscillator that produces a temperature-dependent frequency because of the temperature characteristics of the quartz crystal, an oscillator which has an output frequency dependent upon the value of the capacitance used in the oscillator circuit, and the Geiger tube which converts level of radioactivity to a pulse repetition rate. An example of a E_A-to-$E_{\Delta t}$ domain converter is a voltage-controlled-

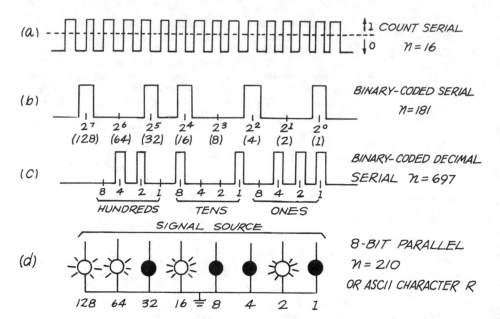

Figure 5.   Serial and parallel digital signals

oscillator or voltage-to-frequency converter which provides an output frequency related to an input voltage.

The $E_{\Delta t}$ domain signal, like the E_A domain signal, is continuously variable since the frequency or pulse width can be varied infinitesimally. However, the $E_{\Delta t}$ signal variable cannot be measured continuously with time or at any instant in time. The minimum time required for conversion of an $E_{\Delta t}$ domain signal to any other domain is one period or one pulse width.

Digital, E_D. In the digital domain, the measurement data are contained in a 2-level signal (HI/LO, 1/0, etc.) which is coded to represent a specific integer (or character) (3). The digital signal may be a coded series of pulses in one channel (serial form) or a coded set of signals on simultaneous multiple channels (parallel form). Representative digital signal waveforms are shown in Figure 5. The count serial waveform (a) is a series of pulses with a clearly defined beginning and end. The number represented is the number of pulses in the series. The count serial waveform of Figure 5 might represent, for instance, the number of photons of a particular energy detected during a single spark excitation. The count serial form is simple but not very efficient. To provide a resolution of one part per thousand, the time required for at least one thousand pulses to occur must be allowed for each series of pulses.

The most efficient serial digital signal is the binary-coded serial signal shown in Figure 5b. In this signal, each pulse time in the series represents a different bit position in a binary number. The appearance of a pulse at a time position indicates a *1;* the absence of a pulse, a *0.* The data are not represented by the exact time of the pulse as in the $E_{\Delta t}$ domain, but by the signal logic level present within a given time range. The binary number represented by the waveform shown is 101101011 which is decimal 181. A series of n pulse times has a resolution of one part in 2^n. Thus a 10-bit series has a resolution of one

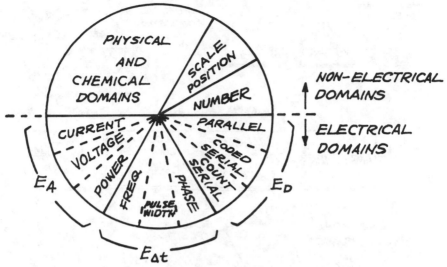

Figure 6. Data domains map

part in $2^{10} = 1024_{10}$, and a 20-bit series has a resolution of better than one part per million.

The binary-coded decimal serial form is somewhat less efficient but very convenient where a decimal numerical output is desired. Each group of four bits represents one decimal digit in a number. Twelve bits can thus represent three decimal digits and provide one part in one thousand resolution.

A parallel digital signal uses a separate wire for each bit position instead of a separate time on a single wire. The principal advantage of parallel digital data connections is speed. An entire "word" (group of bits) can be conveyed from one circuit to another in the time required for the transmission of one bit in a serial connection. An 8-bit parallel data source is shown in Figure 5d connected to indicator lights to show the simultaneous appearance of the data logic levels on all eight data lines. Binary coding (shown), binary-coded decimal coding, and others are used for parallel digital data. Parallel data connections are used in all modern, fast computers. Serial data connections are often used for telemetry and slow computer peripherals such as teletypes.

Mapping Domain Conversions

It has been pointed out that electronic instruments making chemical or physical measurements use no fewer than two data domain conversions. In fact, modern laboratory instruments frequently use three or more domain conversions to perform the desired measurement. Knowing the data domains involved in a particular instrument's operation can help in understanding its operation, applications, limitations, and adaptability as part of a larger measurement system. When analyzing an instrument it is helpful to use the data domains "map" shown in Figure 6. The path of the signal can be traced out on the map as it is followed through the instrument. This process will be illustrated for several chemical instruments of various types.

pH Meter. The block diagram of a conventional pH meter is shown in Figure 7. The combination glass/calomel electrode converts the hydrogen ion activity (chemical composition domain) to an electrical potential (E_A domain). This signal is amplified and converted to a current amplitude that is used to deflect the meter pointer. A

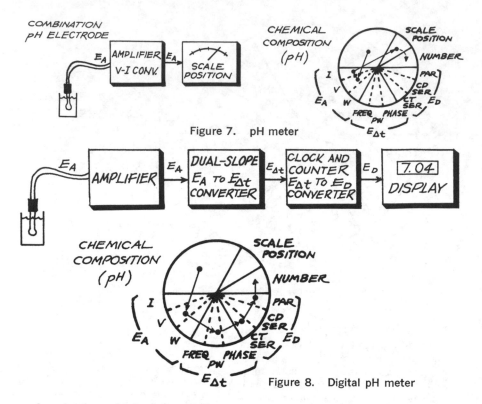

Figure 7. pH meter

Figure 8. Digital pH meter

number is then obtained by reading the position of the pointer against the calibrated scale. The signal path on the data domains map for the pH meter is also shown in Figure 7. Note that there are two instrumental interdomain conversions, one intradomain conversion, and one "manual" interdomain conversion. It will be shown later that interdomain conversions are more complex and error-prone than intradomain conversions. Recording and "servodigital" pH meters have essentially the same block diagram and domains path except that the servo system can be used to convert the E_A voltage signal from the combination electrode directly into the pen position in the case of the recorder or the position of the turns-counting dial (a rotary scale) in the case of the servodigital meter. Notice that even in the latter case the data are never in the digital domain.

Digital pH Meter. A digital pH meter (Figure 8) differs from an ordinary pH meter in that the meter is replaced by an analog-to-digital (A/D) converter and a digital display. A frequently used A/D converter for this application is the dual slope converter. As is often the case, this A/D converter does not convert directly from the E_A to E_D domains. The dual-slope circuit produces a pulse which has a duration proportional to the input signal voltage—i.e., a $E_{\Delta t}$ signal. The pulse width is converted to a digital signal using the pulse to turn an oscillator on and off, generating a count serial digital signal. The count serial signal is in turn converted to parallel digital for the display by a counter. From the domains map it is seen that four interdomain and one intradomain conversions are involved in the measurement.

Digital Temperature Measurement. Two approaches to a digital thermom-

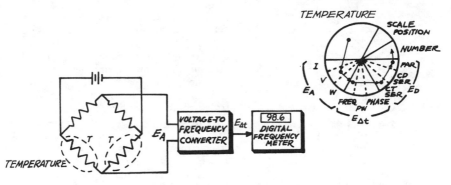

Figure 9. Thermistor-digital temperature measurement

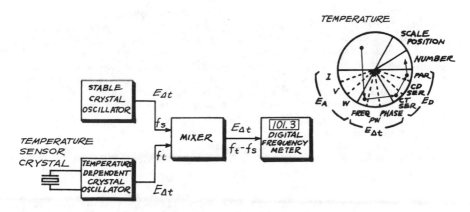

Figure 10. Crystal oscillator temperature measurement

eter are compared here. The first is a thermistor bridge connected to a digital voltmeter (E_D) as shown in Figure 9. The thermistor bridge circuit converts temperature to voltage. In this case the E_A-to-E_D (A/D) conversion is accomplished by a voltage-to-frequency converter and a frequency meter. The digital frequency meter operates by counting the number of cycles of an input signal that occur in a specific time. The resulting domain path is shown. There are four interdomain and one intradomain conversions.

The second approach is the use of a quartz crystal which has a temperature-dependent resonant frequency. An oscillator is used to convert the resonant frequency to an electrical signal in the $E_{\Delta t}$ domain. The block diagram is shown in Figure 10. A mixer is used to subtract a standard frequency, f_s, from the temperature-dependent frequency, f_t, to obtain a signal for which the frequency and temperature are related directly. This is an example of signal modification occurring in the $E_{\Delta t}$ domain. Note that this digital thermometer requires one less interdomain conversion than that of Figure 9. Whether this simplification would result in greater accuracy, however, depends upon the accuracy of the converters involved in each case.

Conversions of Varying Quantities

The examples used in the previous section were measurements of steady-

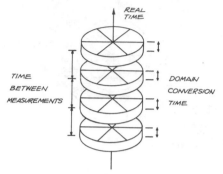

Figure 11. Successive domain conversions

state quantities which were not expected to vary perceptibly over the interval of measurement. When the time variation of quantities in the various data domains is considered, a third dimension (time) needs to be added to the data domain map of Figure 6, as shown in Figure 11. Here each inter-

domain conversion is shown as a slice across the real-time continuum. If a quantity that varies continuously with time is to be converted to the digital domain, the resulting number can only be true for a specific instant in time. It is not possible, therefore, to make a truly continuous digital record of a varying quantity. What can be done is to measure the varying quantity at successive instants in time. The numerical result of each measurement is then stored in order in memory registers, or recorded on punched cards or paper tape, or by magnetic recording devices. If the measurements are made frequently enough for the varying quantity to change only slightly between each time, the digital record can quite accurately represent the amplitude *vs.* time behavior of the measured quantity. Of course, the maximum frequency of measurement is limited by the time re-

CONCEPTS

Data Domains

1. Measurement data are represented in an instrument at any instant by a physical quantity, a chemical quantity, or some property of an electrical signal. The characteristics or properties used to represent the measurement data can be categorized in groups called "data domains."

2. As the data proceed through the instrument, a change in the characteristic or property used to represent the measured data is called a "data domain conversion."

3. All electronic measurement systems can be described as a sequence of two or more data domain converters, each of which can be analyzed separately.

4. Methods of using electrical signals to represent measurement data fall into three major categories or domains: analog, time, and digital.

Measurement Devices

5. All measurement devices employ both a difference detector and a reference standard quantity.

6. Either the difference detector or the reference standard can affect the accuracy of the measurement.

7. The reference standard quantity is the same property or characteristic as that which is being measured.

8. Interdomain converters have the characteristics of measurement devices.

quired to convert the measured quantity into the digital domain and record it. Successive data domain conversions will be illustrated by three examples of digitized measurement systems.

Gas Chromatography. A digital recording gas chromatograph is shown in Figure 12. The components in the sample mixture are separated by the column resulting in a flow of gas of varying composition through the flame ionization detector. The flame in the detector converts the hydrocarbon concentration into an ion concentration (an intradomain conversion) which is then converted by the electric field in the detector to an electrical current (an interdomain conversion). The current signal is amplified and converted to a voltage amplitude suitable for A/D conversion and/or recording. The desired data are a record of detector current *vs.* time. The time relationship of the printed or punched values of the current amplitudes is generally obtained by having the successive A/D conversions performed at precise and regular time intervals. This is accomplished by the timer shown in the block diagram. The domain path for a single conversion (one time slice) is also shown.

Absorption Spectrophotometry. The block diagram of a digitized recording double-beam spectrophotometer is shown in Figure 13. A narrow wavelength range of light from the light source is selected by the monochromator and passed on to the beam switcher and cell compartment. The beam switcher alternately directs the monochromatic beam through the reference and sample cells to the photomultiplier tube detector. This produces an electrical current (E_A domain) which has an amplitude alternating between sample and reference beam intensities P and P_o. The desired output signal for the recorder is absorbance $A = \log_{10} (P_o/P)$, which is accomplished in the log-ratio circuit. This circuit performs a correlation between signal levels measured at two different times. It must,

therefore, have a memory and a synchronizing connection to the beam switcher. The recorder is to plot absorbance *vs.* the wavelength of light from the monochromator. The recorder chart drive thus has a synchronizing connection to the monochromator wavelength drive mechanism.

This standard spectrophotometer was later digitized by putting a retransmitting pot assembly on the servorecorder. This converts the recorder pen position to a voltage amplitude which is connected to an A/D converter and printer or punch. Since the absorbance value recorded for precise *wavelength* (rather than time) intervals is desired, the A/D converter and printer are synchronized to the wavelength drive mechanism rather than to a timer.

The data domains map for the resulting instrument is shown in Figure 13. It contains nine conversions; seven interdomain and two intradomain. The excursion into the scale position domain is unnecessary to the digitizing process and suggests that the A/D converter would have been better connected to the log-ratio circuit output, if possible. It is interesting to note that if photon counting is used to measure the relative intensities, the number of interdomain conversions is reduced to three and a digital log-ratio circuit is required.

Fourier Transform Spectroscopy. A Fourier transform spectrometer is an example of a conceptually very simple data acquisition system connected to a complex data correlating and processing system. The block diagram is shown in Figure 14. To obtain the interference pattern, the A/D converter converts the detector output signal at constant increments of movement of the reference beam mirror. Each piece of data is stored for use in the Fourier transform calculation. The data domains map as shown is complete for each piece of data as it is acquired and stored. A successive approximations type of A/D converter has been assumed in this map. There are four interdomain and two intradomain con-

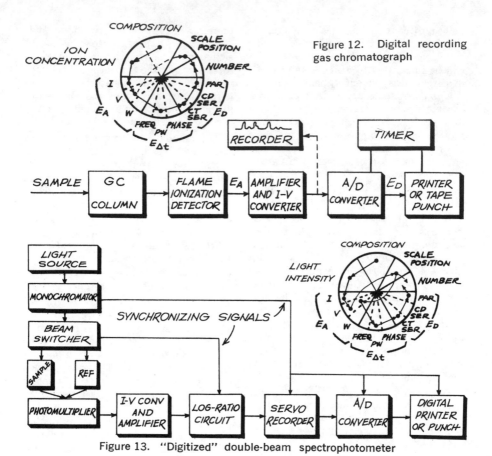

Figure 12. Digital recording gas chromatograph

Figure 13. "Digitized" double-beam spectrophotometer

Figure 14. Fourier transform spectrometer

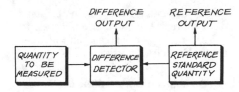

Figure 15. Basic measurement system

versions in all. After the transform calculation (which involves an intercorrelation of all the measured data points) is complete, a plot of absorbance *vs.* wavelength can be made.

This example demonstrates that Fourier transformation from amplitude *vs.* time or space to amplitude *vs.* frequency (or the reverse) is really an intercorrelation of analog signals which have already been "measured." It is essential to distinguish between the data domains involved in the methods of acquiring each data point and the methods of correlating and displaying groups of data points.

Errors in Domain Conversions

To understand the sources of error in domain conversion, it is helpful to review briefly the basic measurement process. Measurement can be defined as: *The determination of a particular characteristic of a sample in terms of a number of standard units for that characteristic.* The comparison of the quantity to be measured with standard units of that quantity is implicit in this definition. The comparison concept in measurement is illustrated by Figure 15. The quantity to be measured is compared with a reference standard quantity. The difference is converted to another form (domain) such as scale position. The quantity measured is then the sum of the standard units in the reference quantity and the difference output calibrated in the same standard units.

All measurement devices involve both a difference detector and a reference standard, although they differ widely in the degree to which one or the other is relied upon in the measurement. As an example, three mass measuring devices can be compared in this regard. With a double pan balance, the unknown mass is compared with standard weights, whole units and fractional, until the difference detector (beam pointer) points to zero. In this case, the accuracy and resolution of the standard weights determines the accuracy of the measurement as long as the difference detector is sufficiently sensitive. No accuracy requirement is placed on the off-null calibrations of the difference detector. The other extreme is a spring-loaded scale, such as a fish or bathroom scale. In this case, the reference standard weights are used to calibrate the scale markings of the manufacturer's original prototype. In use, the measurement accuracy depends entirely upon the off-null markings on the scale. The reference standard quantity compared by the scale in this case is zero weight. In between these two extremes is the single pan balance with an optical scale for fractional weights. Balances of this type rely upon accurate standard weights for the most significant figures and upon off-null calibrations for the less significant figures.

Similar comparisons and analyses can be made for other types of measurement devices. A potentiometric voltage measurement depends much more upon the slidewire (standard voltage unit adjustment) calibration than upon the galvanometer null detector, while an electrical meter depends much more upon the difference detector calibrations than upon the standard. This kind of analysis is helpful in assessing the sources of errors in measurement devices and in choosing among available devices for a particular application. After considering a variety of measurement systems in this way, some basic concepts concerning measurement devices evolve which can be added to the four data domains concepts listed earlier.

(5) All measurement devices employ both a difference detector and a reference standard quantity.

(6) Either the difference detector *or* the reference standard can affect the accuracy of the measurement.

(7) The reference standard quantity is the same property or characteristic as that which is being measured.

A data domain conversion is the conversion of a number of units of some physical, chemical, or electrical characteristic into a related number of units of a different characteristic; for instance, the conversion of units of pH into Nernst factor potential units by a combination pH electrode. Devices for converting data from one domain to another are "measuring" one characteristic in terms of another. Therefore,

(8) Interdomain converters have the characteristics of measurement devices.

Using a combination pH electrode as an example of an interdomain converter to illustrate concepts 5–7: 5) The combination pH electrode itself is the difference detector; the reference standard is the standard buffer solution used to "standardize" the voltage output at a given pH. 6) The conversion error (difference between the predicted and actual potential/pH relationship) depends upon the accuracy of the standard solution and upon the accuracy of the electrode response. The greater the pH difference between the standard and unknown solutions, the more the conversion accuracy depends upon the electrode's characteristics. 7) The reference standard is pH, the units which are being converted to electrical potential.

Once one is accustomed to looking for the difference detector, reference standard, and accuracy dependence of interdomain converters, the basis of the conversion and the sources of error are easier to uncover. Every A/D converter contains a standard voltage or current source and every $\Delta t/D$ converter contains a standard clock oscillator, as expected from concept **7** above. In both cases the conversion accuracy depends directly upon the standard sources and, for various types, to a greater or lesser degree upon the other converter characteristics.

Domain Converter Classification

A classification scheme for data conversion devices would seem desirable for two purposes: to categorize by function the great many devices available, and to provide a way to organize these devices into complete measurement systems. It is natural and useful to classify converter devices according to the domains which the device converts from and to. The domains map shows three electrical domains with three subdomains each, the scale position domain, the number domain, and the physical and chemical domains. For this classification, concentrating on electronic instrumentation, only the electrical, scale position, and number domains will be detailed. There are 110 possible interdomain and intradomain conversions among these 11 domains and subdomains. However, direct converters for most of these transitions are rare or unknown. Thus the number of categories required to encompass the common converters is not unwieldy.

The domains map is a very convenient means of organizing domain converter categories. Figure 16 shows 21 categories of converters arranged by input and output domains. Examples of devices for each listed domain transition are given in the accompanying table. This map and table clearly show what direct transitions are possible and which specific devices will do them. In addition, Figure 16 can be used to obtain and compare many possible combinations of devices to achieve a given transition by following connecting paths. Thus voltage-to-parallel digital converters could be made by paths 3–6–12, 7–12, 4–9–12, 11–13, and 18–15. These five types of A/D converters are all currently marketed.

For any required conversion, that path which has the fewest conversions should also be apparent from tracing the possible routes shown in Figure 16. However, the shortest path is not always the path of choice. For example, to go from scale position to parallel digital, the direct path is 15. However, absolute shaft rotation encoders with a high accuracy and ruggedness requirement could cost much more than the devices needed to take route 19–7–12.

Summary

The data domains and measurement concepts discussed here can be used to

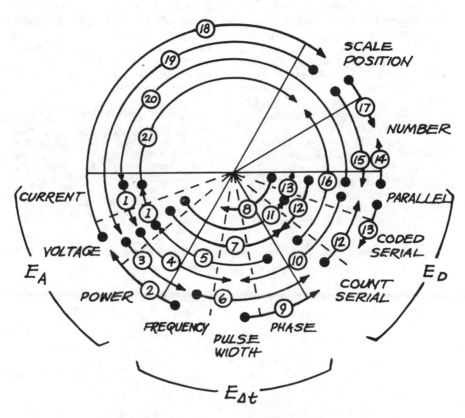

Figure 16. Domain converter classification map

1. Resistor, op amp circuit
2. Count rate meter
3. V-F converter
4. Dual slope A/D converter, ramp A/D converter
5. Phase angle meter
6. Counting gate timer
7. Staircase A/D converter
8. Programmable frequency divider
9. Gated oscillator
10. Preset digital timer
11. Successive approximation A/D converter
12. Counter
13. Shift register
14. Nixie tubes, printer
15. Shaft rotation encoder
16. Stepper motor
17. Mechanical shaft turns counter
18. Recorder
19. Retransmitting potentiometer
20. D/A converter (ladder or weighted sum)
21. Meter

great advantage to analyze and describe available analog/digital instruments, to design or modify measurement systems, and to determine the sources of measurement errors. The data domains map can serve to show the data path from a block diagram or to devise a possible block diagram for instruments or modules, knowing only the input and output domains.

To analyze or describe an instrument, use the instrument description and block diagram to carefully follow the measurement data step-by-step through the instrument. Trace out the path on a domains map as shown in Figures 7–10 and 12–14. Now each converter corresponds to a line segment on the domains map. The conversion errors can be assessed by identifying and studying the reference source and difference detector for each converter. Modifications to instruments can be made by exchanging equivalent converters or by adding appropriate line segments to the instrument's domains map where new domains are to be included. New systems can be designed by completing a chart like Figure 16 for the devices available and comparing all the possible routes between the desired input and output domains.

Readers' comments, criticisms, and discussion on the concepts described in this article are welcomed. The author gratefully acknowledges the many helpful discussions he had with Dr. Howard Malmstadt, Dr. Stanley Crouch, Jim Ingle, and his graduate students during the evolution of these ideas.

References

(1) H. V. Malmstadt and C. G. Enke, "Digital Electronics for Scientists," W. A. Benjamin, New York, N. Y., 1969.
(2) D. Hoeschele Jr., "Analog-to-Digital-to-Analog Conversion Techniques," Wiley, New York, N. Y., 1968.
(3) H. V. Malmstadt and C. G. Enke, "Computer Logic," W. A. Benjamin, New York, N. Y., 1970.

When the Computer Becomes a *Part* of the Instrument

MARVIN MARGOSHES
Technicon Instrument Corp.
Tarrytown, N.Y. 10591

Computers have revolutionized analytical instrumentation and the problems of instrument-computer interfacing are continually confronting analytical chemists. Making the computer a component part of the instrument should provide the most feasible means of realizing the maximum information content of an analytical method

COMPUTERS ARE CHANGING analytical chemistry. Their influence on our instrumentation and methods will be as dramatic and pervasive as the earlier effect of electronics. Those who recognize and make intelligent use of these new developments will reap considerable benefit. The field of X-ray crystallography provides an example of the likely effects. With computerized techniques, the same group of crystallographers was able to increase its number of publications sixfold (1). The accuracy and precision of the data are also improved. As a result, entirely new types of crystallographic studies are being made.

In analytical chemistry, computers are used mainly to automate existing instruments and conventional computations. These applications have been useful, but the real benefits of computerizing will come with the development of new measurement methods that may be possible only with computerized instruments, and from techniques which take advantage of more of the information content of an analytical method. Kaiser (2) recently discussed, in this journal, the information content of an analytical method. Although Kaiser stated that he was not necessarily referring to computers, it is apparent that routine realization of his suggestions requires the use of these machines at least for the computations.

In this article, I will stress the instrumentation aspect, though one really should not isolate the measurement from how the data are used. In particular, I will stress types of measurements that, perhaps, can be accomplished without a computer, but which are most feasible when the computer is made a part of the instrument.

Attaching the Computer to an Instrument

Before going into the main theme of this article, it is worthwhile to take a brief look of the use of computers with conventional analytical instruments. In the first applications, the computer was used off-line only, and analog computers often were used because the digital machines were still expensive and unreliable. Despite these limitations in early work, one of the first uses of computers in analytical chemistry is a good illustration of a technique that is greatly improved by being able to carry out fairly extensive computations.

The application was the analysis of petroleum fractions by infrared spectrometry. These fractions frequently are mixtures of similar compounds, and it is nearly impossible to find an infrared frequency which allows measurement of the concentration of one component specifically, without interference from the other components. Quantitative concentration data were computed by equating the absorbances at each of several wavelengths with the sum of the absorbances of the individual compounds. For a ten-component mixture, measurements were made at ten wavelengths and ten simultaneous equations were solved on an analog computer. In principle, the analytical computations could have been made by hand or with a desk calculator, but the greater speed of the computer made the method much more attractive for routine use.

The gas chromatograph eventually took over this type of analysis. It has the advantage of giving a physical, rather than mathematical, separation of the compounds in a complex mixture. Here, too, the computer has become prominent. The original hardware integrating circuits have been replaced by computer attachments of varying capability, providing for one or many chromatographs and for different data reduction techniques. The instrument industry has developed these many types of systems in response to the widely differing needs of analytical laboratories. The chemist is faced with the task of selecting both the best chromatograph and the best data system for his present and projected needs. This cost-benefit analysis requires time, but there is an obvious advantage in having flexibility of choice.

Numerous types of computation devices also have been used in emission spectrometry over the past two decades. These include analog and digital devices; the digital computers were used in the dedicated, time-sharing, and batch-loading modes. Mostly, the computers were programmed to automate the usual types of computations. It has been proved (3) that improvements in analytical accuracy and precision are possible when more complex calculations are made and/or when more of the information in an emission spectrum is used. Until quite recently, there were no instruments which could make use of more than a small fraction of this information. Automatic, high-speed microphotometers now have been developed (4) to read all of the information on a photographic plate and convert it into machine-readable form. A spectrometer capable of measuring photoelectrically at up to 2048 wavelengths is under development (5). This spectrometer *requires* that a computer be designed in; it cannot be operated in any other way. The automatic microphotometer does not require a computer as a component, but it produces more data than can be handled manually. It is safe to predict that these instrumental developments, together with computational improvements, will have a dramatic impact on methods of emission spectrochemical analysis far beyond mere automation of existing methods.

Pulse-height analysis is done routinely for collection of data in such analytical methods as nuclear activation and X-ray spectrometry. Signal averagers are less common, but they are used in several methods where the signal measured in a short time is accompanied by excessive noise. It is most common to use special pulse-height analyzers and signal averagers, but, in fact, these are forms of computers. The data collected by these instruments are often fed to a larger computer, and it is logical to use a general-purpose computer to combine the tasks of data collection and analysis. The general-purpose computer is usually less expensive than the special-purpose device, though savings here must be balanced against programming cost. The great advantage of the general-purpose computer in these applications is flexibility in data collection and processing through changes in software. The major disadvantage is that general-purpose computers are normally slower in carrying out a function than are machines wired for the specific operation. It is still necessary to use special-purpose signal averagers and pulse-height analyzers when the data input rates are high.

Building the Computer In

When it is recognized that the instrument will always be used with a computer, it becomes logical to design the instrument with the computer as a component rather than as an add-on attachment. Two major advantages come from this approach: (1) It may be possible to make new types of measurements when the computer is used to operate the instrument and/or to convert raw instrument output into a useful form. (2) The computer can serve several purposes and can replace costly specialized circuits.

Fourier transform spectrometers illustrate both advantages. Both nmr and infrared spectra are now being measured by Fourier transform methods. The ir measurements are probably eas-

ier to explain, so they will be described first.

Infrared Fourier Transform Spectrometry. The essential components of an infrared Fourier transform spectrometer are the interferometer and the data system. The latter consists of a computer, interface circuits, a plotter or other device for displaying the spectrum, and a Teletype or similar unit for input and output of nonspectral information.

Figure 1 is a schematic representation of a Michelson interferometer. The beam splitter sends beams of the incoming light to the two mirrors. Light reflected from these mirrors is recombined at the beam splitter and emerges at right angles to the incident beam.

We may now consider what happens when the incoming light is monochromatic and one of the mirrors is moved toward the beam splitter at a constant velocity. At the starting point, the two mirrors are equidistant from the beam splitter, so that both light beams travel the same distance and arrive back at the beam splitter in phase. These beams interfere constructively and the intensity of the beam emerging from the interferometer is at a maximum. When the mirror has moved a distance equal to one-fourth of the wavelength of the incident light, the distance traveled by the beam in that path is one-half wavelength shorter than the path to and from the fixed mirror. The two beams then interfere destructively and give a minimum in the output intensity. When a detector senses the intensity of the beam from the interferometer and its signal is plotted against the distance moved by the mirror, the curve is a cosine function. This is the *interferogram* for monochromatic light.

When monochromatic light of another wavelength is used, the interferogram is identical except that it has a different period. The interferogram for polychromatic light is the sum of the individual cosine functions which result from all of the discrete wavelengths. Each cosine function is weighted by the

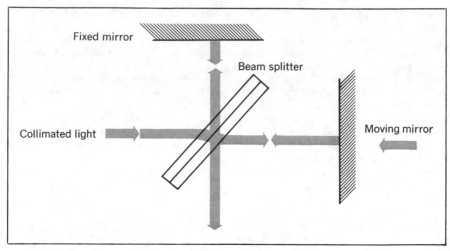

Figure 1.   Michelson interferometer

intensity at that wavelength. This interferogram is simply the Fourier transform of the spectrum, and computing the Fourier transform of the interferogram gives the desired spectrum.

The basic functions of the data system in this case are to store measurements made at equal intervals of mirror movement, to compute the spectrum from the interferogram, and to plot the spectrum. The computation step includes phase correction and apodization as well as the Fourier transform. Phase correction compensates for the probability that the two mirrors are not initially at the same distance from the beam splitter. Apodization removes certain artifacts from the spectrum which arise from the fact that the measured interferogram is only of finite length.

Improved signal-to-noise in the spectrum can be obtained by signal-averaging repeated scans of the interferometer. This becomes another valuable function of the data system. In addition, the computer can be programmed to perform other useful functions. One is to control the resolution by changing the total distance moved by the mirror. Another is to control the form of the spectral plot. Depending on how the

computer is programmed, the size of the plot may be varied and wavelength or wavenumber can be plotted *vs.* transmittance, absorbance, or log absorbance. In addition, the data system can continuously monitor certain instrument functions and alert the operator if certain key components are not operating properly. When the computer controls the instrument, less skilled operators are needed than with conventional instruments. The instrument operation is changed by typing commands and data into the keyboard of the Teletype, rather than by adjustment of knobs.

Automation of the measurement process is only a fringe benefit in this case. The key reason for computerizing is that it is the only practical way to do Fourier transform spectroscopy, and this method has important advantages compared to scanning through the spectrum with a grating or prism monochromator (*6*). In the Fourier transform method the detector senses signals for all wavelengths simultaneously instead of one at a time. This is called the *multiplex advantage* or *Fellgett's advantage*. Also, an interferometer is easily made with mirrors perhaps five centimeters in diameter and thus, the input aperture is much larger than the slit of a monochromator. This is called the *throughput advantage* or *Jacquinot's advantage*. The combination of the two advantages gives a Fourier transform spectrometer the ability to record a spectrum in a much shorter time than can a prism or grating spectrometer. Alternatively, signal-averaging can be used to give a much better signal-to-noise in the spectrum for a given measurement time. The throughput and multiplex advantages are more than large enough to overcome the extra time required in the Fourier transform spectrometer to carry out the computations and to plot the spectrum.

Pulsed Nmr Spectrometry. Pulsed nmr spectrometry is a more recent application of Fourier transforms. The sample in a magnetic field is excited by a radiofrequency pulse of short duration. After the pulse, the resonance signal emitted by the sample is recorded as a function of time. This signal is the Fourier transform of the nmr spectrum.

An nmr measurement made this way has the multiplex advantage, but not the aperture advantage. The signal-to-noise in a spectrum from a single pulse may be quite poor, and signal averaging of repeated pulse signals is usually needed. Various pulse signal sequences may be used, depending on particular measurement requirements. One of the functions of the computer may be to control the pulse sequence. This allows more flexibility than hardware control of the pulses.

than hardware control of the pulses.

A typical application of pulsed nmr is to record spectra from ^{13}C nuclei in samples which were not enriched in this isotope. Such measurements are effectively impossible by the older nmr methods. Commercial computer systems for this purpose have only recently become available, but it is clear that the Fourier transform method will have a major effect on analysis by nmr spectrometry.

The Economics of Computerized Instruments

A change in the design of an instrument to improve one aspect of its performance nearly always involves some drawback. In the case of computerized instruments, the drawback is cost. Computers and interface components are declining in cost, but it is expensive to design the interface and program the computer. The design and programming costs are minimized when a manufacturer can spread them over many instruments, but computerized instruments are generally much more expensive than conventional instruments. For example, a grating infrared spectrometer can be purchased for one-half to less than one-tenth the price of a Fourier transform spectrometer.

The extra cost is justified if it allows new types of measurements which have high economic or scientific value, or if the usual measurements can be made for a total lower expense. It is difficult to make generalizations about the value of new types of measurements, but we can examine costs of making experimental measurements of the usual type with conventional and computerized instruments.

Computerized instruments can be more economical if they significantly reduce the time needed for measurement and computation. When a computer is attached to a conventional instrument, measurement times are not reduced, and only the time for computation is affected. When the instrument is designed for computer operation, significantly shorter measurement times may be achieved. This is true for the Fourier transform infrared spectrometer, where the multiplex and throughput advantages result in much shorter times for recording a spectrum.

Both advantages are greatest in recording high-resolution spectra. Then there are many resolution elements and in conventional instruments it is necessary to use very narrow slits. If a spectrum is measured from 400 to 3800 cm⁻¹ at 0.5-cm⁻¹ resolution, there are 6800 resolution elements, and the multiplex advantage is 6800:1. A typical value for the aperture advantage is 100:1; this is less than the ratio of the area of the interferometer aperture to the area of a slit, but the interferometer must be illuminated by parallel light and a grating or prism spectrometer gains back some of the throughput in the external optics.

A valid comparison of dispersive and interferometer instruments must take into account all time factors, not just the measurement time. In practice, a Fourier transform spectrometer will record a complete spectrum from 400 to 3800 cm⁻¹ at better than 0.5-cm⁻¹ resolution in less than one-half hour. Conservative estimates show that the best commercial grating infrared spectrometer will take more than nine months, at a minimum, to record this spectrum.

The cost of running this one spectrum with the Fourier transform spectrometer is only about $15, including amortization of the instrument, cost of the operator, laboratory space, etc. Amortization alone for the grating spectrometer is about $5000 for nine months. The economics clearly favor the Fourier transform spectrometer even if only a few high-resolution spectra are needed.

Most laboratories have only a limited need for high-resolution spectra. A typical infrared spectrum is recorded in 20 min on a $9000 spectrometer. At this rate, the spectrometer will produce 24 spectra in an eight-hour day, 120 per five-day week, and 6000 per 50-week year. The five-year amortized cost of the spectrometer is only $1800 per year, but the technician cost (with overhead) will be about $15,000 per year. The total cost for the 6000 spectra is about $16,800, neglecting such items as laboratory space and chart paper which are essentially the same for the two types of spectrometer being considered.

The amortized cost of the Fourier transform spectrometer is $13,000 per year, and adding the cost of the technician brings the cost of operation to $28,000 per year. When the work load is only 24 samples per day, it costs about $2.00 more per spectrum to do the work on the interferometer spectrometer.

These cost estimates favor the dispersive spectrometer, but it is necessary to keep in mind several items which can swing the balance. First of all, the Fourier transform spectrometer can record a spectrum in three minutes which is at least as good as that recorded by the dispersive spectrometer in 20 min. The interferometer instrument can run many more than 24 samples per day with almost no increase in cost. When the spectrometer is being used to control a process stream, the ability to record spectra more often can have considerable economic impact. When the spectrometer is being used to support a research scientist, the cost calculations should include the value of his time while he is waiting for the spectrum. Ideally, the scientist will be doing other useful work in this time, but in practice he may use the delay for an impromptu coffee break.

For the one extreme case of high-resolution spectra, it is easy to show that the computerized system is actually much less expensive than the nominally "cheaper" dispersive spectrometer. In the other extreme case, the economic advantage may go one way or the other, depending on individual circumstances. In this latter example, the cost of manpower becomes the key factor, and historically this has been increasing faster than instrument costs. Economic factors thus favor increasing use of computerized instruments in the future.

For the Future
These first steps toward fully computerized instruments will certainly be followed by the development of many

new types of instruments, some of which will be more complex. This prediction is based on advantages of these instruments which have already been cited, and on other benefits of computerizing. For example, the following benefits come from having the analytical information in digital form in a computer:

(1) Higher accuracy and precision in the output compared to analog representation.

(2) Facile control of output size and form.

(3) Easy transfer of data to a larger computer.

(4) Computer correction of data for experimental error.

(5) Computer interpretation of data.

(6) Digital storage of data for later retrieval, interpretation, and comparison with new results.

At least two trends in computerized analytical chemistry are already discernible. New types of research studies are being directed to methods of making use of more information in the analytical signal. When such research was done earlier, it was chiefly of academic interest because the measurements and computations would have required too much time for most applications. New instruments and computer methods eliminate the time restriction (in fact, less time may be needed to measure more data and do more calculations). Acceptance of the new methods requires only that advantages be demonstrated to justify the extra costs. Such demonstrations already have been made in several cases.

A second trend is toward more sophisticated control of the instrument by the computer. Presently, the computer is programmed to operate the instrument by a fixed scheme. The computer potentially can be programmed to alter the experiment in response to analysis of the data being recorded. When this is done, the analytical chemist will specify the desired results to the instrument, rather than stating how the measurements are to be made. At the same time, the computer will be checking the instrument operation. It will monitor several components directly, as is now being done in a few instances, and also will analyze the output for other signs of malfunction. When these tests indicate an instrument malfunction, the computer can terminate the measurement process and indicate to the operator the nature of the disorder. The extra cost for these operations would be regained by more efficient instrument operation.

References

(1) H. Cole, *IBM J. Res. Develop.* **13,** 5 (1969).

(2) Heinrich Kaiser, Anal. Chem. **42** (2), 24A (1970).

(3) Marvin Margoshes, *Spectrochim. Acta* **25B,** 113 (1970).

(4) A. W. Helz, F. G. Walthall, and Sol Berman, *Appl. Spectrosc.* **23,** 508 (1969).

(5) Marvin Margoshes, Pittsburgh Conference on Analytical Chemistry and Applied Spectroscopy, Cleveland, Ohio, March 1970.

(6) M. J. D. Low, Anal. Chem. **41** (6), 97A (1969).

Additional References

(7) T. C. Farrar and E. D. Becker, "Pulse and Fourier Transform NMR," Academic Press, New York, N.Y., 1971.

(8) T. C. Farrar, *Amer. Lab.,* **4** (3), 22 (1972).

The Hall Effect
Devices and Applications

E. D. SISSON[1]

F. W. Bell, Inc., 4949 Freeway Dr., East
Columbus, Ohio 43229

Modern semiconductor technology has produced Hall effect devices capable of very accurately measuring magnetic field flux densities and gradients. Applications of the Hall effect include field monitoring in spectrometers, beam deflection coils, cyclotron or linear accelerator magnets, and magnetic resonance measurements.

ADVANCES in semiconductor technology have produced an amazing variety of new devices. Transistors, diodes, silicon-controlled rectifiers, and many more components obtain their special properties as a result of "junctions"—regions where dissimilar materials are joined together. The electrical properties of the junctions determine the device characteristics. A smaller class of devices utilize only the properties of the bulk semiconductor material itself, and have no junctions. The Hall effect device is an example in which the galvanomagnetic properties of the material provide a sensitive electrical response to an applied magnetic field.

Background

Discovery of the Hall effect is attributed to Dr. Edwin H. Hall, who, in 1879, made the first experimental demonstration of the action of a magnetic field upon the current flowing in a conductor. He showed that the field acted directly on the current, tending to crowd the charges to one side of the conductor, and that a measurable potential, therefore, was generated across the conductor width. Using strips of gold foil, Hall determined that the voltage developed across the strip was proportional to the current flowing and to the magnetic field strength, as shown in Equation 1:

$$V_H = kI_cB \qquad (1)$$

where V_H is the Hall voltage, k is a constant of proportionality, I_c is the strip current, and B is the magnetic field strength. Hall published his work in the *American Journal of Mathematics* in 1879 in a paper entitled "On a New Action of the Magnet on Electric Currents."

The angle at which the magnetic flux lines intersect the Hall plate affected the Hall voltage such that only the component of field perpendicular to the plate was effective. Thus, the expression for the Hall voltage, including the angle, Θ, to the perpendicular, is the following:

$$V_H = kI_cB \cos \Theta \qquad (2)$$

Figure 1 shows these relationships. In addition, Hall theorized that the resistance of the conducting strip should increase in a magnetic field because of

---

[1] New affiliation to be announced.

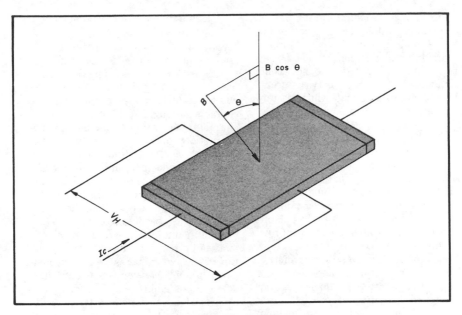

Figure 1. The Hall Plate

Magnetic field density vector, *B*, is shown entering the Hall plate at angle ⊝ to the perpendicular

the modified flow of the current carriers. This effect is now called transverse magnetoresistance.

Unfortunately, the Hall voltages obtained with the simple metals were so extremely small that the Hall effect remained merely a scientific curiosity for many years. It was not until the development of semiconductor compounds with high mobilities that Hall devices gained sufficient output to be practical (*1*). As an example, the compound indium antimonide can produce Hall voltages 100,000 times greater than those obtainable from the pure metals.

The constant of proportionality used in the expression for V_H is referred to as the sensitivity constant and is represented by the symbol γ_{IB}. This is the product sensitivity and is given in units of volts per ampere-kilogauss. One kilogauss is equal to 1/10 tesla, or weber per square meter. Strictly speaking, γ_{IB} is not a true constant because V_H is not in exact linear relationship to B because of temperature, magnetoresistance, and other effects. Very

nearly linear proportionality can be obtained by proper fabrication.

Figure 1 depicts a typical Hall plate, showing a conducting band across each end of the semiconductor plate to help obtain uniform current flow through the plate. Hall connections are centrally positioned along each edge and aligned as accurately as possible on an equipotential line. High-alumina ceramic is often used as a mounting substrate because of the excellent strength and the thermal and dimensional stability of the material. A precision miniature circuit pattern, metallized directly on the substrate, serves to bring the Hall voltage out with a minimum of stray coupling. The final package will vary in size typically from less than 0.01 in. to 1/16 in. in thickness and from 0.1 to 3/4 in. in width or length.

Hall Materials

The most commonly used materials for Hall generators are indium antimonide (InSb) and indium arsenide (InAs). Such materials are made by combining atoms of indium (group III of the periodic table) and antimony or arsenic (group V), and they are termed binary intermetallic compounds. They are not alloys, but true chemical compounds with a 1:1 ratio between group III and group V atoms which occupy alternate sites in the crystal lattice. Their special properties are high electron mobility and small energy gap. The electron mobility of InSb, for example, can be 50 times higher than that of silicon, the material used in junction transistors. Doping, which is the intentional but controlled addition of selected impurities, is used to stabilize carrier concentration and improve thermal stability.

Indium arsenide is often preferred over indium antimonide because of lower temperature influence, even though sensitivity is somewhat lower. Other materials, including germanium, silicon, and gallium arsenide, have been used to advantage in specific applications.

Hall Generator Characteristics

The Hall generator (Figure 2) enjoys several features which distinguish it from semiconductor devices such as transistors, which use junctions as a basis for their operation. Every precaution is taken to assure purely ohmic contacts throughout fabrication. Hall characteristics result from the galvanomagnetic properties of the bulk semiconductor material, and majority carriers determine the primary characteristics. The geometry of the Hall plate and the method of lead connections play secondary roles. Thus, many of the qualities peculiar to junction devices and minority carriers are absent in a Hall generator. For example, junction threshold voltage, with its accompanying nonlinear voltage-current characteristic, is eliminated. Reverse voltage breakdown, junction temperature drift, capacity effects, leakage, and noise associated with minority carriers are absent. Hall generators have inherently linear response through zero field.

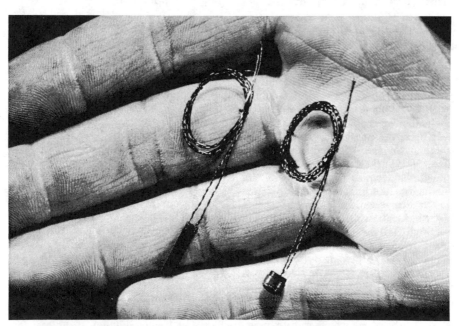

Figure 2.   Transverse Hall generator (rectangular) and axial Hall generator

The active part of the Hall generator, the Hall plate, must be made quite thin (1 to 5 mils in thickness), and it is, therefore, fragile and strain sensitive. However, when properly supported on its substrate and encapsulated, the Hall generator package can be extremely rugged and reliable.

The ability of the Hall generator to directly convert magnetic flux density to output voltage is unique. A typical bulk material InAs unit will produce about 10 mV output per kilogauss of flux density with 100 mA of current flowing. The 10 mV per kG figure is called the magnetic sensitivity and its value depends on the current. Increased current, within the limits of self-heating, will result in correspondingly higher output voltage. Pulsed-current operation can allow considerable increase in output, as can good heat sinking. Devices using InSb material have about three times higher

output, but suffer from greater temperature influence.

The linearity of output voltage with flux density is affected by the dimensional accuracy of the Hall plate, by length-width ratio, doping levels, contacts and other factors, all of which can be well controlled in a modern manufacturing process. Various trimming and circuit resistance adjustments can be made, if needed to reduce further any remaining errors. One percent of reading errors are not uncommon over the range of ±20,000 gauss. Hall generators are now available having only one part in 1000 error over a range of ±30,000 gauss.

Elevated temperatures generally reduce mobility, with a consequent reduction in sensitivity. The loss in output is usually stated as a percentage of error which is a mean value over a specified temperature range. Typically, the error is one percent per °C (−40° to

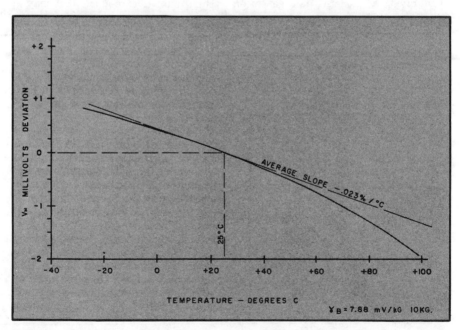

Figure 3. Temperature characteristic of V_H for typical bulk material Hall generator (InAs)

F. W. Bell, Inc., Type JB-802

+80° C range) for indium antimonide. Indium arsenide has less than 0.1 percent per °C error for the same range. Figure 3 shows a typical temperature plot for indium arsenide units. Recently, Hall generators have become available with special material formulation having less than 50 ppm per °C (0.005% per °C) error over the range of −40° to +100°C. No thermal compensation or correcting circuits are used. Probes for cryogenic insertion are also available with typically one percent error at 4.2° K. Temperature also influences the zero field stability, an important factor when reading low Hall voltages. The effect is usually less than one microvolt per °C for the better bulk material InAs Hall generators. A small dc thermal voltage is also present, usually of even smaller magnitude.

To obtain direct readout of magnetic flux density, the Hall voltage must be calibrated using a known field. Once calibrated, a modern Hall generator

will retain calibration almost indefinitely if not abused. In a room-temperature continuous-operation test over a five-year period, test points fell within the limits of confidence of measurement accuracy, using a nuclear magnetic resonance measurement for comparison. Factors that might cause calibration change are mechanical strain, excessive temperature, or thermal shock. Each Hall generator must be individually calibrated because of variations in the Hall coefficient of the semiconductor material.

A characteristic of Hall generators is that they do not possess a true zero, that is, the output may not be zero volts with zero magnetic field. This is because of extremely small positioning errors on the Hall contacts and a resulting voltage offset when current is applied. This residual offset usually measures in the microvolt range and can be easily "zeroed" out by a simple external resistance adjustment. Similarly, small inductive residuals that may ap-

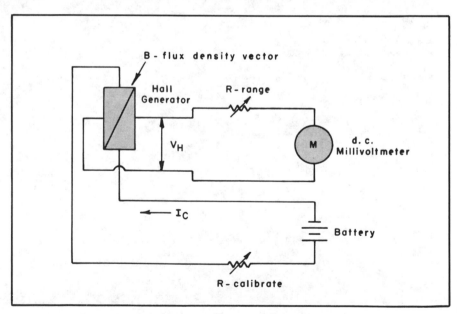

Figure 4.   A simple gauss meter

The millivoltmeter is usually a 0- to 10-mv dc full-scale instrument.   The battery may be a 1.5-V cell capable of delivering 1/4 amp continuously

pear in high-sensitivity measurements can be balanced out without difficulty.

Bulk material Hall generators have extremely low electrical noise output when properly fabricated, and the limits of sensitivity are usually set by amplifier noise or thermal drift. Noise is essentially Johnson (thermal) (2) in the material resistance. Current noise has not been isolated in InAs or InSb units of this type. The resistance of these units ranges from one ohm to 10 ohms in value. Thin-film Hall generators have additional noise arising within the film structure and resistance values are higher, typically 30 to 1000 ohms.

The frequency bandwidth of the Hall generator is not generally limited by the semiconductor material itself, but by the connecting lead wires which have inductive impedance at very high fre-

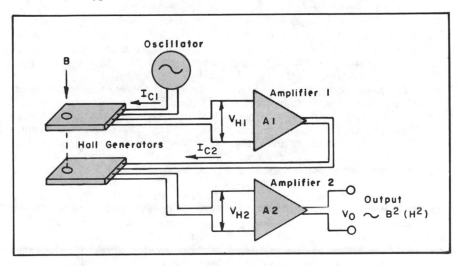

Figure 5. Using two Hall generators to generate a B^2 Output

B = flux density, H = field strength. These quantities are equivalent in air

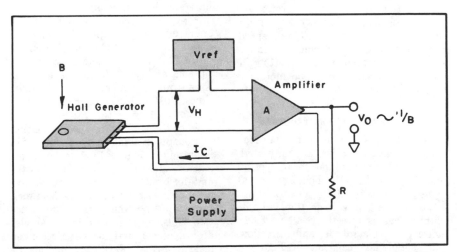

Figure 6. Generation of the inverse function output, using a high-gain amplifier

373

quencies. The semiconductor material, limited by carrier relaxation time, is usable into the gigahertz range, whereas an upper limit of one megahertz is more typical for the Hall package, including leads.

A valuable feature of Hall generators is that they are made entirely from nonmagnetic materials and do not disturb the field being measured. This is in contrast to other types of probes which are made of magnetic metals. Another characteristic which is unique to the Hall generator is that its size can be reduced with no loss in sensitivity. This is in contrast to a coil pickup in which output is related directly to coil area. The small sensing area can be an advantage, since it provides better dimensional resolution and improved accuracy in field gradient measurements. Hall output is a measure of the average flux density over the active area. Typical units have active areas of 1/16-in. diam, but have been made as small as 5×10^{-4} cm^2, and much smaller in thin-film units.

A feature of the Hall generator is that Hall output does not depend on rate-of-change of flux as in a coil, but only on instantaneous flux value. In mechanical sensing applications it provides an output independent of speed, acceleration, frequency, or rate, and works equally well in a static or dynamic field, and down to zero speed. This feature is of particular advantage in the measurement and plotting of static fields, since no probe motion is required.

Putting the Hall Effect to Work

Although a surprising variety of Hall effect applications exists, the most obvious is the sensing and measuring of magnetic fields. By holding current constant, usually at a value that will calibrate the Hall voltage, the Hall output voltage is directly related to flux density in magnitude and direction. The simple arrangement of Figure 4 forms the basis for a gauss meter capable of reading fields from a few gauss to 10,000 gauss or more. Rotation of the Hall probe in the field will provide field direction information in space.

Two Hall generators can measure field gradients by taking the difference output. Two Hall generators can also generate an H^2 output, useful in connection with the mass spectrometer where mass number is related to the square of the field strength (3). This is illustrated in Figure 5 where two units, both mounted in the deflection field, are connected in tandem to obtain an output of H multiplied by H. This permits plotting on a linear H^2 scale.

In another interesting variation the Hall generator is placed in the feedback loop of a high-gain amplifier to obtain a reciprocal, or $1/H$, output. An application where the measured variable depends upon the inverse function of H is the recording of de Haas-van Alphen oscillations (4). The arrangement of Figure 6 will permit plotting directly on a linear $1/H$ scale. In this circuit the input current to the Hall generator is caused to vary inversely with H by the amplifier, which will maintain a fixed $V_H = V_{ref}$ condition required at the input.

The circuits of Figures 5 and 6 both illustrate the ability of the Hall generator to make analog computations by operating on the input flux signal to obtain a desired function of flux output. This ability results from the multiplying ability of the Hall effect; that is, output is proportional to current times flux. Many other applications take advantage of this capability.

Since the Hall probe yields a continuous output reading as long as the field is present, it is useful for driving the H axis on an X-Y plotter in field plotting and in tests using laboratory electromagnets. Other uses include field monitoring in spectrometers, beam deflection coils, cyclotron or linear accelerator magnets, and magnetic resonance measurements. Field programming and control of electromagnets have been carried out using Hall probes.

Hall instruments have advantages in the study of flux transients in solenoids, motors, relays, etc., since thin probes will fit into narrow gaps in confined spaces and can feed signals directly to an oscilloscope for display and analysis of the field waveforms. Magnetic fields within solid material cannot be directly measured with a Hall probe for obvious reasons, but a proportional field value can often be indicated if a suitable air gap is available. A Hall generator positioned to sense the tangential field alongside a flux-carrying member can read the value of surface H and is used for this purpose in B-H plotters, permeameters, and coercimeters. Magnetic susceptibility and saturation magnetization (5) of small samples have been measured using Hall devices.

Industrial applications include the control and adjustment of field strength in the manufacture of dc motors, relays, and similar devices, and in the nondestructive testing of parts by magnetization or by eddy current methods. Noncontacting measurement of electric current is accomplished using Hall generators to sense the field around the conductor. Electric power is measured by Hall multipliers designed for use in power circuits.

A host of other applications exist beyond the scope of this article, many of which have become feasible as a result of recent advances in Hall device technology.

References

(1) "The Hall Effect and Its Applications," F. W. Bell Inc., Columbus, Ohio, 1962.
(2) M. Epstein, L. J. Greenstein, H. M. Sachs, *Proc. Nat. Electron. Conf.,* **15,** 241 (1959).
(3) A. J. Monks, U. S. Patent 3,469,092 (September 23, 1969).
(4) R. J. Higgins, *Rev. Sci. Instrum.,* **36** (11), 1536 (1965).
(5) W. Viehmann, *ibid.,* **33** (5), 537 (1962).

Instrumentation for the Study of Rapid Reactions in Solution

RICHARD M. REICH

American Instrument Co., Division of Travenol Laboratories, Inc.
8030 Georgia Ave., Silver Spring, Md. 20910

Continuous-flow, accelerated-flow, or stopped-flow apparatus, coupled with equipment for relaxation methods such as temperature jump or pressure jump, comprise much of the instrumentation available today for the study of rapid reactions in solution. With continued development of systems having automatic data reduction, such instrumentation is likely to become a standard tool found in most laboratories

Before the early 1920's, the term instantaneous was often applied to reacting systems with half-lives of less than a few seconds. Methods for disturbing a chemical system from equilibrium in a time period which was short with respect to its rate of reaction and methods to monitor and record the properties of a rapidly changing system were not yet developed. Visual, comparative determinations of opacity and color, successive titrations, and measurement of temperature were among the methods used to determine a system's properties. These methods were slow and limited the possible time resolution in the detection of the progress of a reaction.

Rapid-mixing techniques were first introduced in 1923 by Hartridge and Roughton (1) with the development of the continuous-flow apparatus. This system allowed studies of the kinetics of reactions with half-lives as short as a few milliseconds and bypassed the need for a rapid detection system. The reversion spectroscope, developed by Hartridge (2), was used as a visual means of detecting changes in absorbance of the reacting species.

As improvements were made in the response time of detection and display devices, new configurations of rapid-mixing systems were developed which reduced reactant consumption and provided real-time data presentation. But until the introduction of relaxation methods in 1954 by Eigen (3), rate studies were restricted by the limitations of mechanical mixing to millisecond kinetics. Relaxation methods have extended the range of kinetic investigation to sub-nanosecond diffusion-limited reactions (4).

Some of the more widely used instrumentation for flow and single perturbation relaxation methods will be discussed, and no attempt has been made

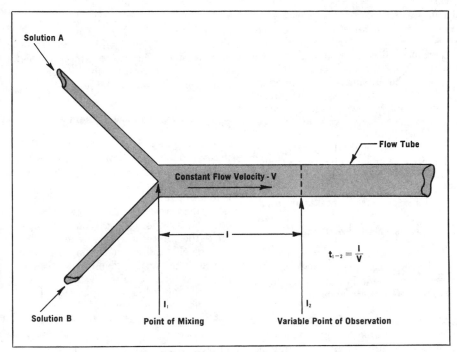

Figure 1. For the continuous-flow apparatus, the age of mixture at the point of observation is proportional to the distance between the point of mixing and the point of observation

to present a comprehensive review of all developments.

Continuous-Flow System

The principles of the continuous-flow apparatus, as illustrated in Figure 1, are based on the assumptions that the time required for complete mixing of solution A and solution B is much smaller than the age of the mixture at the point of observation and that the solutions flow at a constant volume flow rate, Q, through an observation tube of uniform geometry. The progress of the reacting mixture can then be monitored by measurement of some property of the system at various points along the length of the observation tube. For the above conditions, the age of the mixture at the point of observation will be directly proportional to the distance between the point of mixing and the

point of observation. Because the extent of progress of the reaction will remain constant at any fixed distance from the point of mixing, a method of rapid detection is not essential.

The minimum time resolution of the complete apparatus is most often labeled dead time and will be dependent on the efficiency of mixing. Mixing efficiency is a measure of the percentages of solution A and solution B which are homogeneously mixed in a given time interval and is a function of mixer configuration, flow velocity, and solution viscosity. Optimum mixing occurs when the critical Reynolds number for the mixer configuration is exceeded and turbulent flow results. The Reynolds number is proportional to the flow velocity, is inversely proportional to the dynamic viscosity, and requires increased flow velocities for an increase in solution viscosity. Various configurations of mixers have been constructed which are capable of mixing two solutions to 99% completion in less than 1 msec.

Turbulent flow is desired also along the length of the observation tube to complete the mixing of any unmixed reagents after they leave the mixer. Again, turbulent flow can be maintained with sufficient flow velocity.

A major disadvantage of the continuous-flow system is the large amount of reactants consumed while a set of measurements is being performed. The more time required by the detection procedure, the more reactants are used.

Accelerated-Flow System

The accelerated-flow apparatus, designed by Chance (5) in 1940, rapidly mixes two reagents in a manner similar to that of the continuous-flow system, but requires much smaller volumes of reactants. The reactants are contained in two hypodermic syringes which are connected to the input of a mixing chamber. A phototube is used as a detector and is fixed in position on the observation tube a short distance from the point of mixing. The syringe plungers are connected to a manually driven pushing block. The contents of both syringes are rapidly discharged by a manual push through the mixing chamber and flow into the observation area. The solutions are, therefore, accelerated from zero velocity to some maximum velocity before the syringes are exhausted. Since the detector position is fixed with respect to the mixer, the age of the reaction, t, at the point of observation, is inversely proportional to the flow velocity down the axis of the tube.

Chance obtained flow velocities of 10 m/sec in a 1-mm diam circular tube. The detector was positioned 7 mm from the mixer, thereby allowing measurement of times on the order of 1 msec. The electrical output of a potentiometer, mechanically coupled to the pushing block, was differentiated to produce a flow velocity output. The extent of

the reaction was then displayed as a function of flow velocity (proportional to $1/t$) on an oscillographic recorder or oscilloscope. The use of a rapid detection system permitted fluid consumption to be as small as 100 μl.

Stopped-Flow System

The stopped-flow apparatus is the most widely used rapid-mixing technique because of its simplicity and real-time data presentation. Reactants flow through a mixing chamber into an observation area and the flow is then abruptly stopped. At the time of stopping, the mixture is only a few milliseconds old, and the remaining progress of the reaction is monitored with a rapid-response detection system.

The stopped-flow apparatus, diagrammatically shown in Figure 2, is similar to the accelerated-flow apparatus discussed previously, but incorporates a stopping device which is capable of stopping the flow very quickly. A simple and effective method to accomplish this is the addition of a third stopping syringe connected to the effluent part of the observation tube. A driving block, either manually or pneumatically driven, rapidly accelerates the fluids to a steady maximum flow velocity, V_{max}; the stopping syringe begins to fill, and its plunger extends. The flow stops abruptly when the motion of the plunger of the stopping syringe is arrested by a mechanical stop. The age of the reacting mixture, t_0, in the observation area at the time of stopping will be l/V_{max}, where l is the distance between the point of mixing and the point of observation, and V_{max} is the maximum steady-state flow velocity obtained before stopping. The oscilloscope trace, shown in Figure 3, illustrates the different phases of flow. A solution of $NaHCO_3$ with bromphenol blue added as an indicator was mixed with HCl. A rapid-response spectrophotometric system was used to monitor the state of the mixed reagents.

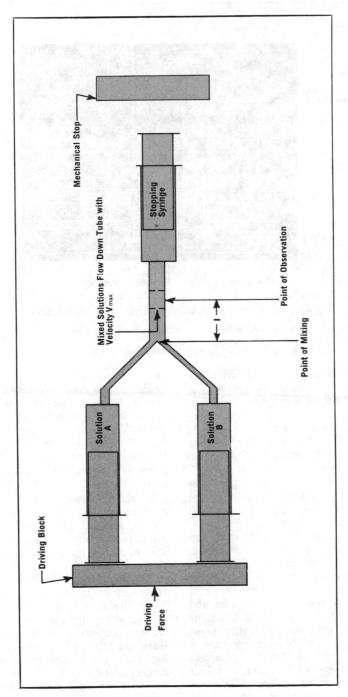

Figure 2. Basic components of stopped-flow apparatus

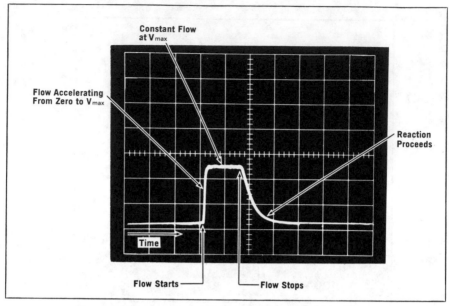

Figure 3. Detection of flow cycle at point of observation

Solution A. NaHCO₃ with indicator bromphenol blue
Solution B. HCl

The acceleration phase clearly illustrates the function of the accelerated-flow apparatus discussed previously. After the solutions reach V_{max}, they flow at constant velocity until stopping occurs. The progress of the resulting reaction (the information of interest) is then monitored.

The dead time of the stopped-flow apparatus is affected by three major factors: efficiency of mixing, transport time, and stopping time.

Mixing efficiency, as explained previously, is dependent on mixer configuration and flow velocity. The transport time is the time required for the mixed solutions to travel from the point of mixing to the observation area and is dependent on flow velocity and the distance between the observation point and the mixer. The stopping time is the time required for the complete closed-fluid system to come to a halt after the stopping syringe makes contact with its mechanical stop. The

dead time of the system will be degraded if the fluids do not come to a stop in a time period which is small in comparison to the mixing and transport times. It is desirable for mixing to reach 98% of completion within the transport time interval. If the minimum time resolution is limited by the transport time, the first observation after the flow stops will be free from artifacts due to incomplete mixing. Mixing artifacts will appear if the solution is transported to the observation area in a time much shorter than that required for complete mixing.

The physical arrangement of one of the many instrumental configurations (6) is shown in Figure 4. To simplify loading of the driving syringes, a valving arrangement is incorporated which seats the tips of a set of loading syringes. The pneumatically actuated driving block is retracted by pressing the reset button, and the reagents contained in the loading syringes are transferred into the drive syringes. To set the apparatus for an experiment, the remains of spent reactants from previous experiments are exhausted from the stopping syringe through a waste valve. A microswitch is mounted on the face of the mechanical stop to signal the detection device when the flow stops. Flow is initiated when the drive button is depressed. The micrometer connected to the mechanical stop will set the amount of distance the stopping plunger must travel before it makes contact. The micrometer will, therefore, allow control of the flow interval. An extended flow interval will increase

the consumption of reactants but allows a thorough flushing of the observation cell before flow stops. For minimum consumption of reactants, the flow interval must be only of sufficient duration for the solutions to travel from the mixer to the observation cell and fill the volume of the cell. Satisfactory results can be obtained with reactant volumes as small as 100 μl. Extended-flow intervals (of any length) will not affect the dead time of the apparatus; the age of the mixture in the observation area will depend on V_{max}, and not how long V_{max} is maintained.

An expanded diagram of the mixer

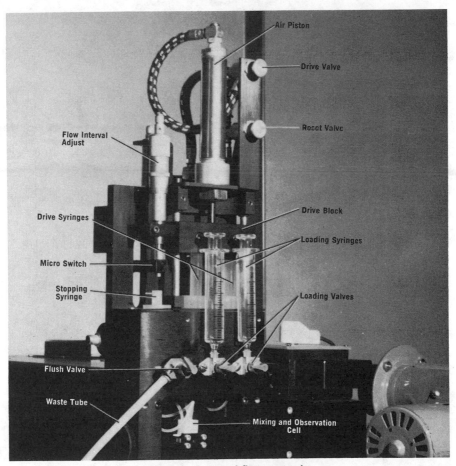

Figure 4. Stopped-flow apparatus

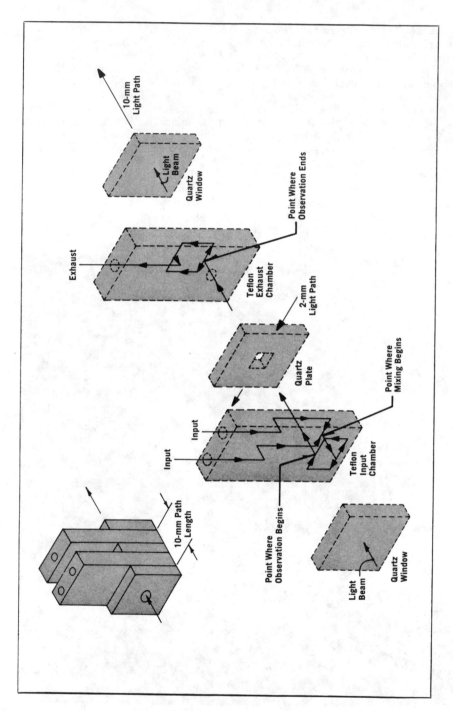

Figure 5. Mixer and observation cell

382

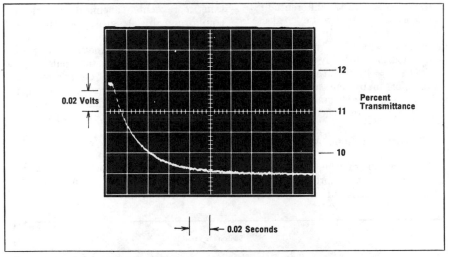

Figure 6. Reaction between aged Ce(IV) and H_2O_2; $\lambda = 420$ nm

Equal amounts of reagents A and B were mixed
Reagent A: 0.01M ceric ammonium nitrate; $(NH_4)_2$ $Ce(NO_3)_6$ in 0.03M HNO_3
Reagent B: 4 x 10^{-4}M H_2O_2 in 0.03M HNO_3
Solution A is aged to promote the formation of a polynuclear Ce(IV) species
Path Length = 10 mm $\lambda = 420$ nm

and observation cell is shown in Figure 5. The efficiency of mixing is greater than 98% in less than 1.5 msec, and the dead time is about 2.5 msec. Figure 6 shows the results of a typical stopped-flow experiment.

Relaxation Methods

A different approach to rate studies, which eliminated the necessity for mixing and extended time resolution into the microsecond range, was taken by Eigen. A solution containing components which are already in an equilibrium state is perturbed by a rapid change of an external parameter such as temperature or pressure. The shift to a new equilibrium state at the new temperature or pressure is observed as a function of time. If an instantaneous change of an external parameter could be accomplished, the resulting shift to a new equilibrium state would not be instantaneous, but its rate would be governed by the reaction mechanism of the chemical system in solution. The time course of the concentration changes of the reactants and products can be characterized by a summation of exponential relaxation terms, each with a characteristic relaxation time constant.

The extent of change in concentrations of the reactants and products obeys the thermodynamic law $d \ln K_p / dT = \Delta H / RT^2$ for a temperature perturbation, and $d \ln K_t / dP = -\Delta V° / RT$ for a pressure perturbation. K_p and K_t are the equilibrium constants at constant pressure and constant temperature, respectively.

To determine the rate-law parameters, a mechanism first must be assumed, the rate-law parameters must be derived in terms of relaxation times, and the proposed mechanism must be verified by experimentation. This may seem like a formidable task, but for small perturbations, where the changes in reactant and product concentrations are less than 10%, second-order concentration effects in the differential rate equations can usually be neglected. Relaxation methods can be applied to all systems which are reversible and have a nonzero enthalpy change of reaction ($\Delta H \neq 0$). Mathematical derivations for many one-step, two-step, and multi-step mechanisms can be found in various references (7–9).

Temperature-Jump Apparatus

The function of the temperature-jump apparatus is to produce a small perturbation in a solution which contains a reversible chemical system in equilibrium, by elevating the temperature 5–10°C within a few microseconds.

The most common method for producing rapid heating is the Joule heating resulting from the discharge of a capacitor through a resistive solution (Figure 7). The magnitude of the temperature rise, ΔT, will be given by

$$\Delta T = W k_t / Q k_s \qquad (1)$$

where W = energy transferred from the capacitor, J; Q = cell volume, cm³; k_t = 0.239 cal/J; and k_s = specific heat

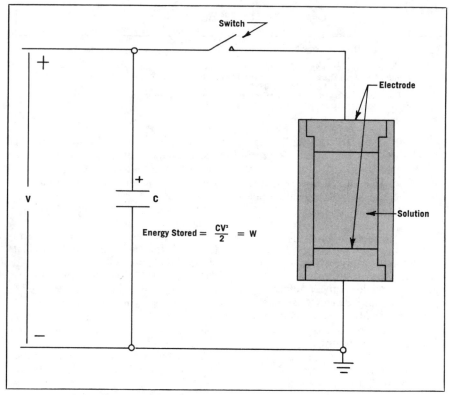

Figure 7. Schematic representation of temperature-jump apparatus

Temperature rise $\Delta T = \dfrac{W}{Q}\dfrac{K_t}{K_s}$

Energy stored $= \dfrac{CV^2}{2} = W$

Heating time constant $\tau_H = \dfrac{RC}{2}$

R = solution resistance, ohms; W = energy, J; Q = volume of cell, cm³; K_t = 0.239 cal/J; K_s = specific heat, cal deg⁻¹ cm⁻³

of the solution, cal deg^{-1} cm^{-3}. The energy, W, stored on the capacitor of capacitance, C (farads), is

$$W = CV^2/2 \qquad (2)$$

where V is the voltage across the capacitor. The simplest way to control ΔT, therefore, will be to adjust V to the required value. For example, a 10,000-V charge on a 0.25-μF capacitor is equivalent to 12.5 J of energy. This would produce approximately a 6°C rise in temperature in a 500-μl cell.

An inert electrolyte is added to the solution to control the resistivity. The resistance between the electrodes is then dictated by the ionic strength of the solution and the geometry of the cell. If stray inductance in the discharge circuit is negligible, the discharge voltage across the electrodes will decrease exponentially with the time constant, RC, where R is the solution resistance in ohms. This results in an exponential temperature rise with a heating-time constant of $RC/2$.

Shock waves of considerable intensity are propagated through the solution when the heating-time constant is reduced below a few microseconds. Behind the high-pressure wavefront exists a low-pressure region which will cause cavitation in the solution. Shock waves of excessive amplitude will cause optical aberrations, if not damage to the cell. The formation of cavitation will cause random artifacts which can degrade the quality of the signal to the point where no useful information can be obtained. To minimize these effects, a number of conditions must be maintained. The maximum ionic strength of the solution must be limited to prevent heating-time constants of much less than 1 μsec. The cell must be ruggedly constructed to withstand the intense shocks produced when the high-voltage capacitor is discharged through a solution containing an air bubble. The shock waves and resulting cavitation will be minimized if an aqueous sample solution is thermostated to 4°C. This is the point of maximum density of water.

The electrodes are usually plated with a thin layer of platinum to prevent attack by acidic or oxidizing solutions. They must be arranged in the sample cell so that thermal gradients, due to nonuniformities in heating, will be minimized in the optical path. Since the refractive index of a solution will be temperature dependent, thermal gradients in the optical path will cause optical distortions.

As mentioned before, the heating-time constant of a temperature-jump apparatus generally will be limited to values greater than 1 μsec. The fastest relaxation process that can be studied, then, will be reactions with relaxation-time constants at least a few times the heating-time constant of the apparatus. The slowest processes that can be observed will depend on the heat-transfer properties of the cell. All measurements should be taken in a time period in which the solution temperature remains relatively constant at the new elevated temperature. The majority of temperature-jump cells will reequilibrate at a rate slow enough to allow measurements to be extended to hundreds of milliseconds.

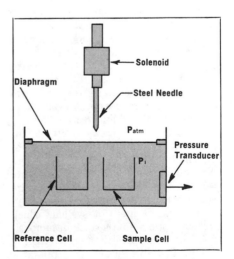

Figure 8. Pressure-jump apparatus

Combination of Stopped-Flow and Temperature-Jump Methods

Stopped-flow methods are limited to the study of reactions with half-lives greater than a few milliseconds by the time required to mix two solutions completely. The temperature-jump method can be used for reactions as short as a few microseconds but is limited to reversible systems in equilibrium. It is possible to combine these two methods (10) to study reactions in which a short-lived steady state appears. The reaction is initiated by rapid mixing and then perturbed during the appearance of this pseudosteady state. The time range of the stopped-flow apparatus is then effectively expanded to microsecond kinetics. In some enzymatic systems, a steady state may be maintained for a few tens of milliseconds. These systems can be quite complex, however, making data reduction very difficult.

Pressure-Jump Apparatus

The pressure-jump apparatus applies a rapid increase or decrease in pressure to a chemical system in equilibrium. The pressure perturbation causes a shift to a new equilibrium state at the new pressure. The relaxation to the new equilibrium is monitored by conductometric or spectrophotometric methods.

The extent of change in equilibrium constant for a change in pressure will be proportional to $\Delta V°$, the standard volume change for the reaction. The pressure-jump technique, therefore, will be applicable to chemical systems in solution which exhibit reasonable volume changes and are not suitable for temperature-jump studies because of the small enthalpy changes. Although the equilibrium constant is usually more sensitive to a change in temperature rather than pressure, pressure-jump studies can be used for systems which are susceptible to thermal decomposition from a temperature perturbation.

The first pressure-jump apparatus was constructed in 1958 by Ljunggren and Lann (11). The pressure in a sealed vessel was raised from 1 atm to 150 atm in 50 msec by opening the valve on a nitrogen tank. Because they were working with ionic species, the relaxation process was monitored by measuring conductance changes. The sample was contained in a conductance cell which formed one arm of an alternating-current Wheatstone bridge.

An improved system was introduced by Strehlow and Becker (12) which caused a pressure drop in a dual conductivity cell, from about 60 atm to 1 atm in 60 μsec. The addition of the reference cell allowed compensation for errors due to temperature and viscosity changes resulting from the pressure drop. The cell compartment, sealed by a metallic disk, was pressurized to 60 atm. A solenoid-controlled steel needle ruptured the disk, causing a rapid drop to atmospheric pressure. A pressure transducer contained within the cell signaled the initiation of the pressure transition. A similar apparatus is illustrated in Figure 8.

Characteristics of a Rapid Photodetection System

The components of a rapid photodetection system, shown in Figure 9, are similar to those found in a standard spectrophotometer but differ greatly in their response characteristics. Noise which is normally filtered out by a slow-response amplifier and recording device can now cause significant degradation in the signal-to-noise ratio of the system. The signal-to-noise ratio, or limit of detection, will improve in direct proportion to the square root of the luminous flux impinging on the photocathode of the photomultiplier tube, and in inverse proportion to the square root of the bandwidth of the detection system.

Because there are no universal sources of radiant energy presently available which will provide both sufficient energy and acceptable stability over the complete ultraviolet, visible, and infrared spectrum, it is necessary to examine the most commonly used sources and

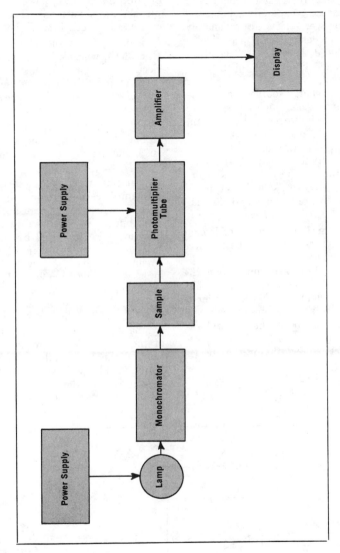

Figure 9. Basic components of rapid photodetection system

evaluate their merits and limitations. For work in the visible and infrared region, a tungsten lamp powered by a low-ripple, highly regulated dc power supply forms a very stable, low-noise source; however, below approximately 300 nm, the energy output is significantly reduced, lowering the signal-to-noise ratio of the system to an unacceptable level.

Ultraviolet energy sources include the mercury and xenon high-pressure arc lamps and the hydrogen and deuterium low-pressure lamps. The intensity of the high-pressure arc lamps is considerably greater than that of the low-pressure lamps. The xenon and mercury lamps, however, suffer from instabilities due to arc wander and fluctuations. In a single-beam spectrophotometric system, the deuterium or hydrogen lamp is, therefore, the best source of ultraviolet energy.

The monochromator must have the largest possible aperture to avoid unnecessary reduction of the luminous flux entering into the observation cell. The slits should be opened as wide as permitted by the specific requirements for spectral bandwidth for the species under study.

In an optimized detection system, the signal-to-noise ratio should be limited by the statistical shot noise inherently produced in the photocathode of a photomultiplier tube and not by the effects of lamp instabilities, power supply ripple, and amplifier noise. The rms noise current produced in the photocathode, $\bar{I}_n$, is described by the relationship $\bar{I}_n = (2\ eI_pB)^{1/2}$, where e is the charge on an electron, C; I_p is the photocathode current, A; and B is the noise bandwidth of the detection system, Hz. The signal-to-noise ratio, I_p/I_n, is, therefore, $(I_p/2\ eB)^{1/2}$. Both I_p and $\bar{I}_n$ are now equally amplified by the n dynodes, producing an anode signal current of magnitude $I_s = (\bar{I}_n + I_p)A^n$, where A is the average amplification per dynode stage. The amplification provided by the dynodes is almost ideal because their noise contribution is considerably less than the shot noise produced by the photocathode. There is a limitation to the amount of radiant energy that may be imposed on the photocathode because of the heat produced by the resulting photocurrent. The maximum amplification of I_p by the dynodes is limited by the maximum anode current rating of the tube. To ensure stable operation of the photomultiplier tube, the power supply, which supplies the high voltage bias to the dynodes, should be about 10 times more stable than the

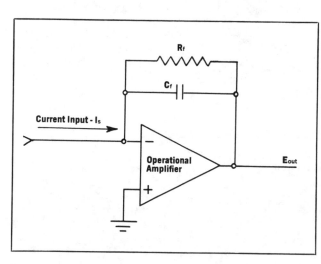

Figure 10. Current to voltage converter

$E_{out} = I_s R_f$

$\tau = R_f C_f = \dfrac{0.16}{B}$

R_f, ohms; C_f, F; I_s, A; E_{out}, V; τ, sec, Hz

desired stability of the signal current.

The amplifier serves the function of transducing the signal current into a proportional voltage and tailoring the system response characteristics to produce a suitable signal level and the best possible signal-to-noise ratio. The expression for shot noise indicates that a decrease in system bandwidth will decrease the proportion of noise in the signal. But an excessive decrease in the bandwidth will affect the response to an event of typically exponential nature containing the information of interest.

The current-to-voltage converter, shown in Figure 10, is a simple and effective circuit which is commonly used, as its name implies, to convert a current to a proportional voltage and to limit the response of the detection system. The anode of the photomultiplier tube is connected directly into the inverting input of the operational amplifier, thereby eliminating a load resistor. This circuit presents a low impedance to the source (the photomultiplier) and, therefore, is much less susceptible to the response-limiting effects of the capacitance of the interconnecting cable. The low frequency gain of the circuit, expressed in $V/\mu A$, A_a, is equal to the resistance of the feedback resistor in megohms. A 1-μA input will produce a 1-V output for a 1-$M\Omega$ feedback resistor. The response to a step input is exponential, with a time constant, τ, of RC seconds, where R is in ohms and C is in farads, and is related to the bandwidth B by $B = 0.16/\tau$. To pass the signal of interest with minimum error, the time constant of the circuit should be at least five times greater than the time constant of the signal.

A storage oscilloscope, used as a display device, will display for periods in excess of 1 hr, an image of a rapid transient event, occurring in microseconds. The time base of the horizontal axis is triggered from an external source, such as a microswitch on a stopped-flow apparatus, signaling the stopping of flow, or a pulse transformer on a temperature-jump apparatus which is initiating the breakdown of a triggered spark gap that causes the energy of the storage capacitor to be transferred to the cell. When the kinetic experiment is complete, the signal trace is stored. The trace may then be examined before making a permanent photographic record for subsequent evaluation and can afford a considerable savings in film usage.

Future Developments

Stopped-flow and temperature-jump instrumentation are the most widely used methods of those discussed. Their level of development is sufficiently mature to produce reliable and repeatable information. The data generated by these techniques can contain a great amount of information of considerable complexity. The optical detection system discussed in the previous section requires manual data reduction from the information presented in a photograph. Enough information can be obtained in a day to require a week of analysis. Analog manipulation, such as logarithmic conversion, will eliminate some of the steps required in analysis but will not improve the limited resolution of a photograph.

Digital data acquisition systems can considerably improve resolution and reduce the time and effort required to perform analysis of rate data. In the past two years, dedicated computer systems have been introduced which digitize the analog rate data, smooth the data by a least squares procedure, and compute various kinetic parameters (13, 14). Future developments in the software for data reduction will significantly increase the efficiency and versatility of these techniques.

Indications are that with the continued development of systems with automatic data reduction, instrumentation for the study of rapid reactions in solutions will become a standard tool found in most laboratories.

389

Literature Cited

(1) H. Hartridge and F. J. W. Roughton, *Proc. Roy. Soc.* (London), **A104**, 376 (1923).

(2) H. Hartridge, *ibid.*, **A102**, 575 (1923).

(3) M. Eigen, *Discuss. Faraday Soc.*, **17**, 194 (1954).

(4) R. G. Pearson, *ibid.*, p 187.

(5) B. Chance, *J. Franklin Inst.*, **229**, 455 (1940).

(6) J. I. Morrow, *Chem. Instrum.*, **2** (4), 375 (1970).

(7) G. W. Castellan, *Ber. Bunsenges. Phys. Chem.*, **67**, 898 (1963).

(8) G. Schwarz, *Rev. Mod. Phys.*, **40** (1), 206 (1968).

(9) A. F. Yapel, Jr. and R. Lumry, "Methods of Biochemical Analysis," Vol 19, Interscience, New York, N.Y. 1970.

(10) J. E. Erman and G. G. Hammes, *Rev. Sci. Instrum.*, **37** (6), 746 (1966).

(11) S. Ljuggren and O. Lamn, *Acta Chem. Scand.*, **12**, 1834 (1958).

(12) H. Strehlow and M. Becker, *Z. Electrochem.*, **63**, 457 (1959).

(13) P. J. De Sa and Q. H. Gibson, *Comput. Biomed. Res.*, **2**, 494 (1969).

(14) B. G. Willis, J. A. Bittikofer, H. L. Pardue, and D. W. Margerum, ANAL. CHEM., **42**, 12 (1970).

Bibliography

(1) E. F. Caldin, "Fast Reactions in Solution," Blackwell Scientific, Oxford, England, 1964.

(2) G. H. Czerlinski, "Chemical Relaxation," Marcel Dekker, New York, N.Y., 1966.

(3) K. Kustin, Ed., "Methods in Enzymology," Vol XVI, Academic Press, New York, N.Y., 1969.

(4) A. Schechter, *Science,* **1970**, 273 (1970).

(5) J. E. Finholt, *J. Chem. Educ.*, **45**, 394 (1968).

(6) L. Faller, *Sci. Amer.*, p 30, May 1969.

Prospects for
Molecular Microscopy

Recent improvements in specimen support films and image
contrast will help achieve the goal of visualizing single
atoms or small molecules by electron microscopy. Other
possible applications in chemical identification
and structure determination, including the detection of
X-rays, fluorescence emission, and secondary electron
emission, are likely to be exploited

J. WENDELL WIGGINS and MICHAEL BEER
Thomas C. Jenkins Department of Biophysics
Johns Hopkins University, Baltimore, Md. 21218

OVER THE LAST 25 YEARS, the electron microscope has emerged as a major device for the determination of the structure of matter. Many recent developments in biology and metallurgy would have been inconceivable without this powerful tool. For all its success, electron microscopy has been beset by a number of fundamental limitations. Particularly, in reaching for the goal of visualizing single atoms or small molecules, there have been four substantial problems: insufficient image contrast, specimen damage, insufficient resolution, and irregularities in the specimen support film.

The problem of resolution has been attacked with a vehemence not afforded the others, perhaps because it was the first recognized. This attack and its successes have been well reported elsewhere (1). Conventional techniques now allow 2–3 Å point-to-point resolution, sufficient for many molecular structure investigations. The problem of specimen damage owing to the electron beam is to a great extent unsolved. Indeed, the extent and nature of the damage are just beginning to be assessed. Primarily, we will deal with progress in the other two problems, specimen support films and image contrast.

Advantages of Graphite Crystallite Specimen Support Films

Specimens are carried into the vacuum system of the electron microscope after mounting on thin support films. These must be as transparent to electrons as possible, stable in the electron beam, and sufficiently good conductors

391

of heat and electricity to avoid local accumulation of charges or excessive temperature during irradiation. Finally, the film should be free of irregular structure to assure that details of the specimen are not obscured by the superimposed structure of the support. Until now the most successful supports were thin films of carbon produced by evaporation in vacuo onto a smooth surface (2). Such a procedure is bound to lead to variable thickness. This is indeed found; these supports have an irregular structure which leads to a granular image. This granularity sets the limit on the minimum mass which can be detected.

Recently, it has been found that evaporated carbon films can be converted to films consisting of crystallites of graphite by subjecting them to a brief heat treatment at 2700°C in an atmosphere of argon (3). These crystallites are easily recognizable, as shown in Figure 1. The high temperatures require that the films be mounted on carbon support grids instead of the more usual copper ones.

On examination in the electron microscope, the graphite crystallites are conspicuously free of the disturbing granularity found in evaporated carbon films. Then the limiting granularity of support films seems largely eliminated in the graphitization. The decrease in granularity is indicated in Figure 2. Here, the intensity of the elastically scattered electrons is sampled repeatedly, showing a scan of a graphite film and a carbon film. Since this intensity of scattered electrons is a measure of the local thickness, the distribution is related to the distribution of thickness —that is, to the granularity. The distribution is clearly narrower for the graphitized film, even though it was, on the average, thicker. This improvement cannot be realized if carbon contamination is introduced during observation. Indeed, if absolutely no contamination had been present in the instrument when the micrographs were taken, the difference between the dis-

Figure 1. Dark field electron micrograph of thin graphite film produced by high temperature heat treatment of carbon film

Crystallites 0.1 to 0.2 μm in size are clearly visible. 50,000×

tributions in Figure 2 would have been even greater. However, one important difficulty remains to be solved. The graphite films have extremely different adsorption properties from the carbon ones; therefore, new specimen mounting procedures must be found.

Improvements in Image Contrast

We shall now turn to the important recent development, by Crewe (4) and his collaborators, of the scanning trans-

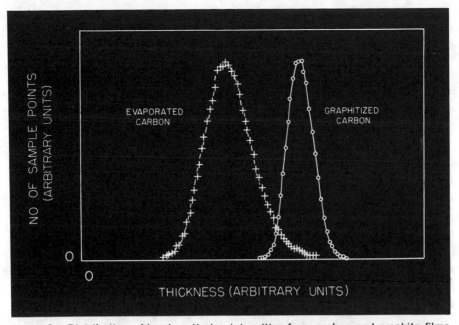

Figure 2.   Distribution of local scattering intensities from carbon and graphite films

Distribution is a measure of the thickness variation.   Though overall thickness of graphite film is considerably greater, thickness variation is considerably less than for carbon film

mission electron microscope (STEM) which has improved the prospects for obtaining better image contrast and some specific chemical information about the components of the specimen. One illustration of the advantage of the STEM is provided by the clear visibility of single heavy atoms (5), as shown in Figure 3. Using similar (though not rigorously identical) specimens, Highton and Beer (6) estimated that the threshold of visibility for the conventional electron microscope is approximately three heavy atoms.

The scanning electron microscope of Crewe differs from the conventional scanning type in two basic respects. The high resolution is made possible by the introduction of a unique and ingenious electron gun combining a field emission source, as developed by Muller et al. (7), and a low aberration accelerating lens (8). At the same time they introduced detection of the elec-

trons after transmission through the specimen, which allows great versatility in assessing the interaction with the specimen. To understand the advantages of the STEM, let us state the mode of image formation in both STEM and the transmission electron microscope (TEM). See Figure 4 for a comparison of the schematic diagrams of the TEM and the STEM.

In the TEM the entire field of view is illuminated continuously. A magnified image of the field of view is produced on a fluorescent screen or, for permanent recording, a photographic plate. If the optical system were ideal, all electrons incident on the field of view would contribute to the final image. The only image contrast would be owing to the fact that when an electron is elastically scattered by the specimen, a phase shift is introduced into its wave function. This causes destructive interference at the corresponding image point. However, the optical system is not perfect. Since the lenses have an unavoidable spherical aberration, optimum spatial resolution is obtained by placing an aperture stop at the back focal plane of one of the lenses, usually the objective. This aperture stop provides additional contrast by completely removing from the image those electrons scattered through an angle greater than that of the aperture. The lenses also have an unavoidable chromatic aberration. Therefore, when an electron loses energy in the scattering process—so-called "inelastic scattering"—it is not focused to the proper image point. Then the three processes—interference, aperture removal, and chromatic defocusing—all remove electrons from the image of a

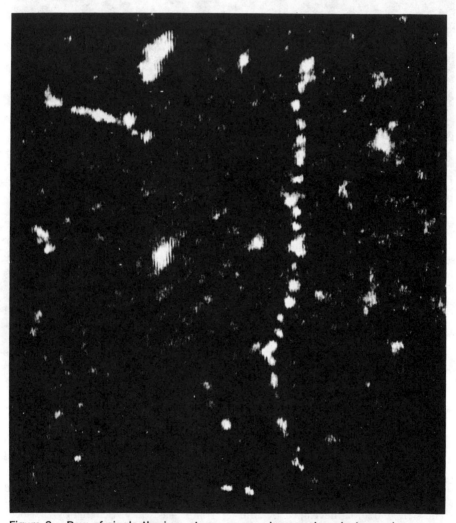

Figure 3.   Row of single thorium atoms as seen in scanning electron microscope

Specimen produced by spraying on carbon films an equimolar mixture of Th(NO₃)₄ and 1, 2, 4, 5-benzenetetracarboxylic acid.   Small spots along chains are single atoms of thorium.   Micrograph is roughly 670 Å to each side

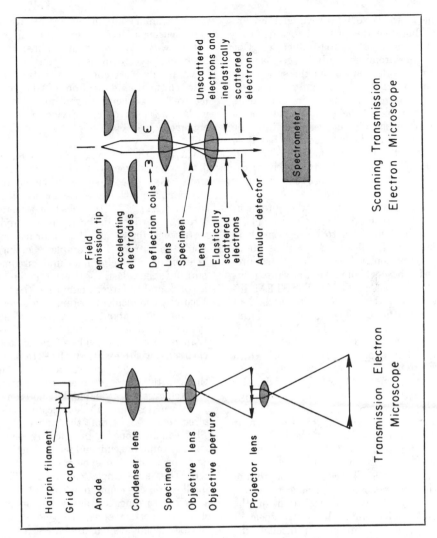

Figure 4. Comparison of schematic diagrams of transmission electron microscope and scanning transmission electron microscope

scattering object.

A calculation including all these effects is difficult. However, several calculations have been made (9). In general, when the calculated contrast of an object of atomic dimensions is compared with substrate noise, its observation seems to be marginally possible.

The TEM can be used in the dark field mode so that the unscattered beam is excluded from the image by an aperture, and only the scattered electrons form the image. In this mode interference contrast effects are considerably reduced. There are both advantages and disadvantages in going from bright field to dark field operation. The main disadvantages are too little illumination for proper focusing, and each of the methods so far proposed for eliminating the unscattered beam introduces some substantial difficulty. One of the prominent advantages is that much of the noise from the supporting film originates from interference effects largely eliminated in dark field. In the bright field mode the photographic plate is darkened before the optimum number of electrons has been collected. In dark field the unscattered electrons which carry no information are absent, and more scattered electrons may be collected.

There are, of course, other problems to be overcome in the use of the TEM, but these are also to be dealt with in the STEM. The considerations above are mentioned because they are precisely those which make it advantageous to work with a STEM. The STEM differs from the TEM in its method of illuminating the field of view and forming the image. In the STEM the illumination is focused on the specimen to a fine spot, the diameter of which determines the spatial resolution limit of the device. Crewe showed that by using the exceedingly bright field emission source, spot sizes down to 5 Å diameter were practicable. In the near future this spot size may be further reduced to perhaps 1 Å. This illumination spot is scanned across the field of view by deflection coils in synchronism with the raster of a cathode ray tube. The intensity of illumination of the CRT is determined by some aspect of the electron transmission through the specimen.

This method of image formation is formally identical to that of the TEM (10, 11). It follows that the resolution limit of the TEM and the STEM depends in the same way on the performance of the objective lens. If comparable lenses are used in the two devices, the resolution limits will be comparable. With instruments using an accelerating voltage of 100 kV, the best resolution attainable today is approximately 2.5 Å. However, in practice there are significant differences.

The primary advantage accrues from the focusing of the beam before interaction with the specimen. Thus, one does not have to focus electrons which have already lost energy in scattering. The chromatic aberration of the objective lens is then no problem. If one wishes, a selective field of view may be achieved by special use of the deflection coils.

Having the spatial information in hand, one is free to process the electrons transmitted by the specimen in any manner which will enhance the contrast (4). If the electrons were sorted according to both energy and scattering angle, all the information would be extracted. Complete sorting is difficult. A helpful natural circumstance is that elastic scattering involves a considerably larger angle than the inelastic. The incident beam has a divergence angle of about 10^{-2} radians. The typical inelastic scattering angle is less than 10^{-3} radians. Therefore, if the electrons with less than 10^{-2} radians emergence angle are separated from those with a greater scattering angle, one achieves a reasonable, but not complete, separation of the elastically scattered electrons from the remainder. This can be done in practice by placing an annular semiconductor junction detector below the specimen to intercept the large angle electrons. The small angle electrons may then be separated according to energy loss in an electron spectrometer. Depending on the particular system used to sort the electrons, one ends up with several electrical signal currents which may be combined in various ways by operational amplifier techniques to produce a variety of types of contrast. For example, division of the total elastic signal by the total inelastic signal produces an image intensity proportional to the atomic number of the scattering object (12).

Possibilities for Chemical Identification

The electron energy loss spectrum contains a wealth of chemical information about the specimen. There is a valuable relationship between the optical absorption spectrum and the electron energy loss spectrum of a specimen. Both spectra are determined by the dielectric constant. The exact relation is complicated and is discussed thoroughly in Ref. *13*. In general, any optical absorption of a certain energy is reproduced by a corresponding loss in the electron spectrum. In addition there are electron energy losses owing to collective excitations of the valence electrons. Therefore, the prospect emerges that chemical identification may be accomplished on the molecular scale just as it is now done for larger samples with a spectrophotometer. Some insight into the specificity of the electron energy loss spectrum may be gained from the data of Crewe et al. (*14, 15*), who have examined adenine and thymine. The spectra are detectably different and so encourage optimism in chemical identification by this approach.

The limit to what can be accomplished by the new techniques for molecular microscopy is probably imposed by the damage to the specimen from the electron beam. The unscattered and elastically scattered electrons do nothing to the specimen. The inelastically scattered electrons cause excitations of electrons, or ionization. Either case may lead to dissociation of the molecules involved. Whether or not sufficient information for identification can be obtained before the character of the specimen is changed is not yet known. There have been both theoretical and experimental studies (*15, 16*). Opinion is divided. The benefits to be gained by successful utilization of the energy loss information are so great that the question must be settled finally by experiment. It is almost certain that energy loss identification can be applied to microscopic structures larger than a few atoms, for example, the organelles of a cell.

Depending on what is indeed the severity of radiation damage, one may conceive of determining the three dimensional structure of molecules along with the subunit identity by this technique.

Finally, it is worth mentioning that the high resolution scanning microscope can be extended so that other manifestations of the interaction of the electron beam with the specimen are detected. Thus, X-rays, fluorescence emission, and secondary electron emission could all be detected.

Acknowledgment

The authors gratefully acknowledge the generosity of A. V. Crewe who provided Figure 3.

References

(1) E. Ruska, *Advan. Opt. Electron Microsc.,* 1, 116 (1966).
(2) D. E. Bradley, *Brit. J. Appl. Phys.,* 5, 65 (1964).
(3) J. R. White, M. Beer, and J. W. Wiggins, *Micron.,* in press.
(4) A. V. Crewe, *Quart Rev. Biophys.,* 3, 137 (1970).
(5) A. V. Crewe, J. Wall, and J. Langmore, *Science,* 168, 1338 (1970).
(6) P. J. Highton and M. Beer, *J. Roy. Microsc. Soc.,* 88, 23 (1968).
(7) E. W. Muller and T. T. Tsong, "Field Ion Microscopy," Elsevier (1969); R. Gomer, "Field Emission and Field Ionization," Harvard University Press (1961).
(8) A. V. Crewe, D. N. Eggenberger, J. Wall, and L. M. Welter, *Rev. Sci. Instrum.,* 39, 576 (1968).
(9) C. B. Eisenhandler and B. M. Siegel, *J. Appl. Phys.,* 37, 1613 (1966).
(10) A. V. Crewe and J. Wall, *Optik,* 30, 461 (1970).
(11) E. Zeitler and M. C. R. Thomson, *ibid.,* 31, 269, 359 (1971).
(12) A. V. Crewe, *Ber. Bunsenges. Phys. Chem.,* 74, 1181 (1970).
(13) P. Nozieres and D. Pines, *Phys. Rev.,* 113, 1254 (1959).
(14) A. V. Crewe, M. Isaacson, and D. Johnson, Proc. EMSA 28th Ann. Mtg., 262 (1970).
(15) A. V. Crewe, M. Isaacson, and D. Johnson, *ibid.,* 264 (1970).
(16) J. R. Breedlove and G. T. Trammell, *Science,* 170, 1310 (1970).

Signal-to-Noise Enhancement Through Instrumental Techniques

Part I. Signals, Noise, and S/N Enhancement in the Frequency Domain

G. M. HIEFTJE,
Department of Chemistry, Indiana University, Bloomington, IN 47401

A consideration of some relatively simple techniques which enable better signal collection can result in improved analytical measurements. Since an increase in measurement time or a loss of resolution can occur simultaneously, a selected trade-off is necessary

Aᴄᴄᴏʀᴅɪɴɢ to communication theory, every measurement which is made involves the observation of a signal. To this way of thinking, all of analytical chemistry, indeed most of physical science, requires the examination of a vast number of signals in widely varying forms.

It is therefore important that we understand as fully as possible the nature of these signals and the way in which we can best observe them. From this unified approach, all signals can be examined collectively, and their common properties exploited to assist us in establishing rules and techniques for their measurement. It is often possible to improve signal measurement but only at some cost, such as an increase in measurement time or a loss of resolution, thereby necessitating a selected trade-off. The following discussion will examine signals, consider problems inherent in observing signals, and consider some relatively simple techniques which enable better signal collection.

Signals and Noise

All signals observed in real life carry with them an undesirable hitchhiker called noise. For our purposes, this noise may be considered to be any portion of the observed signal which we do not want. In many cases a signal can carry a great deal of information, much of which is useful. Nonetheless, any portion of the signal we wish to ignore or would rather not observe is noise, regardless of how useful it might be under different considerations. An example is the radiation emitted by a flame in flame spectrometry. Although the emission signal may contain information about a number of species present in a sample, if our only interest is in sodium, all other species can contribute to unwanted variations in the sodium emission and thereby constitute noise.

From this brief discussion, to most effectively observe any phenomenon of interest, it would be desirable to extract the signal from the noise. Although this is indeed important, it must be realized that noise can never be completely removed, because the devices used to reduce noise will themselves introduce a small but finite amount of noise onto the signal. Thus, it is necessary to employ a figure of merit, the signal-to-noise ratio (S/N), to define the relative freedom from noise in a system or to define the quality of a measurement.

Signal-to-Noise Ratio. Let us examine the signal-to-noise ratio more closely. To obtain the S/N ratio, we must first have ratioable quantities. Because a power ratio is generally most meaningful, we must therefore determine the power available in both the S and N waveforms. If the waveforms are dc, the power will be simply proportional to the respective dc value. If, however, the waveforms are ac, the power will be proportional to the root-mean-square (rms) value of the waveform. The rms value can be found from the following relationship:

$$A_{rms} = \left[\frac{\sum\limits_{i}(\bar{A} - \Delta A_i)^2}{i} \right]^{1/2} \quad (1)$$

where $\bar{A}$ is the mean value of the waveform, and ΔA_i is the deviation of the ith point from the mean.

For a dc signal where the major noise is often a time variation of the signal, Equation 1 takes on added significance. In this case, the signal is the mean value ($\bar{A}$), and noise is just the excursion from this mean, so that Equation 1 will define the noise in the following way:

$$N = \left[\frac{\sum\limits_{i}(S - \Delta S_i)^2}{i} \right]^{1/2} \quad (2)$$

Note that Equation 2 is the same as that used in computing the standard deviation (N) of a series of samples whose mean would be S. Therefore, N is equivalent to the standard deviation, and S is equivalent to the mean. Thus,

$$\frac{S}{N} = \frac{\text{mean}}{\text{standard deviation}} = \frac{1}{\text{relative standard deviation}} \quad (3)$$

This relationship emphasizes the importance of S/N in determining the precision of a measurement. Large S/N values obviously provide better precision (lower relative standard deviation). In general practice, the relative standard deviation is used to describe variations between discrete signal measurements made over a period of time; the signal-to-noise ratio is commonly used to express the variation in a single signal or waveform. However, Equation 3 shows the two to be related if proper evaluation is performed.

From this standpoint, S/N may be conveniently estimated in the common case of a dc signal trace on a recorder, oscilloscope, or other readout device. If the variation in the recorded signal is truly random, that is, if the noise is white, any excursion from the mean value (S) will be no greater than 2.5 times the standard deviation (N) within a 99% confidence limit (1). Because these excursions will be in both positive and negative directions, the standard deviation (N) can therefore be found by taking $1/5$ the peak-to-peak variations in the signal. From this, S/N can be easily calculated. In this estimation of S/N, it is important to recognize that a sufficiently long record of the signal must be obtained to provide a statistically reliable estimate of the peak-to-peak excursions (1). For example, the signal from an instrument having a time constant of 1 sec should be observed for about 2 min.

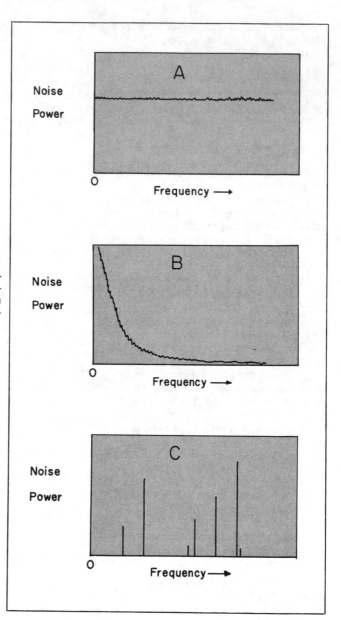

Figure 1. Noise-power spectra commonly found in chemical instrumentation

A. "White" noise
B. Flicker (1/f) noise
C. Interference noise

For ac signals, it is necessary to employ other methods to measure S/N. It is possible, of course, to convert the ac signal to dc by use of a conventional rectification technique and to measure S/N as described. However, during conversion some ac noise will also be rectified and can cause error in the measurement of S. A more accurate measurement of S/N for an ac signal can be made by employing a technique known as autocorrelation. This technique will be described in the second of the two articles in this series. When it is available, a special voltmeter designed to respond to the rms value of a signal can also be used to evaluate S/N; this method is subject to the same error as the ac to dc conversion process but generally involves a simpler measurement.

Noise-Power Spectra. Although signals can usually be characterized in terms of their waveforms (e.g., dc, sine wave), it is impossible to look at noise in the same way. Being inherently unpredictable, noise has no perceivably periodic character so that a noise waveform is almost meaningless. To characterize noise, it is more instructive to examine the frequency components in the noise waveform. This can be done with a noise-power spectrum, which is a plot of the noise power as a function of frequency.

For white noise, as its name implies, a flat power spectrum is found (Figure 1A). White noise is therefore a "hash" of all frequencies, the components of which are random in phase and amplitude. Examples of white noise are shot noise in photomultiplier tubes and Johnson noise caused by the random movement of electrons in a resistor. Johnson, Nyquist, or thermal noise, as it is variously termed, is particularly important in that it arises from the inherently random motion of charged species and therefore occurs in all real measurement situations. This fundamental thermodynamic noise has a flat power spectrum governed by the relationship:

$$E = (4\,RkT\Delta F)^{1/2} \qquad (4)$$

E is the root-mean-square (rms) noise voltage contained in a bandwidth ΔF produced in a resistor of resistance R, k is the Boltzmann constant, and T is the resistor temperature in °K. The noise power, which is proportional to the rms voltage, increases as the square root of both the observed bandwidth and magnitude of the resistance. This characteristic is important in the selection of low-impedance filters to be discussed in the following section.

In contrast to white noise, another common form of noise, called "flicker noise," has a spectrum in which the power is approximately proportional to the reciprocal of the frequency. For this reason, it is often called $1/f$ (one-over-f) noise and has the power spectrum in Figure 1B. Flicker noise is common in most amplifiers and, in fact, in almost all instrumental systems, where it is usually termed "drift".

Another form taken by noise is "interference," often characterized by a line spectrum as shown in Figure 1C. Probably the most common form of interference noise is that arising from the 60-Hz power line. Therefore, a great deal of noise will usually be found at 60 Hz and all its harmonics (120 Hz, 180 Hz, etc.). However, other sources of interference noise, such as high-voltage spark sources, radio transmitters, or microwave devices, must be recognized. These high-frequency noise sources can be serious because interference noise is often added to a signal by transmission through the air, and noise-transmission efficiency increases with frequency.

Perhaps the most troublesome form noise can take is the form of an impulse. Impulse noise occurs erratically from such sources as the start-up of a large motor or a lightning flash, and because of its fast rise-and-fall times, has a broad frequency spectrum. The high peak power inherent in such impulse noise and its broad spectrum (similar to white noise) often make it difficult to eliminate.

Another type of noise which complicates certain measurements can be roughly termed distortion. Distortion is essentially the introduction of foreign frequency components onto the desired signal and can arise from a number of sources such as instrumental nonlinearities, transmission losses and reflections, and the interaction of the signal with various disturbing elements. Because of the unpredictability of some distortion sources, this form of noise can be quite vexing, especially in pulse-measurement systems where it is often encountered.

Most noise encountered in real situations will be a mixture of the noise types described. In general, the power spectrum will have a power proportional to $1/f^n$ where $0 < n < 1$. Additionally, the spectrum will often have line (interference) noise superimposed on it. With this in mind, we can investigate ways of eliminating or minimizing noise to enhance S/N.

Signal/Noise Enhancement by Filtering

To extract a signal from any form of noise, some property of the signal must be employed to aid in its identification and separation from noise. Because signals possess an identifiable waveform and noise does not, all properties a waveform can possess—amplitude, frequency, and phase—can be used in this identification and extraction process. Perhaps the most commonly used property is frequency.

Dc Signal Filtering. For example, a dc signal can often be "smoothed" merely by using a long readout time-constant (damping) or by shunting a capacitor across an instrument's voltage output terminals. Note that this "smoothing" necessitates a trade-off between the response time of the device to a signal change and an improved signal-to-noise ratio. This procedure is nothing more than filtering, wherein all frequencies other than the frequency at which the signal is found are eliminated. In this case, the signal frequency is zero (dc) so that all higher frequencies are attenuated. This is analogous to the somewhat more versatile techniques of analog or digital integration of a signal over a fixed-time interval.

To carry information, the signal must have a finite bandwidth or frequency range (Δf), which means that the dc signal must be permitted to change in value to indicate a signal variation. Thus, the signal is not strictly dc (f_0) but has some additional, low-frequency (slowly varying) components (Δf) which are used to carry information.

This necessary bandwidth makes perfect filtering impossible, since the filter must pass a finite band of frequencies to include all of the signal information. Because noise is found over a broad range of frequencies, some noise is almost certain to be included in this bandwidth and passed by the filter. For this reason, it is best to have the signal located in a frequency region which has little noise power.

Because most noise has a strong $1/f$ component, that is, because drift is generally significant, it is usually advantageous to locate the desired signal away from zero frequency (i.e., away from dc where $1/f$ noise is greatest). To do this, the signal information is impressed on a carrier wave at the desired frequency. This process is called modulation.

Modulation and Tuned Amplifiers. The process of modulation is employed both in everyday devices such as the radio and also in sophisticated electronic signal processing systems. All applications of this technique, regardless of their sophistication, depend upon the impression of one signal upon another. Usually the signal (the desired information) is impressed on another higher frequency wave (called the carrier) by modifying some property of the carrier in accordance with the characteristics of the signal. For example, in the common AM radio, a radio-frequency carrier wave is modulated in amplitude according to the frequency and amplitude of the audio signal which is impressed upon it. In a similar fashion, any property of a carrier wave can be modified in a way such that information from another signal is impressed on it.

Properties of a waveform are amplitude, frequency, phase (position with respect to a reference), and duty cycle (on-time compared to total duration). These properties can be varied on the following typical waveforms: sine wave, square wave, triangular wave, constant-level (dc) wave, and pulse train. Because the square and triangular waves are merely combinations of the fundamental and harmonics of a specific sine wave, these can be thought of as special cases of the sine wave and will therefore not be considered individually. By modulating various characteristics of the other waveforms, specific types of modulation can be effected. When this is done, the resulting modulated wave can be classified as shown in Table I.

From Table I, the carrier wave does not have to be at common electronically detectable frequencies but can range from dc to optical frequencies (10^{15} Hz). Nonetheless, when the signal information (perhaps chemical information about a sample) is forced upon the carrier, the process is called modulation.

Modulation is almost always used to improve the efficiency of transmission of a signal. In instrumental improvement in S/N, this is also the case; by impressing a signal on a carrier at the desired frequency, the information can be brought to a low-noise frequency

Table I. Modulation Forms

Carrier-wave shape	Information-carrying property	Common name	Abbrev	Example
Dc	Amplitude	Analog		Meter reading
Sine wave	Amplitude	Amplitude modulation	SAM	Chopped-photometric signal
Sine wave	Frequency	Frequency modulation	SFM	Voltage-controlled oscillator Wavelength-modulation spectroscopy (2, 3)
Sine wave	Phase	Phase modulation	SPM	Photoelastic effect GC ultrasonic detector (4)
Pulse train	Amplitude	Pulse-amplitude modulation	PAM	Optical chopper Boxcar integrator
Pulse train	Repetition rate	Pulse-frequency modulation	PFM	Photon-counting (5) TV pictures from Mariner satellites
Pulse train	Pulse delay	Pulse-position modulation	PPM	Radar Remote Raman spectroscopy

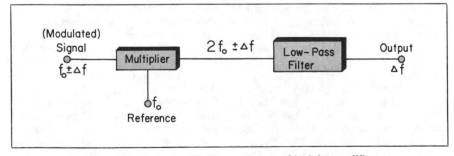

Figure 2.   Instrumental components of lock-in amplifier

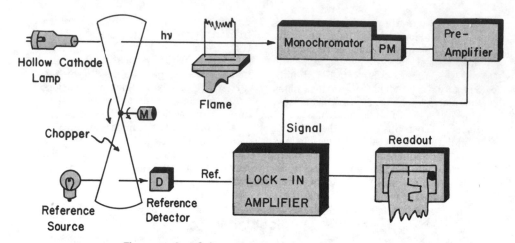

Figure 3. Schematic diagram of atomic absorption spectrophotometer employing lock-in amplification for signal-to-noise enhancement

region so that transmission efficiency and signal recognition are enhanced. In measurements on chemical systems, the modulation methods usually employed are sine-wave amplitude modulation and sine-wave frequency modulation, the former being the most common.

For example, in atomic absorption flame spectrometry if the intensity of a hollow cathode lamp is modulated by a light chopper at a frequency f_0, the lamp signal can be detected at f_0 to minimize the $1/f$ noise present in the system. This increases the effectiveness of filtering, since less noise will now be included in the necessary filter bandwidth. The filters which can be used to detect information at f_0 will necessarily be somewhat more complex than the simple capacitor shunt employed at dc. For example, specially designed "active filters" containing operational amplifiers provide narrow bandwidths having characteristics which can be mathematically defined. The Butterworth filter is such a device (6). Also, band-reject or "notch" filters can be

used to reject specific frequencies (e.g., 60 Hz) where strong interference noise is known to exist.

Often, a filter is combined with an amplifier by use of appropriate feedback so that only a narrow band of frequencies is amplified. This system, called a tuned or frequency-selective amplifier, is often employed instead of a filter because of its ability to increase signal levels through amplification while actually reducing noise by appropriate filtering. Generally, this type of device provides an increase in S/N proportional to the reciprocal of the square root of the bandpass (the range of frequencies passed) of the amplifier. Therefore, the bandpass is made as narrow as is consistent with the necessary signal bandwidth discussed earlier. Unfortunately, this requires that the amplifier and the carrier frequency be stable so that the carrier and amplifier bands match exactly.

Practical experience has shown that bandwidths less than 1 Hz become troublesome in their frequency drift so that this provides a limit to the S/N enhancement attainable with a conventional tuned amplifier. Crystal filters provide one exception to this observation. The naturally stable resonance frequency of an oscillating crystal (e.g., quartz) can be used in a tuned system to furnish a stable but expensive narrow bandwidth amplifier. This unfortunately requires that the signal frequency remain stable at the crystal frequency. To solve the problem of obtaining a narrow bandwidth while maintaining a frequency match between the signal and filter bandpass, a device called a lock-in amplifier is often used.

Lock-In Amplifier

To eliminate the problem of frequency drift in the tuned amplifier and to further decrease the effective amplifier bandpass, a device called a lock-in or phase-sensitive amplifier was developed. This device discriminates against noise, not only on the basis of frequency as does the tuned amplifier, but also according to phase. Thus, the only signals (and noise) amplified by a lock-in amplifier are those having a specified frequency and phase with respect to a reference waveform.

Because the reference waveform can generate the signal (by modulation) or be obtained from the signal (during modulation), it can be phase-related to the signal. By suitable processing, only the signal and noise components having this phase relationship to the reference will be amplified, all others being rejected. This obviously provides even better noise rejection than the tuned amplifier and also, by using a reference waveform, eliminates the problem of frequency drift.

The components of a typical lock-in amplifier are shown in Figure 2. In this system the signal information (Δf) has been impressed on a carrier f_0 by a suitable modulation technique (e.g., optical chopper) and is phase-related to a reference waveform, also at f_0. When the signal (+ noise) is multiplied by the reference, a waveform at $2 f_0$ carrying the signal information (Δf) results. Because Δf is generally low-frequency information, it can be efficiently extracted from the $2 f_0$ carrier by a low-pass filter so that it alone appears at the output. The characteristics of this

system are as follows:

The spectrum of signal frequencies (Δf) centered at f_0 are transformed to the same spectrum centered at dc.

If the signal and reference are of the same frequency, the output is dc, the dc value being largest when the signal and reference are in phase and smallest when they are 90° out of phase.

If the phase of the signal reverses, the sign of the dc output reverses.

Note that the lock-in amplifier performs essentially the reverse function of modulation. For this reason it is one member of a family of devices called demodulators. As an example of the use of a lock-in amplifier, consider an atomic absorption system such as shown in Figure 3.

In this system the reference is phase-locked to the chopper-modulated signal, which is a square wave of amplitude proportional to the hollow cathode emission intensity. Following this signal through the lock-in amplifier of Figure 2, we must first consider the multiplier output. In Figure 4 the multiplier output is shown for signals (or noise) which are in phase with or are 90° or 180° out of phase with respect to the reference wave. Peak-to-peak amplitudes of $2 x$ and $2 y$ are arbitrarily assigned to the signal and reference waveforms, respectively.

From Figure 4 the averaged values of the multiplier output are $+xy$ for the in-phase signal, $-xy$ for 180° phase difference, and 0 for 90° phase difference between the signal and reference. These are the values which would appear at the output of the low-pass filter, as predicted earlier. The $2 f_0$ component discussed earlier is especially obvious in the 90° situation.

As variations on this system, it is common to find both sine-wave signals and/or references, the considerations being the same as above. Note that any noise without a specified frequency and phase relationship to the signal will contribute little to the lock-in output. This provides an enhancement in S/N

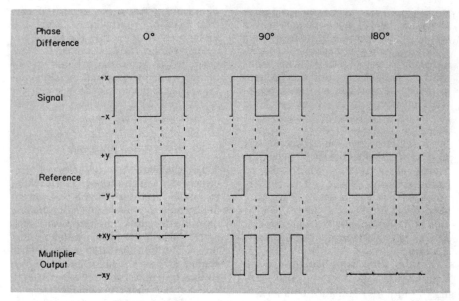

Phase Difference   0°   90°   180°

Signal

Reference

Multiplier Output

Figure 4. Waveforms illustrating multiplier output of lock-in amplifier for various phase differences between signal and reference inputs

which, as with the tuned amplifier, will be approximately equal to the reciprocal of the square root of the amplifier bandpass. For the lock-in amplifier, the effective bandpass (Δf) can be approximated by

$$\Delta f = \frac{1}{4\,RC} \qquad (5)$$

where RC is the time constant of the low pass filter of Figure 2. This argues that the longest time constant should be selected which is consistent with the observation of signal variations (i.e., with the signal bandwidth). For properly designed lock-in amplifiers, it is therefore quite possible to obtain an effective bandpass of less than 0.01 Hz, much smaller than the value of 1 Hz obtainable with a tuned amplifier.

Obviously, neither lock-in nor tuned amplification should be utilized at a frequency which contains a strong interference noise component. If the signal naturally lies at such a frequency, appropriate modulation should be em-

ployed to move the signal information to another frequency region. Thus, one should never, for example, chop a light beam at 60 Hz or any of its harmonics or subharmonics. In a situation where considerable impulse noise is present, tuned or lock-in amplification may be ineffective in improving the signal-to-noise ratio. In these cases, the sporadic appearance and broad spectrum of impulse noise will cause it to be partially passed through these devices rather than eliminated.

Other names which have been used for the lock-in amplifier are synchronous detector (because of the necessary synchronization of signal and reference), phase-lock amplifier, coherent detector, and heterodyne detector. A limitation of these systems, as is apparent from the above discussion, is that the signal to be detected must be periodic or modulated in such a way as to be made periodic. When this is difficult or impossible, other signal enhancement techniques must be used.

These techniques will be considered in the second article of this series.

In the second article, simple filtering and lock-in amplification will be compared to the more versatile techniques of signal averaging, boxcar integration, and correlation. With this information it should be possible to select the most appropriate method of signal-to-noise enhancement for any given measurement situation.

References

(1) V. D. Landon, *Proc. IRE,* 50 (February 1941).

(2) V. Svoboda, *Anal. Chem.,* 40, 1384 (1968).

(3) W. Snelleman, *Spectrochim. Acta,* 23B, 403 (1968).

(4) F. W. Noble, K. Able, and P. W. Cook, *Anal. Chem.,* 36, 1421 (1964).

(5) M. L. Franklin, G. Horlick, and H. V. Malmstadt, *ibid.,* 41, 2 (1969).

(6) Philbrick Applications Manual, Nimrod Press, Boston, MA, 1968. Available from Philbrick/Nexus Research, Dedham, MA 02026.

The author gratefully acknowledges the support of National Science Foundation grant NSF GP 24531.

Signal-to-Noise Enhancement Through Instrumental Techniques

Part II. Signal Averaging, Boxcar Integration, and Correlation Techniques

Techniques are described which can provide signal-to-noise enhancement for nonperiodic or irregular waveforms or for signals which have no synchronizing or reference wave

G. M. HIEFTJE
Department of Chemistry
Indiana University
Bloomington, IN 47401

IN A PREVIOUS ARTICLE (1), it was indicated that noise is found in every experimental signal which can be measured. However, because signals have predictable behavior (i.e., because they are coherent), they can be partially extracted from the noise by use of suitable techniques, provided the necessary increase in measurement time can be tolerated. One such technique, filtering, is most effective if the signal information is located at a frequency or group of frequencies in which little noise is found. Modulation can be used to place the signal information at the desired frequency. However, even when the signal is located in a low-noise frequency region, some noise will always be passed by any filter because of the necessary bandwidth of the signal and of the difficulty involved in matching and maintaining a match between the signal frequencies and the filter (or tuned amplifier) bandpass. A lock-in amplifier was useful in solving these problems but required periodic signals or those which can be made periodic.

In this article, techniques will be described which can provide signal-to-noise enhancement for nonperiodic or irregular waveforms or for signals which have no synchronizing or reference wave. These techniques, signal averaging, boxcar integration, and correlation, will be discussed independently.

Signal Averaging—S/N Enhancement in Time Domain

If a signal is not periodic or repetitive, but is *repeatable*, it can be extracted from noise with the technique of ensemble averaging or signal averaging as it is often termed. This technique uses an averaging or integration procedure similar to that employed when multiple recorder or oscilloscope traces are superimposed to obtain an average or effective value.

Basically, signal averaging involves the instrumental superposition of a number of signal traces by sampling each signal record in the same way and storing the samples in either a digital or analog register. Because the records are each sampled in the same way and at the same corresponding times, the signals add coherently in register while the noise, being random, averages to zero. Averaging the ensemble of records thus provides increased S/N.

More quantitatively, the signal will add in register directly as the number of records sampled (n), whereas the noise will only add as the square root of this number ($\sqrt{n}$), assuming a $1/f^n$ noise spectrum (1, 2). Therefore, the S/N will increase as $n/\sqrt{n} = \sqrt{n}$.

To enjoy this advantage, the sampling of the signal records must meet certain criteria (3). For example, to extract all the information from the signal, it must be sampled at a frequency at least twice as great as its highest frequency component. However, the sampling frequency should not be significantly greater than twice the highest signal frequency. If this occurs, no additional information will be obtained, but more noise will be included than is desirable or necessary merely because the bandwidth is increased by faster sampling. Also, to prevent "aliasing" high-frequency noise down into the sampling frequency region, a high-frequency cut-off filter should be used at the signal input (4). Finally, if any interference (e.g., narrow band or line) noise (1) is expected to be present in the signal, it is important that the signal records not be scanned or sampled at a rate which is either a multiple or submultiple of the interference

frequency. If this latter condition is not fulfilled, no improvement in S/N will occur for the interference noise.

In signal averaging, it is important to be able to properly sample and add the sampled records reproducibly. This requires a synchronizing signal or pulse which can serve to actuate the sampling system. Generally, this "sync pulse" can be derived from the signal or can be used to initiate the signal record. The synchronizing signal must obviously be reliable in predicting the start of the signal for the records to add exactly and for S/N enhancement to be realized.

Because the signal averaging technique operates in the time domain, it is difficult to compare with the lock-in and frequency-selective amplifiers (1) for its capability to increase S/N. For a comparison, it is necessary to transform [via a Fourier transform (5)] the effect of the ensemble averager on a selected signal. When this is done, the averager acts like a comb filter (such as that portrayed in Figure 1) whose teeth are centered at the frequency components of the signal and whose teeth have a bandpass (Δf) of

$$\Delta f = \frac{0.886}{n\tau} \qquad (1)$$

Here, τ is the time taken for the sampling of one record, and n is the number of scans. This type of filter, of course, passes the signal efficiently while removing the noise. For example, with a sampling time for each record of 20 msec and 10^6 scans, a tooth bandwidth of 5×10^{-5} Hz is obtainable—far better than would be possible with a lock-in amplifier (1).

In addition, the signal averager can provide signal-to-noise enhancement for nonrepetitive or nonperiodic signals, a feat which is impossible for tuned or lock-in amplifiers. Because the averager passes *all* the frequency components of the signal *in the proper phase*, the original signal is extracted from noise with its original waveform.

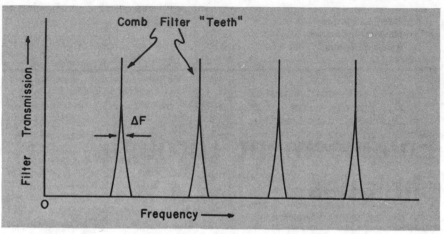

Figure 1. Frequency domain representation of signal averager characteristics. Teeth of comb filter located at frequency components of signal being measured

$$\Delta F = \text{tooth width} = \frac{0.886}{n\tau} \text{ (Hz)} \quad \text{(See text)}$$

Table I. List of Some Commercially Available Hardware Signal Averaging Instruments

Name	Manufacturer
Waveform eductor	Princeton Applied Research Corp.
Enhancetron	Nuclear Data, Inc.
Signal averager	Nicolet Instrument Co., Inc.
Time averaging computer	Varian Associates, Inc.
Spectrum accumulator	JEOLCO, Inc.
LAB-8	Digital Equipment Corp.

This is important in many applications, such as nuclear magnetic resonance (nmr) spectroscopy (2), where the waveform of a signal conveys more information than its frequency components alone.

In nmr, as in a number of applications of signal averaging, it is interesting to note that the signal averager is often preceded by a lock-in amplifier. This combination is used because in situations where $1/f$ noise (i.e., drift) is significant, it is better to sweep a

number of signals rapidly and average them, than to slowly sweep a single signal over the same time period by use of a correspondingly longer filtering time constant on the lock-in amplifier (2). Also, impulse noise will not be as troublesome to signal averaging as it was for lock-in or tuned amplification because for each impulse, only a single deviant point will occur in the final averaged signal. Generally, for a signal which is a smoothly varying time function, the deviant points can be easily discarded if simple but carefully chosen testing criteria are used.

A number of hard-wired signal averagers which are commercially available under various names are listed in Table I. However, note that unlike tuned or lock-in amplification, signal averaging can be conveniently performed under software control by a small laboratory computer. The increasing availability of these computers can be expected to render signal averaging the technique of choice for many routine signal enhancement chores.

From a practical standpoint, hardware ensemble averagers can be used for several purposes other than signal-to-noise enhancement. They can be employed to compute the average value or a more representative value of a signal. Also, because the storage register of the averager can be read at a faster or slower rate than that at which the signal was sampled, the device can be used to slow down or speed up the transmission of data.

Boxcar Integration—Poor Man's Signal Averager

A boxcar integrator is somewhat similar in design to both the lock-in amplifier and the signal averager and performs similar functions, depending on its application. The generalized schematic of a boxcar integrator, shown in Figure 2, indicates that the device is essentially a single-channel signal averager with a sampling gate which can be opened for a selected time during the passage of the signal to be measured. At a fixed delay the gate is synchronized to the signal through the reference channel, so that the same portion of the signal is sampled for each signal passage. The gated samples taken from each passage are then averaged by the low-pass (RC) filter to provide signal-to-noise enhancement *for the portion of the signal which is sampled.*

Clearly, operation of the boxcar integrator in this fashion provides no information about the signal except its average amplitude at a specified time. For complex signals, this is of limited utility. However, for a signal consisting of a square wave or a train of pulses, the average amplitude of the square wave or pulse can be of considerable importance and in many cases is the only information sought. Several examples of pulsed signal detection important to chemical analysis can be cited (*6, 7*), and because of the high signal-to-noise enhancement attainable with pulsed or low duty-cycle signals (*8*), these applications can be expected to rapidly increase in number. Also, the attractiveness of pulsed lasers as sources for many spectrochemical techniques can be expected to make a gated detector, such as the boxcar integrator, increasingly useful as more and better laser sources are developed.

Operation of the boxcar integrator as a detector for pulsed signals requires only the optimization of the gate width and delay for the specific signal sought. As with the signal averager, the pulsed signal need not be periodic but must be repeatable and have a synchronizing reference which is time-locked to the signal. Any lack of synchronization between the signal and reference waveforms will cause "jitter" in the gate opening and resultant error or noise at the output. For this reason the gate width and delay in a commercial boxcar detector must be extremely precise and jitter-free, usually to within a few nanoseconds.

Compared to the measurement of a single pulse, the boxcar detector provides a signal-to-noise enhancement equal to the square root of the number of pulses integrated. However, if the boxcar integrator is compared to the commonly measured "average value" of a pulsed signal, an even greater S/N enhancement is realized. The degree of enhancement depends quantitatively, among other things, on the duty cycle of the signal (on-time compared to total duration) but can be qualitatively evaluated by considering the amount of noise which is detected in *each* case.

In measuring the average value, the detector must be continuously turned on and will therefore detect noise for the full duration of the signal. By contrast, the boxcar detector is "on" only during the gate pulse so that noise is detected only for that short time. Because both measurements include the total signal pulse, the reduction in detected noise in the boxcar integrator produces a dramatic signal-to-noise improvement over that of the average value measurement.

The similarity of the boxcar integrator of Figure 2 to a lock-in amplifier (1) is immediately apparent. In fact, the boxcar detector can serve as a lock-in amplifier simply by setting the gate width to pass the signal at half-cycle

intervals. The added ability to gate low duty-cycle signals makes the boxcar integrator more versatile, however, and more sensitive in the detection of these signals. Like a lock-in amplifier, the boxcar integrator is especially susceptible to impulse noise and to interference noise which is at a multiple or submultiple of the sampling frequency. The sampling frequency should therefore be carefully chosen.

A second mode of operation of the boxcar integrator enables the enhancement of signals having a complex waveform. In this mode the boxcar integrator can again be thought of as a single-channel signal averager. Instead of being locked to a specific time channel, however, the boxcar gate is moved from channel to channel so that eventually the entire complex waveform is sampled. This can be done by automatically stepping or scanning the variable gate delay of Figure 2 so that the signal segments are scanned in se-

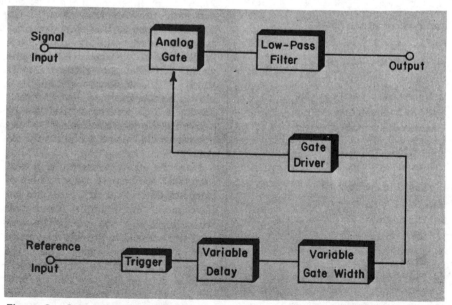

Figure 2. Schematic block diagram of boxcar integrator. Variable delay can be operated in either fixed, stepped, or scan mode

quence from zero delay to a delay which is greater than the signal duration.

In the scanning mode of operation the boxcar integrator provides signal-to-noise enhancement by sampling the signal a number of times at each value of time delay. In a stepped delay system this is accomplished by incrementing the delay after a desired number of signals have been sampled. A continuously scanned delay can be similarly employed, provided that the delay scan rate is significantly slower than the repetition rate of the signal. For irregular or nonperiodic signals the stepped delay is generally preferable, to ensure that the same number of signal waveforms are read at each delay position.

Compared to a signal averager, the boxcar integrator is a rather inefficient instrument for enhancing complex signals. Because the signal is sampled only briefly for each of its appearances, a great deal of information is thrown away. In fact, for a signal averager having n channels, signal enhancement will proceed n times faster than for a boxcar integrator with similar gate characteristics. From this, seemingly, instrumental simplicity is the only advantage of the boxcar integrator. However, because the boxcar integrator has only a single channel (gate), the gate can be made far more sophisticated than would be practical for the large number of channels in the signal averager. Generally, the gate in a boxcar integrator can operate much faster and more precisely than those in the signal averager, so that the boxcar detector is most useful in applications involving fast signals. Also, because signal improvement proceeds more slowly with the boxcar system, it is most advantageously applied to signals having a high repetition rate, usually above 100 Hz where suitable signal averagers would beome prohibitively expensive.

Like the signal averager, a boxcar integrator can be used for a number of purposes other than signal-to-noise enhancement. For example, fast signals with a high repetition rate can be sampled and scanned to provide readout on a much slower time scale. This is the principle on which the sampling oscilloscope operates and provides the basis for an even slower readout of fast signals on a device such as a servo recorder. Time resolved spectroscopy can also be performed with a stroboscopic sampling unit similar in design to a boxcar integrator (9).

Correlation Techniques

The correlation techniques for S/N enhancement are an outgrowth from information theory and depend on the relationship between a signal and a delayed version of itself (autocorrelation) or of another signal (cross-correlation). Mathematically, if $V_1(t)$ and $V_2(t)$ are two functions (signals), the cross-correlation function $[C_{1,2}(\tau)]$ is

$$C_{1,2}(\tau) = \tau \xrightarrow{\lim} \infty \frac{1}{2T} \int_{-T}^{+T} V_1(t) V_2(t - \tau) dt \quad (2)$$

$$\tau = \text{delay}$$

$$t = \text{time}$$

For the special case of $V_1(t) = V_2(t)$, a similar expression gives the autocorrelation function $[C_{1,1}(\tau)]$

$$C_{1,1}(\tau) = \tau \xrightarrow{\lim} \infty \frac{1}{2T} \int_{-T}^{+T} V_1(t) V_1(t - \tau) dt \quad (3)$$

From these equations, the correlation functions indicate whether coherence exists between two signals or within a signal with respect to a natural or artificial variable delay (τ). As an ex-

413

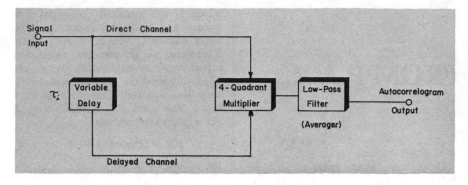

Figure 3. Functional diagram of autocorrelation computer. Conversion to
cross-correlator possible by changing
variable delay input to reference signal.
4-Quadrant multiplier accepts and outputs both positive and negative waveforms

Figure 4. Autocorrelograms of vari-
ous input signals. $C_{1,1}(\tau) =$ autocorrelation function

A. Autocorrelogram of sine-wave signal
B. Autocorrelogram of random signal (noise)
C. Autocorrelogram of band-limited noise
   showing both positive and negative delay
   (τ)
D. Autocorrelogram of extremely noisy sine-
   wave signal (S/N < 1)

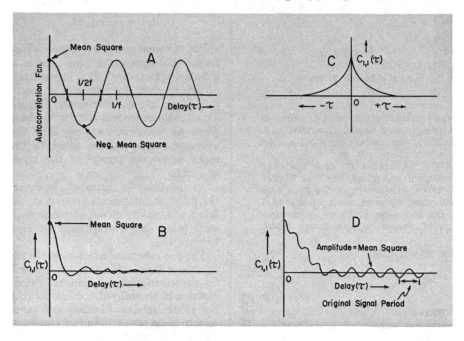

ample of this, a strong cross-correlation exists between the New York blackout a few years ago and the later birth rate, with a natural delay time (τ) of nine months.

To implement and further elucidate the autocorrelation function, the hardware or software model of Figure 3 can be used. Figures 3 and 2 of Reference 1 show that the correlator is similar to a lock-in amplifier which incorporates a variable delay.

In the device of Figure 3 the signal is multiplied by a version of itself delayed by a fixed amount (τ), and the mean of this product is taken as one point of the correlation function $[C_{1,1}(\tau_i)]$. By taking more points in this fashion and plotting them as a function of the delay τ_i, the complete autocorrelation function or autocorrelogram can be traced out. As a simple example of this, refer to Figure 4A and assume a sine-wave input to the instrument in Figure 3.

When the delay (τ) is zero, the sine wave of frequency f_1 is fed directly to the multiplier along both channels so that the multiplier output will be a sine^2 wave. The averager will then output the mean of this wave which will be the mean square value of the input sine wave. Next, when τ is increased by a time equal to a quarter period of the sine wave, the sine wave will be 90° out of phase with its delayed version. From Figure 3 and the discussion of the phase-lock amplifier (1), the output of the averager will be zero under these conditions.

For the case of a delay $\tau = 1/2 f_1$, e.g., ½ cycle or 180°, the sine waves will be exactly out of phase at the multiplier, so that a negative sine^2 wave will result. The averager will then output a value equal to the negative mean square of the input sine wave. From these points and Figure 4A, the correlogram in this case will be a cosine wave of frequency $1/\tau = f_1$ which is the same as the input frequency and of amplitude equal to the mean square of the input sine wave. Therefore, all information about the input signal is maintained except the phase information, since here a cosine wave was obtained whereas a sine-wave input was used. Mathematically, if the input waveform was

$$f(t) = A \cos (\omega t + \theta) \qquad (4)$$

the output would be

$$C_{1,1}(\tau) = \frac{A^2}{2} \cos \omega t \qquad (5)$$

For a more complex input waveform, the correlogram has quite a different appearance. If a broad band of frequencies (e.g., noise) is introduced into the autocorrelator (Figure 3), all frequencies will still be in phase when no delay is used, so that the averager output will be equal to the mean square, or (rms)^2 of the input at $\tau = 0$.

As the delay is increased, however, the different frequencies present at the input will rapidly go out of phase with each other and average out to zero (Figure 4B).

The less coherence which exists in the correlator input, of course, the less will be the correlation between the input signal and its delayed replica so that the autocorrelation function will damp to zero more rapidly. In fact, for completely random, white noise, no time correlation exists. Therefore, for perfectly white noise, the autocorrelation function is a delta function with a value of the (rms)^2 at $\tau = 0$ and with a value everywhere else equal to zero. For real, band-limited noise, such as found at an amplifier output, however, a function similar to Figure 4C is found. Referring to Figure 4C, the autocorrelogram is an even function, i.e., is symmetrical about $\tau = 0$. This can be appreciated by considering the negative delay ($\tau < 0$) to be effected by inserting the variable delay of Figure 3 into the direct channel rather than the delayed channel. Obviously, the mirror image correlogram will be produced.

If to a random waveform is added a coherent signal, e.g., a sine wave, the (one-sided) autocorrelation function will appear, as in Figure 4D where the random waveform (noise) contribution rapidly damps to zero, as in Figure 4C. The coherent signal, however, continues to contribute at much greater delays. This enables the signal (the sine wave) to be extracted *very* efficiently from the noise (random wave). This technique is especially useful when no means is available for synchronizing a sampling system to the signal, which is necessary with the tuned and phase-lock amplifier, with the signal averager, and with the boxcar integrator. Also, autocorrelation techniques are useful in situations where $S/N \ll 1$, i.e., when the signal is often unrecognizable or impossible to detect apart from the noise.

Signal-to-noise ratios can be conveniently measured from an autocorrelogram such as Figure 4D. The point at $\tau = 0$ is the $(rms)^2$ value of $S + N$, but the peak value of the ac signal after the noise contribution has damped to zero is the $(rms)^2$ value of the signal alone. From these quantities, the ratio of the rms values of S and N can be easily computed.

Cross-correlation is similar to autocorrelation except that the delayed signal arises from a second source, probably from the modulating function of the signal or as in phase-lock amplification, from a reference coupled to the signal modulator. In this case, the frequency components common to both waveforms will appear in the correlogram. For example, if a sine wave and square wave of the same frequency are cross-correlated, the correlogram would be a sine wave of that frequency and of amplitude equal to the average value of the product of the sine and square waves. This is because the sine wave fundamental of the square wave is the only frequency component common to both waveforms.

Although cross-correlation requires a reference as does phase-lock amplification, note that the reference and signal waveforms need not be sine or square waves but can take any form. Therefore, a signal can be modulated with any desired waveform, a capability which has advantage in certain applications (*10*). Like signal averaging, the correlation techniques are effective in reducing impulse noise by a simple elimination of obviously deviant points in the correlogram; interference noise effects can also be minimized with correlation.

Because the correlation techniques produce a phase-related representation of the common frequencies between two signals or within a signal, the correlogram may appear quite different from the original signal. As an example, the autocorrelation function of a square wave is a triangular wave of the same fundamental frequency. The triangular and square waves, of course, contain the same frequency components, although in the triangular wave the components are phase-related (i.e., are all cosine waves) and begin at $\tau = 0$. This indicates once more how the amplitude and frequency information is maintained in computing the correlation functions, although the phase information is lost.

Although the correlation techniques have not yet been extensively employed in chemical analysis, it is likely that their use will increase. Previously, it was necessary to calculate point-by-point correlograms or to use expensive hardware correlation computers. Now, with the increasing introduction of small digital computers into the chemical laboratory, it will become increasingly convenient to compute auto or cross-correlation functions with inexpensive software-based systems.

Summary

In this and in the previous article on instrumental methods of signal-to-noise enhancement (1), a number of old and new concepts have been introduced in a hopefully coherent form. No attempt has been made to maintain mathematical rigor in the treatment, and no technique has been treated exhaustively. Although several examples of the use of each technique were cited, these were not considered in detail and, of course, many additional applications could be listed. Rather, it is hoped that the reader with little background in communication theory has gained some familiarity with the subject and its instrumental implementation. Readers with a more extensive background have hopefully had their memories jogged and also benefited by a somewhat different treatment of the subject.

References

(1) G. M. Hieftje, *Anal. Chem.*, **44** (6) 81 (1972).

(2) R. R. Ernst, *Rev. Sci. Instrum.*, **36**, 1689 (1965).

(3) "Threshold Signals," J. L. Lawson, G. E. Uhlenbeck, Eds., Dover Publ., New York, NY, 1965, Chap. 3.

(4) G. Horlick, H. V. Malmstadt, *Anal. Chem.*, **42**, 1361 (1970).

(5) R. Bracewell, "The Fourier Transform and Its Application," McGraw-Hill, New York, NY, 1965, Chap. 2.

(6) G. M. Hieftje and H. V. Malmstadt, *Anal. Chem.*, **41**, 1735 (1969).

(7) L. M. Fraser and J. D. Winefordner, *ibid.*, **43**, 1693 (1971).

(8) V. G. Mossotti, Abstracts, 161st National Meeting of the American Chemical Society, Los Angeles, CA, April 1971, No. Anal. 27.

(9) M. L. Bhaumik, G. L. Clark, J. Snell, and L. Ferder, *Rev. Sci. Instrum.*, **36**, 37 (1965).

(10) R. R. Ernst, *J. Mag. Resonance*, **3**, 10 (1970).

The author gratefully acknowledges the support of National Science Foundation grant NSF GP 24531.

INDEX

INDEX